Editor
Ajay S. Kalamdhad
Department of Civil Engineering
Indian Institute of Technology Guwahati
Guwahati, Assam, India

Jointly published with Capital Publishing Company, New Delhi, India

The print edition is not for sale in SAARC countries. Customers from SAARC countries – Afghanistan, Bangladesh, Bhutan, India, the Maldives, Nepal, Pakistan and Sri Lanka, please order the book from:

Capital Publishing Company, 7/28, Mahaveer Street, Ansari Road, Daryaganj, New Delhi 110 002, India.

ISBN 978-3-030-70462-9 ISBN 978-3-030-70463-6 (eBook)
https://doi.org/10.1007/978-3-030-70463-6

This Springer imprint is published by the registered company Springer Nature Switzerland AG
The registered company address is: Gewerbestrasse 11, 6330 Cham, Switzerland

Ajay S. Kalamdhad
Editor

Integrated Approaches Towards Solid Waste Management

Springer

Integrated Approaches Towards Solid Waste Management

Preface

Increasing world population has put enormous pressure on the planet when it comes to the amount of wastes generated. Such is the case that in a few more years, with this alarmingly increasing rate of waste production, the earth would reach its assimilative capacity. This calls for efficient measures of waste treatment and disposal. Financially developing countries of the world have limited access to vital resources such as funding, human resources etc. for overcoming the problems of waste management. Therefore, it becomes highly essential to develop new strategies and economically viable technologies for effective management of wastes. This book presents various new concepts and strategies of waste management, coupled with treatment technologies and recycling of waste products. It also focuses on the treatment of wastewaters, groundwater and the application of GIS techniques for waste management. Special focus has also been made to the recycling of waste products and converting them to useful products or services. Various treatment technologies such as composting, anaerobic digestion and conversion of waste to energy like the use of microbial fuel cells have been inducted in this book. This book is intended to be of substantial help to the students and researchers, industrialists and academicians working in the field of environmental engineering and sciences. This can also be of use to many universities/institutes worldwide which have developed new prospectuses and offer specialized courses on environmental and sustainable development. Finally, this book will also be of use to all the policy makers; government or private organizations for proper design and operation of solid waste management programmes.

Guwahati, Assam, India Ajay S. Kalamdhad

Contents

Chapter 1
Utilisation of Recycled Slaughterhouse Wastes for Vegetable Cultivation in Rural India

Shantanu Bhunia, Ankita Bhowmik, Rambilash Mallick, and Joydeep Mukherjee

1 Introduction

In recent years, per capita consumption of meat in India is increasing with the growing population and urbanisation. India ranks as the fourth largest consumer of beef meat worldwide according to the USDA FAS Livestock and Poultry: World Markets and Trade Report (2019). The data from surveys on slaughterhouses revealed that there are approximately 3600 authorised (legal) and 32,000 unlicensed (illegal) slaughterhouses in India, most of these slaughterhouses grew in the rural areas over the last 72 years, generating large amounts of organic wastes due to the increasing demand for buffalo meat (Kennedy et al., 2018). These hazardous wastes are generally disposed off without any treatment directly to the landfill sites, which may affect the environment as pollutants and poses a health risk.

The safe disposal of slaughterhouse residues mainly bovine blood and rumen digesta has become an economical as well as an environmental problem in many developing countries. The wastes are characterised by the presence of high organic content usually composed of protein, lipid and small amounts of carbohydrate that can lead to a high methane potential. Such organic wastes are a potential reservoir of pathogenic microorganisms infecting both humans and animals, and their improper disposal through open dumping or landfilling creates environmental as well as human health hazards (Franke-Whittle and Insam, 2013). Different treatment methods for the safe disposal of hazardous and highly polluting slaughterhouse wastes exist including anaerobic digestion, acid treatment, burning, composting, rendering and incineration. These sophisticated and expensive waste

S. Bhunia · A. Bhowmik · J. Mukherjee (✉)
School of Environmental Studies, Jadavpur University, Kolkata, India
e-mail: joydeep.mukherjee@jadavpuruniversity.in

R. Mallick
Department of Agronomy, University of Calcutta, Kolkata, India

A. S. Kalamdhad (ed.), *Integrated Approaches Towards Solid Waste Management*, https://doi.org/10.1007/978-3-030-70463-6_1

management technologies are difficult to apply in rural abattoirs of developing countries like India due to poor infrastructural facilities and policies. Proper treatments may improve waste quality, reduce its phytotoxicity and could be used as an organic nitrogen fertiliser in agriculture (Salminen et al., 2001).

After sufficient heat treatment, highly polluted slaughterhouse wastes (blood and rumen content of bovine animals) were converted to a non-hazardous organic product, which provided nitrogen (49,440.64 mg N kg^{-1}), phosphorus (1628.75 mg P kg^{-1}) and potassium (9367.94 mg K kg^{-1}) as a major source of nutrients for the cultivation of hybrid vegetables (tomato, brinjal and chili) and fruits in India (Roy et al., 2013). Application of dried slaughterhouse wastes (68 kg N ha^{-1} of soil) during the field cultivation of tomato in India showed better growth and higher productivity (33 t ha^{-1}) as compared to the commercially available diammonium phosphate (DAP) (Roy et al., 2016). These agronomic practices may improve the structure, fertility and water-holding capability of soil for the better growth of plants as well as reduce landfill waste disposal and uses of chemical fertilisers. In few studies, composted slaughterhouse wastes were utilised as an organic manure for the cultivation of bell pepper in Mexico (Llaven et al., 2008); maize, mustard and triticale in Hungary (Ragalyi and Kadar, 2012) and soybean in Brazil (Nunes et al., 2015), which showed higher productivity and better fruit characteristics. The residual effects of composted slaughterhouse wastes had also been studied by Ragalyi and Kadar (2012) who observed a significant residual effect even after three to four years of treatment. In this study, 'bovine-blood-rumen-digesta-mixture' (BBRDM) has been utilised as an organic nitrogen fertiliser for the successive cultivations of cowpea and amaranth. The residual fertility of BBRDM treated soils was determined following similar approaches by conducting successive pot cultivation of amaranth after 90 days of bell pepper study, which was the subject of our previous communication.

Composting is an organic waste disposal method that allows biological decomposition of organic materials to conserve and recycle nutrients essential for the plant growth and make an application in soils possible. Composting is a time-consuming and tedious process with several difficulties (particularly stated for windrow composting) such as high moisture problem, reduced nutrient availability of the composted product, foul smell and greenhouse gas emissions (Tritt and Schuchardt, 1992). On the other hand, drying has no such negative impacts on the atmosphere. Recycling of bovine blood and rumen digesta through sun drying to fertilise vegetable crops reported by Roy et al. (2013) maybe an environment-friendly and cost-effective waste management technology for the rural slaughterhouses of South 24 Parganas district, West Bengal. Sufficient heat treatment can destroy the major slaughterhouse pathogens such as *Staphylococcus*, *Bacillus*, *Brucella, Salmonella, Clostridium*, *Mycobacterium*, *Escherichia*, *Erysipelothrix*, etc. (Roy et al., 2015), as well as reduce CO_2, N_2O and CH_4 emissions from pharmaceutical as well as other organic wastes due to removal of moisture content (Majumdar et al., 2006).

Very few studies on recycling and reuse of slaughterhouse wastes have been reported to date. For this reason, the present study was conducted to carry out recycling of waste blood and rumen content of bovine animals into an organic nitrogen fertiliser using a designed tray-dryer system and its application in agriculture as an alternative to commercial fertilisers for the cultivation of cowpea and amaranth in India as well as effects of BBRDM on soil health and climate change.

2 Materials and Methods

2.1 Recycling of Bovine Blood and Rumen Content

Waste blood of cattle and rumen digesta were taken immediately after slaughtering of bovine animals from the rural abattoirs of Magrahat village, South 24 Parganas (India) for further processing. A mixture of bovine blood and rumen waste was prepared in a ratio of 3:1 according to Roy et al. (2013) and dried using 10 trays (having a surface area of 7680 cm^2 for each tray), in a modelled tray dryer at 120 °C for 6–8 hours to obtain BBRDM. The heating load of the dryer was 8 kW.

2.2 Quality Assessment of BBRDM

The following were measured for BBRDM: pH by electrochemical method, total Kjeldahl nitrogen (TKN) by conventional Kjeldahl method, total phosphorus (TP) by vanado-molybdate method and potassium (K) by flame photometric method (Radojevic and Bashkin, 1999).

The surface morphology and elemental composition of BBRDM were studied by ZEISS EVO 18 scanning electron microscopy coupled with energy-dispersive X-ray spectroscopy (EDS, Bruker, XFlash 6I30) to determine the quality of BBRDM as fertiliser (Sharma et al., 2019).

2.3 Performance Evaluation of BBRDM

The fertilising potential of BBRDM was evaluated by conducting a pot experiment during the last monsoon (June to September 2019). This experiment was performed under a shed at the rooftop of the Oceanography building, Jadavpur University (22°N and 88°E) following a completely randomised block design with four replications.

Four sets of six pots, a total of 24 pots (16 cm diameter × 18 cm depth) were filled with (1) soil of Heria village (near Magrahat II block), (2) soil + DAP (N/P/K = 18:46:0), (3) soil + vermicompost (N/P/K = 4:1:1), (4) soil + LDB (low dose of BBRDM), (5) soil + RDB (recommended dose of BBRDM) and (6) soil + HDB (high dose of BBRDM) and were used for the cultivation of cowpea (*Vigna unguiculata* L.). The soil was taken from Heria village of Magrahat II block, South 24 Parganas district (India) and dried at room temperature for 10–15 days. The soil type was loamy (brownish-grey; composed of 42% of sand, 38% of silt and 20% of clay; 1.2% organic matter; pH 6.4). Seedlings of cowpea obtained from a local nursery located in Santoshpur (Kolkata) were planted within the top 3–5 cm of soil, one seedling in each pot. The applied doses of BBRDM were 3, 6 and 9 g kg^{-1} of soil determined according to 45:75:75 kg NPK ha^{-1} as recommended for hybrid cowpea production, where concentration of available N was maintained invariable for all the fertilisers applied during our cultivation. The minimum and maximum mean rooftop temperature during the cultivation season ranged from 26 °C to 33 °C.

The residual effects of dried slaughterhouse wastes were analysed by conducting another pot cultivation of amaranth following similar block design method with four replications and yield characteristics were assessed after 45 days of the seed sowing.

2.4 *Soil Nutrient Availability During Cultivation*

Available N (NH_4^+), inorganic phosphorus (PO_4^{3-}) and organic carbon (C) of soil were measured every 2 weeks after fertilisation according to Subbiah and Asija (1956), Bray and Kurtz (1945) and Walkely and Black (1934), respectively.

2.5 *Quantification of Emitted Methanes*

Soil samples from six different treatments were taken after few days of fertilisation, filled upto 45% in 40 mL borosilicate screw cap vials, and kept at laboratory temperature under dark for 3 days (Chan and Parkin, 2001). In 5 mL aerobic headspace, emitted methane from the soils of cultivation pots and waste dumping sites was measured using SYSTRONICS GC-8205 gas chromatograph equipped with flame ionisation detector (GC-FID) (column temperature 50 °C, detector temperature 140 °C, carrier gas H_2 and flow rate 30 mL/min). Methane fluxes from cultivation and waste dumping sites were determined according to Khoiyangbam et al. (2004) with a few modifications. All the experiments were made in duplicate sets.

2.6 *Study of Soil Microbial Communities*

Soil samples (approximately 20 g fresh weight) for microbiological analysis, were taken from the 3–5 cm top layer of six different soil treatments (four rows and six columns) before and during the cultivation of hybrid cowpea (4th, 8th, 12th, 16th and 20th weeks), sieved (<5 mm) and stored at 4 °C. The abundance (in terms of cfu

mL^{-1}) of actinomycetes, bacteria, free living nitrogen-fixing soil *Azotobacter*, phosphate-solubilizing bacteria (PSB) and fungi was measured according to the serial dilution agar plate method of Jett et al. (1997) using nutrient agar (NA), Ashbys mannitol agar (AM), Pikovskaya's agar (PVK), Actinomycete isolation agar (AIA) and Cooke Rose Bengal (CRB), respectively. Cyanobacteria were cultivated using BG-11 broth under the light intensity of 50 μmol photons/m^2/s (12 h/day for 30 days), methanolic extracts were prepared after centrifugation of cyanobacterial mass and concentration of chlorophyll-a (μg g^{-1} of soil) was determined at 663 nm according to Pramanik et al. (2011).

2.7 Statistics

All the statistical analyses were carried out using a one-way analysis of variance (ANOVA) in SPSS (Chicago, USA) for Windows version 16.0 combined with Tukey's *post-hoc* test. Different soil treatments were significant at a 5% cut-off level.

3 Results and Discussion

3.1 Process Description, Technology Upgradation and BBRDM Characteristics

Globally, blood and undigested food of ruminants are considered as major slaughterhouse wastes. Such organic wastes are highly polluted and soggy and by their nature require a careful drying to reduce moisture content, growth of pathogenic microorganisms and other deteriorative reactions, which may prolong the shelf-life of recycled products (Mujumdar and Law, 2010).

Previously, our group (Roy et al., 2013) established a cost-effective recycling method for rural slaughterhouse wastes, where the mixture of cattle blood and rumen content (in a ratio of 3:1) was boiled using a wood-burning stove for 90 min followed by sun drying for 3 successive days to obtain BBRDM. Several problems such as high moisture content, foul smell, massive loss of available nutrients, infestation of insects and birds and an abundance of airborne pathogens stated for the sun drying of agricultural as well as other organic wastes (Misha et al., 2013). The traditional sun drying is open, climate dependent and time-consuming method generally practiced in rural areas of developing countries, and also occupies a large space. On the other hand, a tray-drying process may solve all the drawbacks and complications observed during the sun drying of slaughterhouse wastes. The process for recycling of blood and rumen contents (Fig. 1) collected after slaughtering of bovine animals from the rural abattoirs of Magrahat village, South 24 Parganas district of West Bengal state using a tray-dryer system (Fig. 2) following a simple principle of

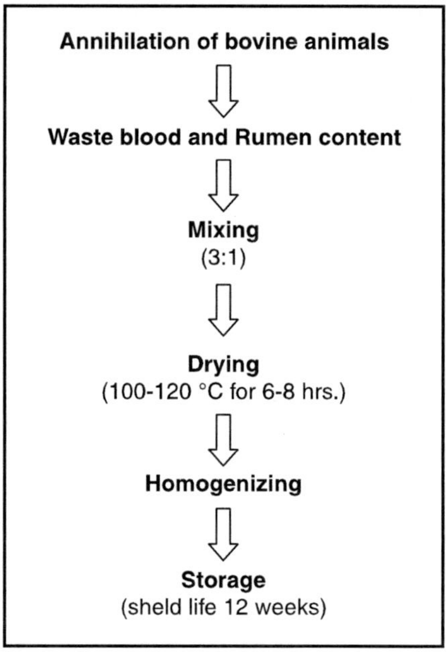

Fig. 1. Production process of bovine-blood-rumen-digesta-mixture (BBRDM).

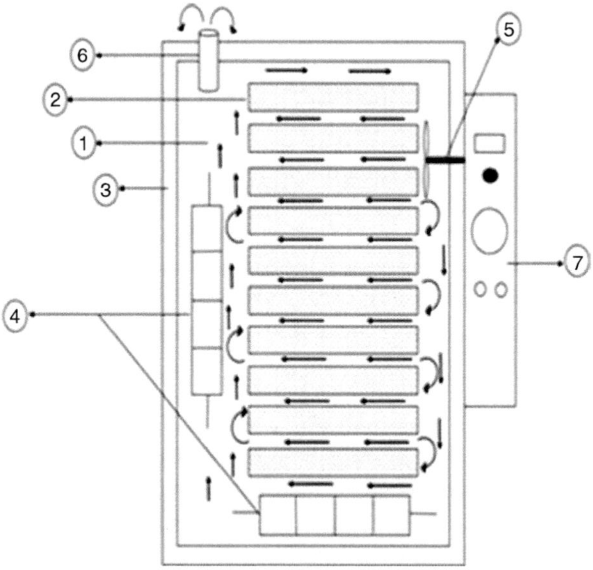

Fig. 2. Schematic diagram of 10 trays modelled tray dryer, where: (1) drying cabinet, (2) tray, (3) insulator, (4) heating coils, (5) fan, (6) exhaust outlet and (7) control panel.

convective heating has already been discussed in sub-section 'Recycling of bovine blood and rumen content' under 'Materials and Methods'.

Recycled bovine-blood-rumen-digesta-mixture was olive-brown (2.5Y 4/3) in colour, pH ranged between 7.0 and 8.0 and contained 16% of moisture after sufficient heat treatment. The N/P/K ratio of BBRDM was about 8:1:2, where the C/N ratio was 4.68 (Fig. 3).

3.2 N Fertilisation, Methane Emissions and Yield Performance

Proper doses and timing of fertilisation may affect the quality and yield of vegetable crops. The application of organic fertiliser improves plant health and maximises yield potential either by supplying nutrients directly to plants or by modifying the microbial status of soil. The consecutive supply of available N and P for the better growth and development of vegetable crops is required. From the previous studies of Roy et al. (2013, 2015, 2016), we can consider BBRDM as an organic nitrogen fertiliser having a 4.68 C/N ratio and contains several micronutrients like calcium, boron, zinc, iron essential for the improvement of vegetable crops. During the cultivation of cowpea, low (3 g BBRDM kg^{-1} of soil) and recommended (6 g BBRDM kg^{-1} of soil) doses of BBRDM fertilisation showed highest yield, although plants fertilised with high dose of BBRDM (9 g BBRDM kg^{-1} of soil) died may be due to higher accumulation of nitrogen especially ammonia (Table 1). We observed similar effects (09 in S, 16 in S + DAP, 21 in S + VC, 32 in S + LDB and 37 in S+ RDB g yield/tub) from residual soils of different fertiliser treatments (DAP, vermicompost and BBRDM) when pot cultivation of amaranth was carried out even after 90 days of cowpea cultivation. Similar results were found by Olayinka et al. (1998), Adediran et al. (2005) and Zhang et al. (2014) according to whom application of animal manures increased the yield potential of cowpea plants and has become an alternative to conventional agronomic practices. The plants fertilised with recycled slaughterhouse wastes (BBRDM) initiated about 2 weeks early flowering and fruiting in comparison with DAP as well as organic vermicompost probably due to the presence of boron. During pot cultivations of solanaceous vegetables, Roy et al. (2013)

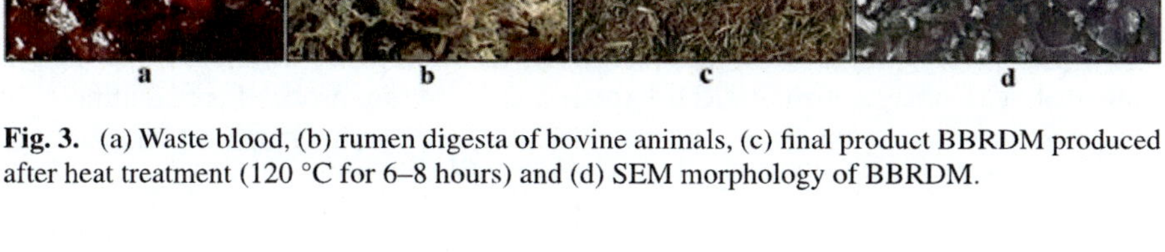

Fig. 3. (a) Waste blood, (b) rumen digesta of bovine animals, (c) final product BBRDM produced after heat treatment (120 °C for 6–8 hours) and (d) SEM morphology of BBRDM.

Table 1. Pot cultivation of cowpea under organic and inorganic fertilisation regimes

	Fertilisation treatments					
Yield parameters	*S*	*DAP*	*VC*	*LDB*	*RDB*	*HDB*
Plant height (cm)	112^{e}	156^{c}	139^{d}	182^{a}	168^{b}	–
Number of leaves	32^{d}	66^{b}	53^{c}	78^{a}	69^{b}	–
Leaf surface area (cm^2)	46^{c}	53^{a}	50^{b}	54^{a}	53^{a}	–
Stem diameter (mm)	4.7^{d}	5.2^{c}	4.9^{d}	6.7^{a}	6.3^{b}	–
Number of fruits	14^{d}	22^{c}	21^{c}	32^{a}	27^{b}	–
Chlorophyll a + b (mg cc^{-1})	0.8^{e}	1.6^{c}	1.2^{d}	2.4^{a}	1.9^{b}	–

The superscripts within each row showing significant differences among different fertilisation treatments at 0.05 cut-off level. S: soil, DAP: diammonium phosphate (CF), VC: vermicompost, LDB: BBRDM (low dose), RDB: BBRDM (recommended dose) and HDB: BBRDM (high dose).

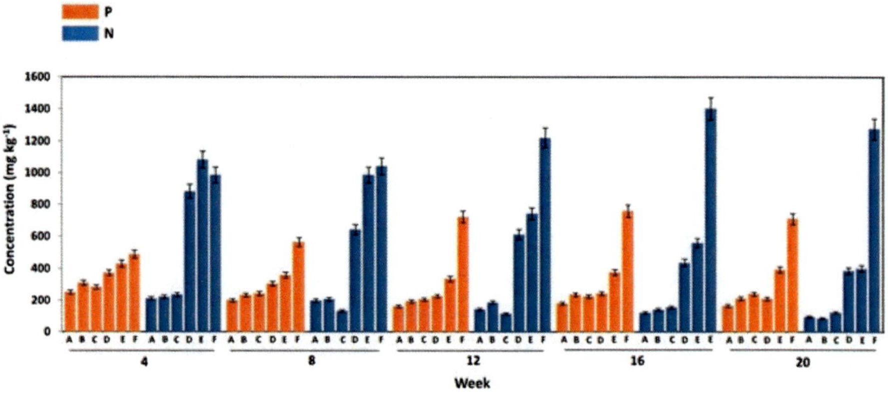

Fig. 4. Rapid utilisation of available soil N during hybrid cowpea (*Vigna unguiculata* L.) cultivation as compared to P, determined according to Subbiah and Asija (1956) and Bray and Kurtz (1945), respectively. A: soil (control), B: soil + diammonium phosphate (DAP), C: soil + vermicompost (VC), D: soil + low dose of BBRDM (LDB), E: soil + recommended dose of BBRDM (RDB) and F: soil + high dose of BBRDM (HDB). Mean ± SD ($n = 6$)

found an early induction of flowering and fruiting in BBRDM fertilised tomato, chili and brinjal plants. In addition, rapid utilisation of nitrogen in organically cultivated soils was observed by Marinari et al. (2010), Galvez et al. (2012) and Roy et al. (2016) during both the pot and field cultivations of solanaceous vegetable crops. Similar trend of N mineralisation was noticed recently in case of our hybrid cowpea production system (Fig. 4). Emissions of methane from the agricultural fields increased significantly with the increasing nitrogen applications in association with a supply of organic C to soils, which encourages methane, nitrous oxide as well as carbon dioxide emissions (Cai et al., 2007). During our investigation, we found a methane flux of 20.84 $\mu g\ g^{-1}\ h^{-1}$ from the open abattoir waste dumping sites of Magrahat village, which was approximately a hundred and fifty-five times higher than that was observed in BBRDM treated soils. After proper heat treatment, a reduction in methane emissions was recorded from organic slaughterhouse wastes probably due to loss of moisture as reported by Majumdar et al. (2006) in case of pharmaceutical and other organic wastes, which makes BBRDM suitable for soil application.

3.3 *Microbial Abundance Under DifferentFertilisation Regimes*

As shown in Table 2, populations (in terms of cfu mL^{-1}) of bacteria, nitrogen-fixing soil *Azotobacter*, phosphobacteria, cyanobacteria and fungi in soils fertilised with BBRDM were obtained higher than the populations from DAP treated soils, while actinobacteria became dominant in chemically amended soils. Similar findings have been reported by Tu et al. (2006), Chaudhry et al. (2012), Roy et al. (2013) and Li et al. (2019), where addition of organic amendments to soils resulted in increase of copiotrophic populations, enzyme activities and availability of soil nutrients essential for plant growth and development.

Soil microbes are an integral part of agricultural ecosystems involved in organic matter decomposition, elemental transformation and nutrient assimilation as well as regulation of plant growth and combating harmful pathogens present in soil. Organic fertilisation can improve the health of plant-soil systems by supplying essential nutrients, protecting soil enzymes and through increasing the numbers of beneficial soil microbes. N appears in both inorganic and organic forms in soil, although plants uptake only biologically available nitrate (NO_3^-) and ammonium (NH_4^+), the inorganic forms of N. The biological transformation of organic N to ammonia followed by the oxidation of nitrite to meet crop's need by the fastidious ammonia and nitrite-oxidising bacteria affiliated to phyla *Proteobacteria*, *Planctomycetes*, *Chloroflexi* and *Nitrospirae* helps the plants to achieve sustainable productivity (Han et al., 2018). Higher abundance of nitrogen-fixing *Azotobacter*, phosphate-solubilising

Table 2. Microbial abundance during hybrid cowpea cultivation

Soil microorganisms	*Fertilisation treatments*					
	S	*DAP*	*VC*	*LDB*	*RDB*	*HDB*
Bacteria ($\times 10^7$ mL^{-1}) in NA	18^d	15^e	24^c	38^b	46^a	39^b
Azotobacter ($\times 10^7$ mL^{-1}) in AM	4^e	2^f	5^d	8^c	12^a	9^b
PSB ($\times 10^7$ mL^{-1}) in PVK	8^d	5^e	12^c	17^b	20^a	19^a
Actinomycetes ($\times 10^7$ mL^{-1}) in AIA	10^e	15^a	12^c	14^b	11^d	11^d
Fungi ($\times 10^4$ mL^{-1}) in CRB	6^e	4^f	9^d	14^c	17^b	18^a
Cyanobacteria (µg chl-a g^{-1} of soil)	0.62^f	1.53^e	2.04^c	2.21^b	2.59^a	1.98^d

The superscripts within each row showing significant differences among different fertilisation treatments at a 5% cut-off level. S: soil, DAP: diammonium phosphate (CF), VC: vermicompost, LDB: BBRDM (low dose), RDB: BBRDM (recommended dose) and HDB: BBRDM (high dose). PSB: phosphate-solubilising bacteria, NA: nutrient agar medium, AM: Ashby's mannitol agar, PVK: Pikovskaya's agar medium, AIA: Actinomycetes isolation agar and CRB: Cooke Rose Bengal.

bacteria and fungi from the rhizosphere of organically cultivated cowpea indicates that such microbial communities are proficient to absorb supplied organic matters. This may provide essential nutrients to soils like nitrogen and phosphorus by increasing urease and phosphatase activities, respectively, as well as producing growth-promoting hormonal substances for plants. These results are in confirmation with the previous findings of Chang et al. (2007), Mikanova et al. (2009) and Roy et al. (2013) who observed an increased abundance of *Azotobacter* and phosphate solubilisers during the cultivation of vegetable crops. On the other hand, growth and abundance of soil *Actinobacteria* were maximum in chemically cultivated soils probably due to low carbon supply. Similar observations have been stated by Wang et al. (2017) and Li et al. (2017) who found a recovery of oligotrophs with the corresponding decrease of organic C and TN in paddy soils.

The abundance, diversity and composition of soil microbial communities affected positively by the long-term application of organic fertilisers derived from animal sources may be due to high organic C and N content of organic manures, which support copiotrophs to proliferate fast (Wang et al., 2017). However, complex responses from microorganisms of agricultural soils to organic fertilisation are very difficult to understand (Morris and Blackwood, 2015).

4 Conclusions

Recycling of bovine blood and rumen digesta using a designed tray-dryer system suggesting an environment-friendly and cost-effective alternative for the safe disposal of hazardous slaughterhouse wastes, which would maintain a clean and healthy environment around the rural abattoirs, and recycled slaughterhouse wastes can be a replacement for chemical fertilisers.

Acknowledgements This research work was financially supported by the Department of Higher Education, Science & Technology and Biotechnology (DHESTBT), Government of West Bengal [ST/P/S&T/1G-81/2017]. Authors are thankful to Dr. Anupam Debsarkar, Associate Professor of Civil Engineering, Jadavpur University and Dr. Subrata Hait, Associate Professor of Civil and Environmental Engineering, Indian Institute of Technology Patna for their valuable suggestions and Mr. Anarul Gazi for providing the slaughterhouse wastes.

References

Adediran, J.A., Taiwo, L.B., Akande, M.O., Sobulo, R.A. and Idowu, O.J. (2005). Application of organic and inorganic fertilizer for sustainable maize and cowpea yields in Nigeria. *J Plant Nutr*, **27(7):** 1163–1181. https://doi.org/10.1081/PLN-120038542.

Bray, R.H. and Kurtz, L.T. (1945). Determination of total, organic, and available forms of phosphorus in soils. *Soil Sci*, **59(1):** 39–46.

Cai, Z., Shan, Y. and Xu, H. (2007). Effects of nitrogen fertilization on CH_4 emissions from rice fields. *Soil Sci Plant Nutr*, **53(4):** 353–361. https://doi.org/10.1111/j.1747-0765.2007.00153.x.

Chan, A.S.K. and Parkin, T.B. (2001). Methane oxidation and production activity in soils from natural and agricultural ecosystems. *J Environ Qual*, **30(6):** 1896–1903. https://doi.org/10.2134/jeq2001.1896.

Chang, E.H., Chung, R.S. and Tsai, Y.H. (2007). Effect of different application rates of organic fertilizer on soil enzyme activity and microbial population. *Soil Sci Plant Nutr*, **53(2):** 132–140. https://doi.org/10.1111/j.1747-0765.2007.00122.x.

Chaudhry, V., Rehman, A., Mishra, A., Chauhan, P.S. and Nautiyal, C.S. (2012). Changes in bacterial community structure of agricultural land due to long-term organic and chemical amendments. *Microb Ecol*, **64(2):** 450–460. https://doi.org/10.1007/s00248-012-0025-y.

Franke-Whittle, I.H. and Insam, H. (2013). Treatment alternatives of slaughterhouse wastes, and their effect on the inactivation of different pathogens: A review. *Crit Rev Microbiol*, **39(2):** 139–151. https://doi.org/10.3109/1040841X.2012.694410.

Galvez, A., Sinicco, T., Cayuela, M.L., Mingorance, M.D., Fornasier, F. and Mondini, C. (2012). Short term effects of bioenergy by-products on soil C and N dynamics, nutrient availability and biochemical properties. *Agric Ecosyst Environ*, **160:**3–14. https://doi.org/10.1016/j.agee.2011.06.015.

Han, S., Li, X., Luo, X., Wen, S., Chen, W. and Huang, Q. (2018). Nitrite-oxidizing bacteria community composition and diversity are influenced by fertilizer regimes, but are independent of the soil aggregate in acidic subtropical red soil. *Front Microbiol*, **9(885):** 1–12. https://doi.org/10.3389/fmicb.2018.00885.

Jett, B.D., Hatter, K.L., Huycke, M.M. and Gilmore, M.S. (1997). Simplified agar plate method for quantifying viable bacteria. *Biotechniques*, **23(4):** 648–650.

Kennedy, U., Sharma, A. and Phillips, C.J. (2018). The Sheltering of unwanted cattle, experiences in India and implications for cattle industries elsewhere. *Animals*, **8(64):** 1–8. https://doi.org/10.3390/ani8050064.

Khoiyangbam, R.S., Kumar, S., Jain, M.C., Gupta, N., Kumar, A. and Kumar, V. (2004). Methane emission from fixed dome biogas plants in hilly and plain regions of northern India. *Bioresour Technol*, **95(1):** 35–39. https://doi.org/10.1016/j.biortech.2004.02.009.

Li, F., Chen, L., Zhang, J., Yin, J. and Huang, S. (2017). Bacterial community structure after long-term organic and inorganic fertilization reveals important associations between soil nutrients and specific taxa involved in nutrient transformations. *Front Microbiol*, **187(8):** 1–12. https://doi.org/10.3389/fmicb.2017.00187.

Li, W., Liu, M., Wu, M., Jiang, C., Kuzyakov, Y., Gavrichkova, O., Feng, Y., Dong, Y. and Li, Z. (2019). Bacterial community succession in paddy soil depending on rice fertilization. *Appl Soil Ecol*, **144:** 92–97. https://doi.org/10.1016/j.apsoil.2019.07.014.

Llaven, M.A.O., Jimenez, J.L.G., Coro, B.I.C., Rincon-Rosales, R., Molina, J.M., Dendooven, L. and Gutierrez-Miceli, F.A. (2008). Fruit characteristics of bell pepper cultivated in sheep manure vermicompost substituted soil. *J Plant Nutr*, **31(9):** 1585–1598. https://doi.org/10.1080/01904160802244738.

Majumdar, D., Patel, J., Bhatt, N. and Desai, P. (2006). Emission of methane and carbon dioxide and earthworm survival during composting of pharmaceutical sludge and spent mycelia. *Bioresour Technol*, **97(4):** 648–658. https://doi.org/10.1016/j.biortech.2005.03.015.

Marinari, S., Lagomarsino, A., Moscatelli, M.C., Tizio, A. and Campiglia, E. (2010). Soil carbon and nitrogen mineralization kinetics in organic and conventional three-year cropping systems. *Soil Tillage Res*, **109(2):** 161–168. https://doi.org/10.1016/j.still.2010.06.002.

Mikanova, O., Friedlova, M. and Simon, T. (2009). The influence of fertilization and crop rotation on soil microbial characteristics in the long-term field experiment. *Plant Soil Environ*, **55(1):** 11–16.

Misha, S., Mat, S., Ruslan, M.H., Sopian, K. and Salleh, E. (2013). Review on the application of a tray dryer system for agricultural products. *World Appl Sci J*, **22(3):** 424–433. https://doi.org/10.5829/idosi.wasj.2013.22.03.343.

Morris, S.J. and Blackwood, C.B. (2015). The ecology of the soil biota and their function. *In:* Paul, E.A. (ed.). Soil Microbiology, Ecology and Biochemistry, 4th edn. Elsevier, Boston (USA), pp. 294–309.

Mujumdar, A.S. and Law, C.L. (2010). Drying technology: Trends and applications in postharvest processing. *Food Bioprocess Tech*, **3(6):** 843–852. https://doi.org/10.1007/s11947-010-0353-1.

Nunes, W.A.G.D.A., Menezes, J.F.S., Benites, V.D.M., Lima Junior, S.A.D. and Oliveira, A.D.S. (2015). Use of organic compost produced from slaughterhouse waste as fertilizer in soybean and corn crops. *Sci Agric*, **72(4):** 343–350. https://doi.org/10.1590/0103-9016-2014-0094.

Olayinka, A., Adetunji, A. and Adebayo, A. (1998). Effect of organic amendments on nodulation and nitrogen fixation by cowpea. *J Plant Nutr*, **21(11):** 2455–2464. https://doi.org/10.1080/01904169809365577.

Pramanik, A., Sundararaman, M., Das, S., Ghosh, U. and Mukherjee, J. (2011). Isolation and characterization of cyanobacteria possessing antimicrobial activity from the Sundarbans, the world's largest tidal mangrove forest. *J Phycol*, **47(4):** 731–743. https://doi.org/10.1111/j.1529-8817.2011.01017.x.

Radojevic, M. and Bashkin, V.N. (1999). Practical Environmental Analysis. Royal Society of Chemistry, Cambridge.

Ragalyi, P. and Kadar, I. (2012). Effect of organic fertilizers made from slaughterhouse wastes on yield of crops. *Arch Agron Soil Sci*, **58(1):** 122–126. https://doi.org/10.1080/03650340.2012.695863.

Roy, M., Das, R., Debsarcar, A., Sen, P.K. and Mukherjee, J. (2016). Conversion of rural abattoir wastes to an organic fertilizer and its application the field cultivation of tomato in India. *Renew Agric Food Syst*, **31(4):** 350–360. https://doi.org/10.1017/S1742170515000289.

Roy, M., Das, R., Kundu, A., Karmakar, S., Das, S., Sen, P., Debsarcar, A. and Mukherjee, J. (2015). Organic cultivation of tomato in India with recycled slaughterhouse wastes: evaluation of fertilizer and fruit safety. *Agriculture*, **5(3):** 826–856. https://doi.org/10.3390/agriculture5030826.

Roy, M., Karmakar, S., Debsarcar, A., Sen, P.K. and Mukherjee, J. (2013). Application of rural slaughterhouse waste as an organic fertilizer for pot cultivation of solanaceous vegetables in India. *Int J Recycl Org Waste Agric*, **2(6):** 1–11. https://doi.org/10.1186/2251-7715-2-6.

Salminen, E., Rintala, J., Harkonen, J., Kuitunen, M., Hogmander, H. and Oikari, A. (2001). Anaerobically digested poultry slaughterhouse wastes as fertiliser in agriculture. *Bioresour Technol*, **78(1):** 81–88. https://doi.org/10.1016/S0960-8524(00)00160-7.

Sharma, A., Ganguly, R. and Gupta, A. (2019). Spectral characterization and quality assessment of organic compost for agricultural purposes. *Int J Recycl Org Waste Agric*, **8(2):** 197–213. https://doi.org/10.1007/s40093-018-0233-7.

Subbiah, B.V. and Asija, G.L. (1956). A rapid procedure for determination of available nitrogen in soils. *Curr Sci*, **25:** 259–260.

Tritt, W.P. and Schuchardt, F. (1992). Materials flow and possibilities of treating liquid and solid wastes from slaughterhouses in Germany: A review. *Bioresour Technol*, **41(3):** 235–245. https://doi.org/10.1016/0960-8524(92)90008-L.

Tu, C., Ristaino, J.B. and Hu, S. (2006). Soil microbial biomass and activity in organic tomato farming systems: Effects of organic inputs and straw mulching. *Soil Biol Biochem*, **38(2):** 247–255. https://doi.org/10.1016/j.soilbio.2005.05.002.

Walkely, A.J. and Black, I.A. (1934). Estimation of soil organic carbon by the chromic acid titration method. *Soil Sci*, **37(1):** 29–38.

Wang, J., Song, Y., Ma, T., Raza, W., Li, J., Howland, J.G., Huang, Q. and Shen, Q. (2017). Impacts of inorganic and organic fertilization treatments on bacterial and fungal communities in a paddy soil. *Appl Soil Ecol*, **112:** 42–50. https://doi.org/10.1016/j.apsoil.2017.01.005.

Zhang, G., Chen, Z., Zhang, A., Chen, L., Wu, Z. and Ma, X. (2014). Phosphorus composition and phosphatase activities in soils affected by long-term application of pig manure and inorganic fertilizers. *Commun Soil Sci Plan*, **45(14):** 1866–1876. https://doi.org/10.1080/00103624.2014.909831.

Chapter 2
Operation and Maintenance of Wash Infrastructure in Residential Schools in Nashik Tribal Belt of Maharashtra: A Case Study

Abhijit Aawari, Bini Samal, Abani Kumar Nayak, and Jitendra Katre

1 Introduction

Water, sanitation and hygiene (WASH) in school has emerged as the key milestone for achieving the Sustainable Development Goals 4 and 6. Apart from providing necessary infrastructure to children and WASH facilities for institutions, behaviour change, communication plays vital role in sustaining the WASH. Schooling and learning are two different aspects, which follow same core values, but the impact and output vary widely. The World Development Report, 2018 highlights a global learning crisis and notes that, while school enrolment has improved, learning outcomes remain poor in many parts of the world. Because of this crisis, 250 million primary school-age children (38%) are currently failing to learn basic numeracy and literacy, and just one in 10 young people in low-income countries are on track to gain basic secondary skills by 2030.

Apart from the lack in quality education insufficient sanitation facilities poses major hurdle to develop a healthy learning environment for children. According to Joint Monitoring Programme, Global Baseline Report, 2018, highlights that in India, one toilet is available for every 54 girls and 63 boys in comparison to the adequacy norm of one toilet per 15 to 20 students. Only 36% of schools have functional incinerators for disposal of sanitary waste and only 69% schools have available drinking water for students considering the disparity in urban and private schools. The availability of WASH facilities for peri-urban and rural areas decreases considerably. The situation worsens in schools for tribal and marginalised communities.

Operation and maintenance (O&M) activities in school, which encompass not only technical issues, but also managerial, social, financial and institutional issues,

A. Aawari (✉) · B. Samal · A. K. Nayak · J. Katre
WaterAid, New Delhi, India
e-mail: AbhijitAawari@wateraid.org

A. S. Kalamdhad (ed.), *Integrated Approaches Towards Solid Waste Management*, https://doi.org/10.1007/978-3-030-70463-6_2

must be directed towards the elimination or reduction of the major constraints, which prevent the achievement of sustainability (Brikké, 2000).

In school, O&M is a crucial element of sustainability, and a frequent cause of failure of water supply and sanitation service facilities in the past. Many failures are not technical ones. They may result from poor planning, inadequate cost recovery or the outreach inadequacies of centralised agencies (WASH Facilities in Schools of West Bengal and Bangladesh, Operation and Maintenance manual).

Well maintained WASH facilities can reduce girls' absenteeism by up to 20% as well as reduce the incidence of diarrhoea by up to 40% in both boys and girls.

With limited available facilities and over burdening, disparity in services and larger load on existing infrastructure leads to frequent wear and tear with immediate need for repair or replacement. Currently, there is no such system available for operation and maintenance of WASH infrastructure apart from conventionally followed 'on-call' basis method. This case study explores the provision for operation and maintenance in residential schools for tribal students in the state of Maharashtra called as 'Ashramshalas'.

In Maharashtra, there are 552 Ashramshalas in 28 districts, with a total of 1,87,392 students studying in 2016–2017. The Government of Maharashtra, along with development partners, is working to ensure that these tribal residential schools have the requisite functional water, sanitation and hygiene facilities, thereby providing children with a conducive learning and living environment. Each Ashramshala should provide free residential facilities, food, school uniforms, bedding, books and stationery items to every child. Additionally, the Ashramshala facilities should also include nutritious meals, access to clean drinking water, sanitation and basic health care. Although these provisions are mandatory, the implementation varies.

'Promoting WASH Compliant Ashramshalas in Maharashtra', a Corporate Social Responsibility (CSR) initiative of the National Stock Exchange Foundation seeks to work in 172 Ashramshalas in Nashik Additional Tribal Commissioner (ATC), Tribal Development Department (TDD), Government of Maharashtra. Over a three-year period, the project's primary objective is to facilitate creation of appropriate and adequate water, sanitation and hygiene (WASH) facilities for approximately 70,000 children in 172 Ashramshalas. This includes functional toilets, safe drinking water, clean surroundings, as well as the adoption of critical hygiene practices such as handwashing and menstrual hygiene management. To ensure operation and maintenance of facilities (O&M) for sustained change, strengthening of institutions is also a key focus.

2 Literature Review

O&M encompasses all the activities needed to run a water supply, sanitation and hygiene facilities, but excludes the construction of new facilities. The overall aim of operation and maintenance is to ensure efficiency, effectiveness and sustainability of water supply and sanitation facilities (Castro et al., 2009). O&M goes beyond mere technical definition and includes all managerial aspects necessary to run WASH facilities, i.e. selecting personnel for WASH committees (school management committees and student committees), purchase and management of spare parts and hygiene material, financial management, management of assets and the planning of extensions.

3 Understanding the Meaning and Importance of Operation and Maintenance

Operation refers to the direct access to the system by the user, to the activities of any operational staff (e.g. operators of water pumps), and to the rules or policies, which may be devised to govern who may access the system, when, and under what conditions. Operation deals with ensuring daily smooth functioning of the water supply system and other activities associated with it, e.g. cost of running the system, water testing, electricity supply for pumps etc. (Carter, 2009).

Maintenance refers to the technical activities that are needed to keep the system working. Maintenance requires skills, tools and spare parts. Maintenance deals with upkeep of the machines, system in good working condition and preventing breakdowns, damages to the machinery. Maintenance can be classified into three sub-categories:

- **Preventive or routine maintenance:** It is a routine check of the WASH infrastructure for minor repairs. Under this type of maintenance, monitoring must be taken as part of precautionary measures. It is essential for upkeep of the system. In Ashramshala O&M, this is part of daily, fortnightly and monthly maintenance.
- **Periodic maintenance:** It covers more major repairs, maintenance work such as replacing the faulty parts, oiling of locomotive parts from skilled personnel. Periodic maintenance may be costly, but it is necessary to improve the longevity of the system. In Ashramshala O&M, this is part of annual maintenance.
- **Emergency maintenance (crisis maintenance):** Repair of unexpected damage or breakdown of the system.

Successful O&M requires that WASH needs in institutions are well described and responsibilities are well defined. Further, staff dealing with O&M must be well trained and have the requisite skills to carry out the necessary tasks.

4 Global Scenario of Wash and O&M in Schools

Among the main actions the WASH sector will emphasise is improved access for children to an adequate number of child-friendly WASH facilities, which will also enhance school attendance, especially for girls. This includes increased access to water for drinking and handwashing. Strengthened collaboration with health agencies and institutions will also be encouraged to ensure minimum standards for WASH infrastructure are met in health centres, hospitals and nutritional centres, and that persons with specific needs have access to adapted sanitation facilities (UNHCR Global Strategy for Public Health 2014–2018).

5 Importance of O&M in Curbing Water-Borne Diseases

Diarrheal disease—nearly 90% of which has been attributed to suboptimal water, sanitation and hygiene (WASH)—is one of the largest causes of morbidity and mortality in children under five years of age in low- and middle-income countries, where it kills more children than HIV, malaria and measles combined. WASH interventions aim to prevent and control transmission routes of bacteria (e.g., *Shigella*, *E. coli*) (Ramesh and Blanchet, 2015).

National review: UNICEF WASH for School Children South Asia Report 2012 states that out of total 13,03,829 public schools in India, 2,82,481 need repair for sanitation facilities and 2,57,171 schools need new facilities. This implies nearly 41% schools are in need of new or retrofitting of WASH facilities.

6 O&M in Residential Schools: Challenges and Opportunities

The major O&M related challenges that have been observed in Ashramshalas are as follows:

- Overburdened WASH infrastructure as the toilets and water points are used by the students as well as the nearby community. As a result, a toilet that is meant to be used by 15 students or a water point meant for 20 students, is catering to five times that capacity.
- Lack of trained personnel to carry out regular minor repairs, particularly related to plumbing and masonry work, results in defunct infrastructure.
- Lapses in cleaning as roles and responsibilities may not be clearly defined or cleaning staff may not be regular, and when present, are tasked with cleaning a large Ashramshala campus.
- Lack of adequate guidance on O&M, in terms of tasks/activities to be undertaken, roles and responsibilities, and oversight mechanisms.
- Lack of separate cleaning staff for boys' and girls' toilets in both schools and hostels.
- Poor quality and maintenance of existing infrastructure.
- Non-handing over of newly created infrastructure is forcing the current infrastructure to become defunct.

- No adherence to the standard norms on WASH.
- No proper mechanism in place for regular cleaning of toilets.

7 Methodology/Approach

During the implementation, it was prominently observed that the inadequate maintenance, lack of knowledge and skilled labour to repair or replace the defect was major reason for infrastructure failing to serve its purpose.

To address the issues of O&M in Ashramshala, consultative workshops were organised with all key stakeholders from Tribal Development Department in Nashik. The workshops were attended by head masters, superintendents and representatives from state government (TDD), UNICEF along with KRC partner. Three models for piloting emerged from these workshops.

In all the models, personnel engaged by the partner will visit Ashramshala at least once in a week and inspect each WASH infrastructure. Their scope of work has been kept very simple as to carry out the minor repairs and maintenance of the WASH infrastructure. The material for the work will be provided by the school administration through funds available with the School Management Committee (SMC). If any major repair is required, the HM shall be informed for immediate action.

8 Results

Three different models were piloted namely—Social Service Provider model, Village Entrepreneurship model and Institutional Strengthening model. The details of each model are given below.

9 Social Service Provider (ITI and Youth)

Social Service Provider model is tweaked version of standard Annual Maintenance Contracts for regular maintenance. It is implmeneted in two ways where the technical resource person is hired from either Industrial Training Institutes (ITI) with traits as plumbing, masonary or local youths are identified and trained for the works related to WASH operation and maintenance. ITI model was implemented in a centralised manner and Youth was done with cluster approach.

Model was designed in a way to provide services of O&M to Ashramshalas as Annual Maintenance Contract. The schedule for visits to Ashramshala was prepared at KRC level and every Ashramshala was visited at least once in a week. During the visits, mandatory checking of all infrastructure was carried out and minor repairs

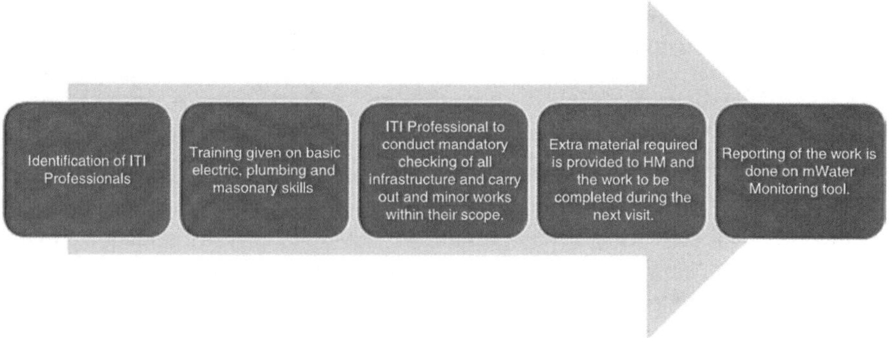

Fig. 1. Methodology for implementing Social Service Provider model.

were carried out as per the scope of work of the personnel engaged. If extra material is required for the replacement such as taps, plumbing material, the assigned personnel will provide list of such required material to Ashramshala and the work will be done during the next visit (Fig. 1).

10 Key Learnings and Observations

- System based model: After establishing the system for regular O&M of WASH infrastructure the model requires minimum handholding support.
- Support to Ashramshalas of skilled technician with knowledge of multiple traits such as masonry, plumbing, tiling, electrification etc.
- Streamline of material procurement from the SMC and utilisation of SMC funds.
- Presence of monitoring agency for logistic support is required.
- Job opportunities for the local youth and ITI students.
- Hands on experience of working in WASH sector for the local youth and ITI.
- Possibility of asking to do work related to other than WASH infrastructure by School Management Committee.

11 Village Entrepreneurship Model

Village Entrepreneurship model depends largely on the integration of ZP schools and Ashramshalas under one umbrella for O&M work on WASH infrastructure. The model comprises of identifying and recruiting a person or an agency from the intervention area comprising of two Gram panchayat with Ashramshalas and Zilla Parishad schools under the Gram panchayat. A few roles and responsibilities of the identified O&M staff are:

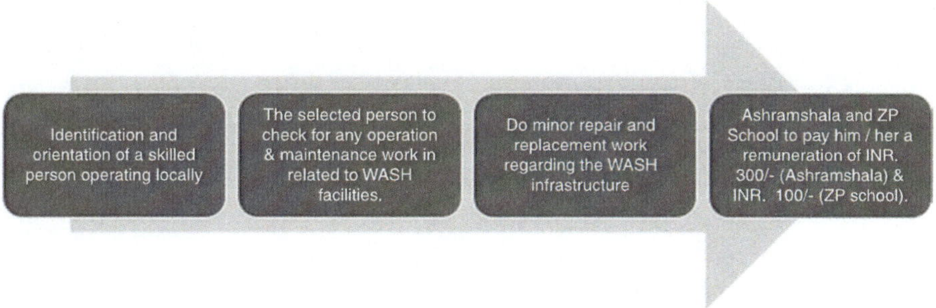

Fig. 2. Methodology for implementing Village Entrepreneurship model.

- To visit twice a week in the Ashramshala.
- To check for any O&M work related to WASH facilities.
- Do minor repair and replacement work regarding the WASH infrastructure.
- After each visit, Ashramshala and ZP school should pay him/her a remuneration of INR 300 (Ashramshala) & INR 100 (ZP school) (Fig. 2).

12 Key Learnings and Observations

- Revenue model for local person doing odd jobs in the nearby village.
- Skilled technician support to Ashramshala as well as ZP schools in the selected GPs.
- O&M of the WASH infrastructure on regular basis through locally engaging the personnel.
- Significant difference is financial capability and infrastructure between ZP schools and Ashramshalas.
- Possibility of local influencing on selection of personnel for the job.
- Lack of willingness by the local personnel to work on anything related to sanitation.
- Empowering local entrepreneurs and developing sustainable business model for them.

13 Institutional Strengthening

Institutional strengthening is the model based on capacity building of the existing sanitary staff in the Ashramshala to carry out the minor repair and maintenance work of WASH infrastructure. The reason for selecting Nandurbar region for this model is the geographical terrain and longer distance between two Ashramshalas, making it difficult to reach out to schools on regular basis (Fig. 3).

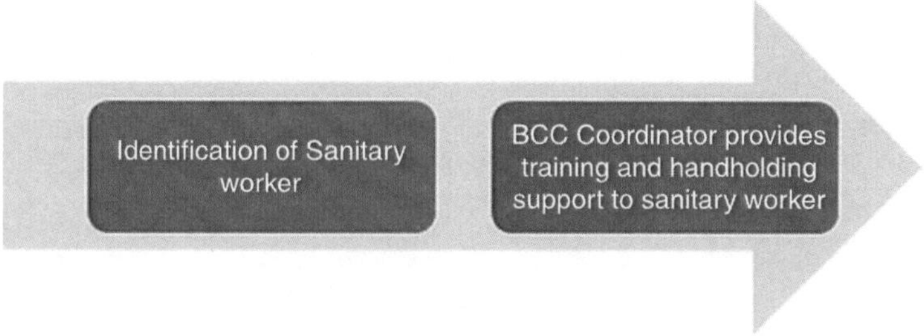

Fig. 3. Methodology for implementing Institutional Strengthening model.

14 Key Learnings and Observations

- Logistic delay can be eliminated as the worker is present in the school every day.
- Knowledge of the local conditions, vendors for material work can be done without time lag.
- Lack of incentive scheme for the sanitary workers to do the extra work.
- Absence of sanitary workers in most of the Ashramshalas due to vacant posts.
- Lack of willingness by sanitary worker to do anything other than tasks described in their job description.
- Resources for the sanitary workers are minimum as well as their skill set, making this model with limited opportunities.

15 Conclusion

Looking at Ashramshala with average strength of 400 students, the number of recurring works are huge. After assessing the visiting pattern of frequency of repair work, feedback from TDD PO, HMs there are different 261 number of recurring works which require skilled technician.

Ashramshala on their level does do some repair work through contracted worker or by themselves, which brings down the work which does not seem of high priority but has a significant impact on WASH. There are 163 number of recurring works, which is often not taken on priority due to which existing structure becomes obsolete.

These works which required skilled technician throughout the year are addressed on weekly basis and significant shift in availability of functional WASH facilities is observed for the children.

Ashramshala poses a different challenge in comparison to existing humongous one with regular schools. Students reside in Ashramshalas along with studying. The average time spent at home by regular students is spent in Ashramshala campus making it a foundation of their behavioural aspects and personality development.

Therefore, the WASH infrastructures are subjected to extensive use by the student resulting in frequent damages.

Inability to carry out repair works from time to time increases the load in existing functional infrastructure posing a larger evil of total system breakdown. Increased pressure increases risk of permanent damage, water logging, which leads to epidemics, water source and water storage pollution directly affecting the health of students.

Dynamics of the Ashramshala works in such way that even when a work is contracted to external personnel or contractor, SMC waits until quantum of work is considerably large as a result minor repair work is prolonged. Social Service Provider Model addresses this key challenge at considerably low cost and has a potential to be scaled up and provide sustainable, regular solution for the O&M issue.

Acknowledgement We would like to thank the National Stock Exchange Foundation (NSEF) for funding this project.

References

Ramesh, A. and Blanchet, K. (2015). Interventions on Health Outcomes in Humanitarian Crises: A Systematic Review. Retrieved from https://www.ncbi.nlm.nih.gov/pmc/articles/PMC4580573/.

Carter, R.C. (2009). Operation and maintenance of rural water supplies. *In:* Perspectives No. 2.Rural Water Supply Network (RWSN), St. Gallen.

Castro, V., Msuya, N. and Makoye, C. (2009). Sustainable Community Management of Urban Water and Sanitation Schemes (A Training Manual).Water and Sanitation Program-Africa, World Bank, Nairobi.

Brikké, F. (2000). Operation and Maintenance of Rural Water Supply and Sanitation Systems. IRC International Water and Sanitation Centre, Delft Netherlands.

Hutton, G. and Haller, L. (2004). Evaluation of the Costs and Benefits of Water and Sanitation Improvements at the Global Level. Retrieved from http://www.who.int/water_sanitation_health/wsh0404.pdf.

IRC International Water and Sanitation Centre. (May 2005). School Sanitation and Hygiene Notes & News.

Joint Monitoring Programme, Global Baseline Report on WASH in Schools (2018). Retrieved from https://washdata.org/sites/default/files/documents/reports/2018-11/JMP%20WASH%20in%20Schozols%20WEB%20final.pdf.

Tribal Development Department, Government of Maharashtra. https://tribal.maharashtra.gov.in/1120/Ashram-Schools.

UNICEF WASH for School Children South Asia Report (2012).

United Nations High Commissioner for Refugees, Global Strategy for Public Health 2014–2018. Public Health, Food, Nutrition and WASH. Retrieved from https://wash.unhcr.org/download/unhcr-global-strategy-for-public-health-2014-2018/.

WASH Facilities in Schools of West Bengal and Bangladesh, Operation and Maintenance manual by UNICEF.

World Development Report (2018). Learning to Realize Educations Promise. Retrieved from https://www.worldbank.org/en/publication/wdr2018.

Chapter 3
Metal Resistant Bacteria in Animal Manure Induces Bacterial Resistance to Antibiotics: Their Co-occurrence in Compost, Soil and Water

Payal Mazumder, Jyoti, Meena Khwairakpam, and Ajay S. Kalamdhad

1 Introduction

Animal manure is a conveyance for the emanation and dispersion of antibiotic resistance. Antibiotics are extensively employed in animal and human medicines and in diet in aquaculture and animal agriculture farms. In the agriculture sphere, they are used as food additives to enhance the growth of animals and for curing diseases. Sulfonamides, tetracyclines and quinolones are the most common and customarily used antibiotics. The embezzlement of antibiotics predominantly escalated the generation and proliferation of ARGs and ARBs (Qiao et al., 2018). ARGs and ARBs have been persistently reported in the environment, viz., air, soil, sediment and water (Philip et al., 2018; Murray et al., 2019). They have turned into a universal human health crisis and also a major environmental concern (WHO, 2015). Additionally to the excessive use and exploitation of antibiotics, metals can promote the occurrence and vast spread of antibiotic resistance (AR). Several researches have proclaimed that metals augment AR in different environments, where the heavy metal concentration is notably correlated with the profusion of ARGs (Zhou et al., 2016). For instance, abundance of copper (Cu) and arsenic (As) were found to be cognate with the abundance of tetracycline (tet) resistance genes collected from swine farm manure samples (Zhu et al., 2013). Furthermore, co-selection of

P. Mazumder (✉)
Centre for the Environment, Indian Institute of Technology Guwahati,
Guwahati, Assam, India
e-mail: payal93@iitg.ac.in

Jyoti · M. Khwairakpam
Centre for Rural Technology, Indian Institute of Technology Guwahati,
Guwahati, Assam, India

A. S. Kalamdhad
Department of Civil Engineering, Indian Institute of Technology Guwahati,
Guwahati, Assam, India

A. S. Kalamdhad (ed.), *Integrated Approaches Towards Solid Waste Management*, https://doi.org/10.1007/978-3-030-70463-6_3

community-level resistance increased for tetracycline and vancomycin was reported upon increase in Cu concentrations in the soil (Berg et al., 2010). The co-occurrence of antibiotics and metals resistance in bacteria occur mainly via mechanisms viz., cross-resistance, i.e., metals and antibiotics prefer the same resistant pathway where they may use same protein/efflux pumps and co-resistance where, the antibiotic and metal resistance genes are located in the vicinity, i.e., same plasmid. Faecal bacteria are prevalent in aquatic ecosystems and albeit they are vulnerable to antibiotic pollution/resistance, it is plausible if they may also play a role on ARG transmission in natural ecosystems.

Soil environments are contemplated predominant pools of ARGs and antibiotics, supporting the selective survival of strains conferring resistance and the licentious exchange of genetic material among bacteria of unrelated origins. In particular, metal-contaminated soil remains the home for a large number and diversity of native microorganisms. On the other hand, fresh surface water constitutes pathogenic bacteria from animals and human derived from discharges of agricultural runoff and industrial wastewater as well as from contaminated soil leaching. ARGs for major antibiotic classes are discovered and studied from naturally occurring bacteria from aquatic ecosystems as well as bacteria of clinical origin (Zhang et al., 2009). Nonetheless, the co-occurrence of MRGs and ARGs from agricultural run offs and wastewater effluents are seldom monitored. According to Dr. Tedros Adhanom Ghebreyesus, Director-General of World Health Organization (WHO), '*Antimicrobial resistance is a global health emergency that will seriously jeopardize progress in modern medicine*'. WHO conducted the world antibiotics week (12–18 November 2018) to create public awareness and encourage health practitioners and policy makers to put effort to avoid antibiotic resistance.

Due to the extensive utilisation of antibiotics in livestock farming, residual antibiotics, ARBs and ARGs thrive in the soil environment when animal faecal matter is applied to the soil. This may further possess threat to contaminate the ground and surface water. Composting at thermophilic temperatures have been reported to eradicate the ARBs. However, the prevalence of ARBs is reported during the aerobic storage period of composts. Salmonella and Campylobacter, two of the many bacteria commonly transmitted through food, cause an estimated 4,10,000 antibiotic-resistant infections in the United States each year (CDC, 2013). Presence of ARGs in our food and drinking water will further reduce the effectiveness of antibiotics in treating infections. Therefore, it is imperative to find the ARB/Gs and MRGs influencing ARB/Gs in the water environment and focus on the probability of their co-occurrence in nature as well as agricultural run offs and surface waters.

2 Origin of ARB/Gs in Manure

Manures are usually a non-uniform mixture of animal urine and faeces. It is rich in nutrient content, and can serve the purpose of fertiliser to be applied in agricultural land and crops. However, it contains a lot of microbes/faecal coliforms and a large community of these microbes can be transferred to the environment where manure is applied. Therefore, understanding the microbial diversity and ecology of cattle faeces is of main significance for manure management and development of biological management tech which, helps us not only to deal with the microbes in the faecal matter, i.e., to be used as fertiliser but also diminishing its potential to impact human health (Durso and Cook, 2014). One of the major concerns with the application of cattle manure to agricultural fields is the risk of ARBs and ARGs to be transferred to the soil and plants thereby, to humans during exposure/consumption. The transfer of faecal microorganisms and pathogens indicate that these microbes reach surface and ground water from compost/manure in the agricultural soil (Thurston-Enriquez et al., 2005), and from the contaminated soil and water to the food crops (Mootian et al., 2009). This suggests that similar pathway may also be followed by various ARBs and ARGs to water and food. Figure 1 shows the probable pathways of ARB/Gs endurance in the environment. ARB/Gs can transfer from manure and manure applied soil to plants, groundwater and surface water bodies thereby, entering the food chain. It is crucial to check the ARG profiles for source tracking in such environmental macrocosms, and quite a few studies have deciphered these gene profiles of AR by employing quantitative real-time polymerase chain reaction (PCR) (Pruden et al., 2006; Storteboom et al., 2010). Further, with metagenomic analysis information about microbial community data, metabolic pathway details

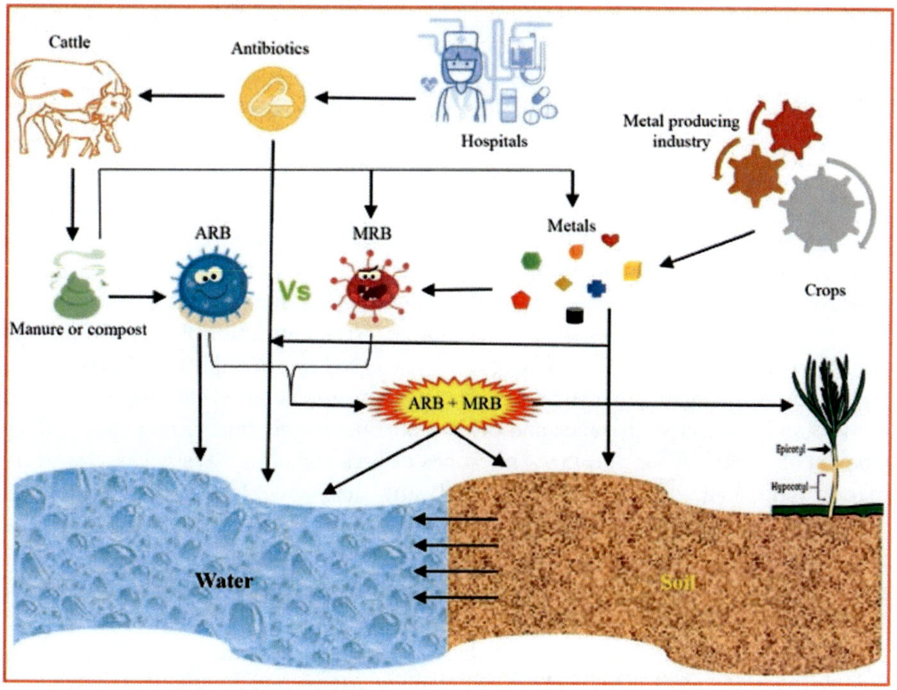

Fig. 1. Schematic diagram of MRBs and ARBs transmission in the environment

and abundance of ARGs can be clubbed together to get the holistic profile of certain sets of samples. The bovine faecal microbial community composition has been reported by several researchers, with the help of culture-independent as well as, culture-based technologies (Wilssens and Buttiaux, 1958; McGarvey et al., 2005; Wu et al., 2007; Dowd et al., 2008; Durso et al., 2010).

3 Persistence of ARB/Gs in the Agroecosystem

Using antibiotics for rearing animals creates environmental, ethical and economical concerns. Antibiotics that are used for disease treatment and curing ailments are vital to animal husbandry and comes under the 'five freedoms' accorded to animal agriculture and livestock farming (Poletto and Horzel, 2012). Antibiotics are also known to enhance the fodder feed efficiency, that is responsible for higher meat generation for the smaller number of animals produced/cultured. Increasing efficiency leads to increased profit and decreased prices due to lesser land required, lesser manure produced so management cost is cut down. However, there is concern that the usage of antibiotics makes the microbiota having ARGs flourish in the faecal matter, and these ARBs then can be transmitted from agricultural ecosystems to humans. Even though the proper pathway by which ARB/Gs are transmitted from soil, water, anthropods, lower level animals to humans is not clearly known and needs to be investigated (Phillips et al., 2004; Gibbs et al., 2006; Allen et al., 2010; Wright, 2010; Ahmad et al., 2011; Hong et al., 2012; Ashbolt et al., 2013). There are particular genes present that are triggered by each antibiotic medicine/drug, which encodes for its resistance. For example, tetracycline resistance can be manifested by mainly three mechanisms, which is coded by ~25 separate genes (Levy et al., 1999). Thus, it is vital to understand what is the magnitude of ARB/Gs in the farm soil, and how much it has/can diminished/increased (in case of horizontal transfer and replication) over time, what type of management techniques can be employed, which type of these genes should be checked, and in which type of microorganisms. Different ARGs can be retained by different strains of an organism (Ibekwe et al., 2011), on the other hand, same ARGs can be present in multiple taxa/family of microbes (Durso et al., 2011a; Durso et al., 2011b; Frye et al., 2011; Durso et al., 2012). Microorganisms that survive in harsh environment has greater probability to possess and transfer such genes and diseases to susceptible host (Zhou et al., 2010; Cook et al., 2014). These group of microbes or taxa can, thus, give necessary information and can be targeted for analysing and tracking clinically related AR. Biological, physico-chemical, spatio-temporal complexes and 'many ecologies' of resistance of natural ecosystems may reflect in the complications of AR measurement in agricultural systems and its impact on human health (Singer et al., 2006; Quintana-Hayashi and Thakur, 2012; Keelara et al., 2013; Mather et al., 2013; Netthisinghe et al., 2013). Various strategies and rule targets are planned and implemented by FAO (2007) to manage and reduce AR in bovine, poultry, dairy and swine farm related operations.

4 Co-occurrence of MRGs with ARGs

It has been reported that besides the over dose of antibiotics, metals can also trigger the genes for AR in microorganisms and aid in its spread (Seiler and Berendonk, 2012; Zhu et al., 2013; Gullberg et al., 2014). Many studies have documented that in certain environments, AR is enriched by metals and there seems to be a direct correlation with metal concentration (Zhou et al., 2016). For instance, mercury (Hg), zinc (Zn) and Cu in soil have been found to be firmly correlated with sul III and sul A (sulfonamide resistance genes) (Baker-Austin et al., 2006). Similarly, investigation of swine manure samples showed that tet (tetracycline resistance) genes were strongly related with heavy metals Cu and As (Ji et al., 2012). Again, vancomycin and tetracycline resistance was found to have occurred in soil enriched with Cu metal (Berg et al., 2010). Berg et al. (2005) also reported that when Cu was applied to agricultural soil media it simultaneously induced resistance for Cu and few antibiotics viz., chloramphenicol, ampicillin and tetracycline. Also, cadmium (Cd) and nickel (Ni) induced resistance to ampicillin or chloramphenicol in bacterial microcosms (Stepanauskas et al., 2006). Generally, there are two mechanisms of co-selection of MRGs and ARGs in bacteria against metals and antibiotics, (1) when metal and AR follows the same pathway e.g., common efflux pumps, called the cross-resistance and (2) when the genes responsible for AR and metal resistance occur in same loci or in plasmids (mobile genetic element), known as co-resistance.

The total microorganisms count and composition in animal faecal matter/manure changes on application of certain treatments or pre-treatments, e.g., composting/ vermicomposting, aerobic digestion, thermal treatment, etc. (Cote et al., 2006; Topp et al., 2009). Such pre-application methods while dealing with animal manure may decline the emergence of ARB/Gs and might also reduce the possibility of horizontal gene transfer (HGT) upon soil application of the processed manure (Chen et al., 2016). Aquatic systems are the major pools in the environment where the occurrence, proliferation and spread of MRGs and ARGs thrive in bacteria (Allen et al., 2010). Reddy and Dubey (2019) reported the occurrence of ARGs and MRGs in the river Ganges with the help of high throughput metagenomics analysis technique. Also, the unchecked application of manure/compost in agricultural land leads to the generation and enhancement of MRGs and ARGs (Guo et al., 2019). Also, Knapp et al. (2011) reported that heavy metals are strongly related to ARGs and significantly take part in contributing to the selection of MRGs and ARGs. It was also found that the relation between metals and ARGs were stronger than antibiotics and ARGs (Ji et al., 2012). Roberto et al. (2019) reported that rapid urbanisation and change in land use patterns changed the co-abundance of MRGs with ARB/Gs. Antibiotics such as tetracycline and Sulfamethoxazole along with heavy metals in the sediments of Changshou Lake were associated with ARGs (Lu et al., 2020).

Similarly, many other reports have documented the occurrence of MRGs and ARGs in the same vicinity or loci or in the same DNA fragments. Microplastics are hotspot zones where human bacterial pathogens present in metal polluted aquatic environment co-selects for multi-drug resistance (Imran et al., 2019). Zhang et al. (2020) reported that As(III) changed bacterial community structure and facilitated co-selection of ARBs and increased ARG abundance. Laffite et al. (2020) studied the river water and sediment of Kinshasa, Democratic Republic of the Congo and found high ARG copies in the sediment along with higher metal and organic matter content. Dickinson et al. (2019) reported Zn contamination in core sediments and its relatedness with cefotaxime, trimethoprim and oxacillin resistant bacteria. Wang et al. (2020) reported Cu induced cell damage and conjugative transfer of ARGs. Mainly, metals are non-degradable in nature and exerts a selective pressure that might facilitate the ARGs further (Huang et al., 2019). Therefore, it is very important to analyse and have the knowledge of relationships between MRGs and ARGs so that their mode of action and mechanism in the environment can be deciphered.

5 Conclusion

The percentage of pharmaceuticals used in the United States is the largest in the world. Developing countries like India, lacks good health care systems as well as efficient wastewater treatment plants, thus, adding a lot to antibiotic related resistance. Today pharmaceuticals (especially antibiotics) have emerged as one of the principal classes of '*contaminants of emerging concern*'. Although the original purpose of these chemicals was and still is to save lives, however, unabated use/misuse, failure to follow guidelines etc. has resulted in a silent epidemic; i.e., the rise of untreatable forms of common diseases caused by the resistant micro-organisms or '*superbugs*'. Moreover, the ill effects of such compounds are not limited to microbial resistance alone, other serious impacts include: adverse effects on the 'early life stages' of aquatic organisms like *Daphnia magna* and *Artemia snail.* Antibiotic resistance (AR) is an emerging global issue for the environment and public health. The application of cattle manure in agriculture has led to the accumulation of ARGs and MRGs in the soil environment and can leach to surface water and also percolate to ground water thus affecting the water environment as well. The focus of this manuscript is to focus on the co-occurrence of MRGs and ARGs, and their relationships within faecal bacteria and their transport from animal rearing farms to water bodies.

The uptake of ARB/Gs and MRGs by plants have proven to be an important indirect pathway of exposure to humans. Most of the pharmaceutical load in surface water has been attributed to urban waste water treatment plants/sewage treatment plants (UWTPs/STPs). However, this is an issue which is mainly common in the urban regions; in rural or agriculture dominated regions there are important alternative nonpoint sources like runoff of land applied manures/composts as well as biosolids. As majority load of antibiotic resistant bacteria is sourced back to human or animal faeces, and the related effluents; the novel technique of microbial source tracking (MST) helps us to discern the original contaminant sources through the use of target microbes or chemicals. In the present times, only the assessment of antibiotic resistance of bacteria in natural water bodies like streams and rivers is not enough. Therefore, it is equally important to research the source, occurrence and fate of the antimicrobial resistance bacteria and genes which are the causative factors for resistance in natural non-resistance microbes; along with the source apportionment of the various hotspots of microbe release to the natural environment.

References

Ahmad, A., Ghosh, A., Schal, C. and Zurek, L. (2011). Insects in confined swine operations carry a large antibiotic resistant and potentially virulent enterococcal community. *BMC Microbiol*, **11:** 23.

Allen, H.K., Donato, J., Wang, H.H., Cloud-Hansen, K.A., Davies, J. and Handelsman, J. (2010) Call of the wild: Antibiotic resistance genes in natural environments. *Nat Rev Microbiol*, **8:** 251–259.

Ashbolt, N.J., Amézquita, A., Backhaus, T., Borriello, P., Brandt, K.K., Collignon, P., Coors, A., Finley, R., Gaze, W. H. and Heberer, T. (2013). Human Health Risk Assessment (HHRA) for environmental development and transfer of antibiotic resistance. *Environ Health Perspect* **121:** 993–1001.

Baker-Austin, C., Wright, M.S., Stepanauskas, R. and McArthur, J.V. (2006). Co-selection of antibiotic and metal resistance. *Trends Microbiol,* **14:** 176–182.

Berg, J., Thorsen, M.K., Holm, P.E., Jensen, J., Nybroe, O. and Brandt, K.K. (2010). Cu exposure under field conditions co-selects for antibiotic resistance as determined by a novel cultivation-independent bacterial community tolerance assay. *Environ Sci Technol*, **44:** 8724–8728.

Berg, J., Tom-Petersen, A. and Nybroe, O. (2005). Copper amendment of agricultural soil selects for bacterial antibiotic resistance in the field. *Lett Appl Microbiol*, **40:** 146–151.

Centers for Disease Control and Prevention (CDC). (2013). CDC Threat Report, http://www.cdc.gov/drugre-sistance/threat-report-2013/

Chen, Q., An, X., Li, H., et al. (2016). Long-term field application of sewage sludge increases the abundance of antibiotic resistance genes in soil. *Environ Int*, **92:** 1–10.

Cook, K.L., Netthisinghe, A.M.P. and Gilfillen, R.A. (2014). Detection of pathogens, indicators, and antibiotic resistance genes following land application of poultry manure. *J Environ Qual*, Accepted for publication.

Cote, C., Masse, D.I. and Quessy, S. (2006). Reduction of indicator and pathogenic microorganisms by psychrophilic anaerobic digestion in swine slurries. *Bioresour Technol*, **97:** 686–691.

Dickinson, A.W., Power, A., Hansen, M.G., Brandt, K.K., Piliposian, G., Applebey, P., O'Neill, P.A., Jones, R.T., Sierocinski, P., Koskella, B. and Vos, M. (2019). Heavy metal pollution and co-selection for antibiotic resistance: A microbial palaeontology approach. *Environ Int*, **132:** 105117. doi:https://doi.org/10.1016/j.envint.2019.105117.

Dowd, S.E., Callaway, T.R., Wolcott, R.D., Sun, Y., McKeehan, T., Hagevoort, R.G. and Edrington, T.S. (2008). Evaluation of the bacterial diversity in the feces of cattle using 16S rDNA bacterial tag-encoded FLX amplicon pyrosequencing (bTEFAP). *BMC Microbiol*, **8:** 125.

Durso, L.M. and Cook, K.L. (2014). Impacts of antibiotic use in agriculture: What are the benefits and risks? *Curr Opin Microbiol*, **19:** 37–44. https://doi.org/10.1016/j.mib.2014.05.019.

Durso, L.M., Harhay, G.P., Smith, T.P., Bono, J.L., DeSantis, T.Z. and Clawson, M.L. (2011a). Bacterial community analysis of beef cattle feedlots reveals that pen surface is distinct from feces. *Foodborne Pathog Dis*, **8:** 647–649.

Durso, L.M., Harhay, G.P., Bono, J.L. and Smith, T.P.L. (2011b). Virulence- associated and antibiotic resistance genes of microbial populations in cattle feces analyzed using a metagenomic approach. *J Microbiol Methods*, **84:** 278–282.

Durso, L.M., Harhay, G.P., Smith, T.P., Bono, J.L., Desantis, T.Z., Harhay, D.M., Andersen, G.L., Keen, J.E., Laegreid, W.W. and Clawson, M.L. (2010). Diversity of beef cattle fecal microbiota varies by individual. *Appl Environ Microbiol*, **14:** 2010 (Epub ahead of print May).

Durso, L.M., Miller, D.N. and Wienhold, B.J. (2012). Distribution and quantification of antibiotic resistant genes and bacteria across agricultural and non-agricultural metagenomes. *PLoS ONE*, **7:** e48325.

FAO/WHO/OIE (November 2007). Expert Meeting on Critically Important Antimicrobials Report, Rome, Italy.

Frye, J.G., Lindsey, R.L., Meinersmann, R.J., Berrang, M.E., Jackson, C.R., Englen, M.D., Turpin, J.B. and Fedorka-Cray, P.J. (2011). Related antimicrobial resistance genes detected in different bacterial species co-isolated from swine fecal samples. *Foodborne Pathog Dis*, **8:** 663–679.

Gibbs, S.G., Green, C.F., Tarwater, P.M., Mota, L.C., Mena, K.D. and Scarpino, P.V. (2006). Isolation of antibiotic-resistant bacteria from the air plume downwind of a swine confined or concentrated animal feeding operation. *Environ Health Perspect*, **114:** 1032–1037.

Gullberg, E., Albrecht, L.M., Karlsson, C., Sandegren, L. and Andersson, D.I. (2014). Selection of a multi drug resistance plasmid by sublethal levels of antibiotics and heavy metals. *mBio*, **5(5):** 01918–01914. doi: 10.1128/mBio.01918-14. PMID: 25293762; PMCID: PMC4196238.

Guo, H., Gu, J., Wang, X., Yu, J., Nasir, M., Peng, H., Zhang, R., Hu, T., Wang, Q. and Ma, J. (2019). Responses of antibiotic and heavy metal resistance genes to bamboo charcoal and bamboo vinegar during aerobic composting. *Environ Pollut*, **252:** 1097–1105.

Hong, P.Y., Li, X., Yang, X., Shinkai, T., Zhang, Y., Wang, X. and Mackie, R.I. (2012). Monitoring airborne biotic contaminants in the indoor environment of pig and poultry confinement buildings. *Environ Microbiol*, **14:** 1420–1431.

Huang, J., Zhang, Y., Arhonditsis, G.B., Gao, J., Chen, Q., Wu, N., Dong, F. and Shi, W. (2019). How successful are the restoration efforts of China's lakes and reservoirs? *Environ Int*, **123:** 96–103.

Ibekwe, A.M., Murinda, S.E. and Graves, A.K. 2011. Genetic diversity and antimicrobial resistance of *Escherichia coli* from human and animal sources uncovers multiple resistances from human sources. *PLoS ONE*, **6:** e20819.

Imran, Md., Das, K.R. and Naik, M.M. (2019). Co-selection of multi-antibiotic resistance in bacterial pathogens in metal and microplastic contaminated environments: An emerging health threat. *Chemosphere*, **215**: 846–857. doi:https://doi.org/10.1016/j.chemosphere.2018.10.114.

Ji, X., Shen, Q., Liu, F., Ma, J., Xu, G., Wang, Y. and Wu, M. (2012). Antibiotic resistance gene abundances associated with antibiotics and heavy metals in animal manures and agricultural soils adjacent to feedlots in Shanghai; China. *J Hazard Mater*, **235–236:** 178–185.

Keelara, S., Scott, H.M., Morrow, W.M., Gebreyes, W.A., Correa, M., Nayak, R., Stefanova, R. and Thakur, S. (2013). Longitudinal study of distributions of similar antimicrobial-resistant *Salmonella serovars* in pigs and their environment in two distinct swine production systems. *Appl Environ Microbiol*, **79:** 5167–5178.

Knapp, C.W., Mc Cluskey, S.M., Singh, B.K., Campbell, C.D., Hudson, G. and Graham, D.W. (2011). Antibiotic resistance gene abundances correlate with metal and geochemical conditions in archived Scottish soils. *PLoS One*, **6(11):** e27300.

Laffite, A., Al Salah, D.M.M., Slaveykova, V.I., Otamonga, J.P. and Poté, J. (2020). Impact of anthropogenic activities on the occurrence and distribution of toxic metals, extending-spectra β-lactamases and carbapenem resistance in sub-Saharan African urban rivers. *Sci Total Environ*. Available online, 138129. In Press. doi:https://doi.org/10.1016/j.scitotenv.2020.138129.

Levy, S.B., McMurry, L.M., Barbosa, T.M., Burdett, V., Courvalin, P., Hillen, W., Roberts, M.C., Rood, J.I. and Taylor, D.E. (1999). Nomenclature for new tetracycline resistance determinants. *Antimicrob Agents Chemother*, **43:** 1523–1524.

Lu, L., Liu, J., Li, Z., Zou, X., Guo, J., Liu, Z., Yang, J. and Zhou, Y. (2020). Antibiotic resistance gene abundances associated with heavy metals and antibiotics in the sediments of Changshou Lake in the three Gorges Reservoir area, China. *Ecol Indic*, **113:** 106275. doi:https://doi.org/10.1016/j.ecolind.2020.106275.

Mather, A.E., Reid, S.W., Maskell, D.J., Parkhill, J., Fookes, M.C., Harris, S.R., Brown, D.J., Coia, J.E., Mulvey, M.R. and Gilmour, M.W. (2013). Distinguishable epidemics of multidrug-resistant *Salmonella typhimurium* DT104 in different hosts. *Science*, **341:** 1514–1517.

McGarvey, J.A., Miller, W.G., Sanchez, S., Silva, C.J. and Whitehand, L.C. (2005). Comparison of bacterial populations and chemical composition of dairy wastewater held in circulated and stagnant lagoons. *J Appl Microbiol*, **99:** 867–877.

Mootian, G., Wu, W.H. and Matthews, K.R. (2009). Transfer of *Escherichia coli* O157:H7 from soil, water and manure contaminated with low numbers of the pathogen to lettuce plants. *J Food Prot*, **72:** 2308–2312.

Murray, R., Tien, Y.C., Scott, A. and Topp, E. (2019). The impact of municipal sewage sludge stabilization processes on the abundance, field persistence, and transmission of antibiotic resistant bacteria and antibiotic resistance genes to vegetables at harvest. *Sci Total Environ*, **651(2):** 1680–1687.

Netthisinghe, A.M., Cook, K.L., Rice, C., Gilfillen, R.A. and Sistani, K.R. (2013). Soil nutrients, bacteria populations, and veterinary pharmaceuticals across a backgrounding beef feedlot. *J Environ Qual*, **42:** 532–544.

Phillips, I., Casewell, M., Cox, T., De Groot, B., Friis, C., Jones, R., Nightingale, C., Preston, R. and Waddell, J. (2004). Does the use of antibiotics in food animals pose a risk to human health? A critical review of published data. *J Antimicrob Chemother*, **53:** 28–52.

Philip, J.M., Aravind, U.K. and Aravinda Kumar, C.T. (2018). Emerging contaminants in Indian environmental matrices-a review. *Chemosphere*, **190:** 307–326.

Poletto, R. and Hotzel, M.J. (2012). The Five Freedoms in the global animal agriculture market: Challenges and achievements as opportunities. *Anim. Front.*, **2:** 22–30.

Pruden, A., Pei, R., Storteboom, H. and Carlson, K.H. (2006). Antibiotic resistance genes as emerging contaminants: Studies in northern Colorado. *Environ Sci Technol*, **40:** 7445–7450.

Qiao, M., Ying, G.G., Singer, A.C. and Zhu, Y.G. (2018). Review of antibiotic resistance in China and its environment. *Environ Int*, **110:** 160–172.

Quintana-Hayashi, M.P. and Thakur, S. (2012). Longitudinal study of the persistence of antimicrobial-resistant Campylobacter strains in distinct swine production systems on farms, at slaughter, and in the environment. *Appl Environ Microbiol*, **78:** 2698–2705.

Reddy, B. and Dubey, S.K. (2019). River Ganges water as reservoir of microbes with antibiotic and metal ion resistance genes: High throughput metagenomic approach. *Environ Pollut*, **246:** 443–451.

Roberto, A.A., Van Gray, J.B., Engohang-Ndong, J. and Leff, L.G. (2019). Distribution and co-occurrence of antibiotic and metal resistance genes in biofilms of an anthropogenically impacted stream. *Sci Total Environ*, **688:** 437–449.

Seiler, C. and Berendonk, T.U. (2012). Heavy metal driven co-selection of antibiotic resistance in soil and water bodies impacted by agriculture and aquaculture. *Front Microbiol*, **3:** 399.

Singer, R.S., Ward, M.P. and Maldonado, G. (2006). Can landscape ecology untangle the complexity of antibiotic resistance? *Nat Rev Microbiol*, **4:** 943–952.

Stepanauskas, R., Glenn, T.C., Jagoe, C.H., Tuckfield, R.C., Lindell, A.H., King, C.J. and McArthur, J.V. (2006). Co-selection for microbial resistance to metals and antibiotics in freshwater microcosms. *Environ Microbiol*, **8:** 1510–1514.

Storteboom, H., Arabi, M., Davis, J.G., Crimi, B. and Pruden, A. (2010). Tracking antibiotic resistance genes in the South Platte River basin using molecular signatures of urban, agricultural and pristine sources. *Environ Sci Technol*, **44:** 7397–7404.

Thurston-Enriquez, J.A., Gilley, J.E. and Eghball, B. (2005). Microbial quality of runoff following land application of cattle manure and swine slurry. *J Water Health*, **3:** 157–171.

Topp, E., Scott, A., Lapen, D.R., et al. (2009). Livestock waste treatment systems for reducing environmental exposure to hazardous enteric pathogens: Some considerations. *Bioresour Technol*, **100:** 5395–5398.

Wang, Q., Liu, L., Hou, Z., Wang, L., Ma, D., Yang, G., Guo, S., Luo, J., Qi, L. and Luo, Y. (2020). Heavy metal copper accelerates the conjugative transfer of antibiotic resistance genes in freshwater microcosms. *Sci Total Environ*, **717**: 137055. doi:https://doi.org/10.1016/j.scitotenv.2020.137055.

WHO (2015). Global Action Plan on Antimicrobial Resistance. https://www.who.int/antimicrobial-resistance/publications/global-action-plan/en/.

Wilssens, A. and Buttiaux, R. (1958). Bacteria in fecal flora of normal cow. *Ann Inst Pasteur (Paris)*, **94:** 332–340.

Wright, G.D. (2010). Antibiotic resistance in the environment: A link to the clinic? *Curr Opin Microbiol*, **13:** 589–594.

Wu, X.Y., Walker, M., Vanselow, B., Chao, R.L. and Chin, J. (2007). Characterization of mesophilic bacilli in faeces of feedlot cattle. *J Appl Microbiol*, **102:** 872–879.

Zhang, M., Wan, K., Zeng, J., Lin, W., Ye, C. and Yu, X. (2020). Co-selection and stability of bacterial antibiotic resistance by arsenic pollution accidents in source water. *Environ Int*, **135**: 105351. doi:https://doi.org/10.1016/j.envint.2019.105351.

Zhang, Y., Marrs, C.F., Simon, C. and Xi, C. (2009). Wastewater treatment contributes to selective increase of antibiotic resistance among *Acinetobacter* spp. *Sci Total Environ*, **407(12):** 3702–3706.

Zhou, Z., Raskin, L. and Zilles, J.L. (2010). Effects of swine manure on macrolide, lincosamide, and streptogramin B antimicrobial resistance in soils. *Appl Environ Microbiol*, **76:** 2218–2224.

Zhou, B., Wang, C., Zhao, Q., Wang, Y., Huo, M., Wang, J. and Wang. S. (2016). Prevalence and dissemination of antibiotic resistance genes and co-selection of heavy metals in Chinese dairy farms. *J Hazard Mater*, **320:** 10–17.

Zhu, Y.G., Johnson, T.A., Su, J.Q., Qiao, M., Guo, G.X., Stedtfeld, R.D., Hashsham, S.A. and Tiedje, J.M. (2013). Diverse and abundant antibiotic resistance genes in Chinese swine farms. *Proc Natl Acad Sci U.S.A.*, **110:** 3435–3440.

Chapter 4
Artificial Neural Network Model for Prediction of Methane Fraction in Landfill Gas from Pretreated Waste in Bioreactor Landfills

Tamru Tesseme Aragaw and Sumedha Chakma

1 Introduction

Landfilling remains to be the principal method of municipal solid waste (MSW) disposal worldwide due to its economic advantage compared to the other alternatives such as incineration and composting (Chakma and Mathur, 2012; Siddiqui et al., 2012; Zhang et al., 2012). However, recently increasing attention has been given to the significant threat to public health and environmental pollution problems related to the landfills (Lou and Nair, 2009). These problems are repeatedly allied with the formation of high organic pollutants (organic carbon and nitrogen) in the leachate and emissions of the greenhouse gases during the anaerobic decomposition of MSW (Amini et al., 2013; Zhang et al., 2012). The methane (50–60%) and carbon dioxide (40–50%) are the predominate components in the landfill gas (LFG) (Chakma and Mathur, 2016). Few trace species namely hydrogen sulphide, nitrous oxide and some other organic compounds are also regularly found plentiful in the landfill gases.

As a result of this, it is advisable to limit the quantity of organic fraction of the MSW that would be disposed to the landfill to reduce the emission of the methane gas into the environment (Bayard et al., 2009; Di Maria et al., 2013). The biological pretreatment of the MSW through simple composting has been found as the most operational and cost-effective technology throughout in the EU member countries in compliance with the Landfill Directive (EC, 1999) to reduce the biodegradable fraction of the MSW disposed to the landfill. The methane (CH_4) released from the biological decay of the landfilled waste is the principal donor to the global warming

T. T. Aragaw (✉)
Faculty of Water Supply and Environmental Engineering, Arba Minch Water Technology Institute, Arba Minch University, Arba Minch, Ethiopia

S. Chakma
Department of Civil Engineering, Indian Institute of Technology, Delhi, New Delhi, India

A. S. Kalamdhad (ed.), *Integrated Approaches Towards Solid Waste Management*, https://doi.org/10.1007/978-3-030-70463-6_4

(De Gioannis et al., 2009). The determination of the methane gas emissions from the landfill is a vital step for appraising the landfill gas control procedures and enlightening evaluations of CH_4 productions for greenhouse gas inventories.

In the last 100 years of time horizon with the association to the potential of the carbon dioxide, the methane gas has been found with 25 times higher heat trapping efficiency within the atmosphere, which could contribute significantly for global warming (Karanjekar et al., 2015). Moreover, the methane is contributing about 20% of the global warming; nevertheless, the annual contribution of the methane emission worldwide is accounted approximately 200 times smaller than the carbon dioxide (Çelekli et al., 2013; Di Maria et al., 2013). The landfill sites are representative of the third largest anthropogenic source of emission for the methane (Jahandideh et al., 2009; Thompson et al., 2009).

The LFG compositions and its rate of production are often ruled by numerous sites-specific variables such as the waste composition, ambiance temperature, pH, moisture contents and availability of nutrient and landfill operations (Chakma and Mathur, 2011; Karanjekar et al., 2015). Under this consideration, the waste composition and ambiance temperature are found as the major influencing factors for LFG production. It is due to the volume of gas production fully governed by the quantity and types of the substrate (organic waste), which are being decomposed. The LFG production consists of several complex processes during the anaerobic digestion of solid wastes, which involves many classes of microbial action as illustrated in Fig. 1.

Quantifying of the methane composition and LFG to energy production potential is the most difficult duty due to the uncertainties occurred in the modelling LFG generation (Chakma and Mathur, 2016). Estimating of the methane composition in

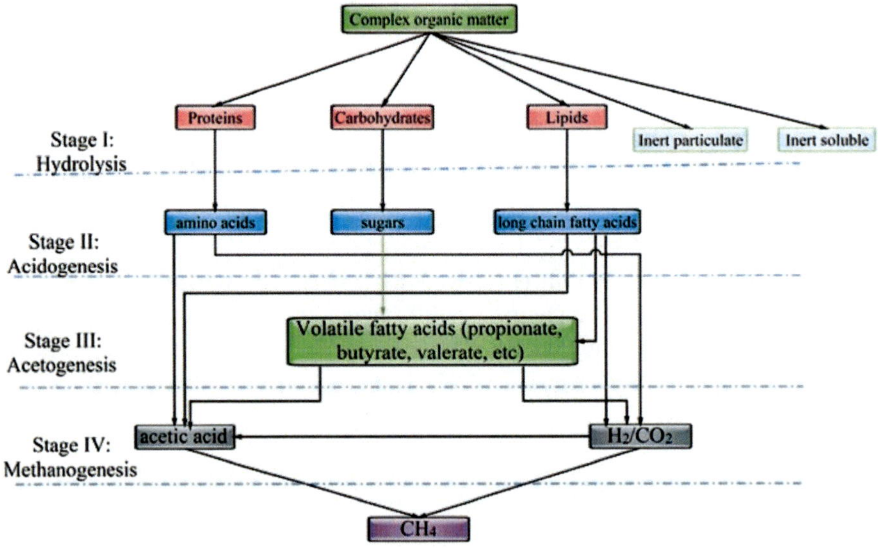

Fig. 1. Flowchart for the methane production in anaerobic bioreactor landfill.

the landfill generation have been recently computed using mathematical models, which are reliant on the amount and composition of waste to be disposed, the moisture content and the landfill cover material as well as the efficiency of the LFG collection system (Amini et al., 2012). However, these factors are commonly poorly defined; resulting in a significant ambiguity in model outcomes. Substantial development in the mathematical modeling of LFG generation from anaerobic digestion process by using mass balance ideologies and reaction kinetics has been reported by numerous scholars (Amini et al., 2013; Karanjekar et al., 2015 and Thompson et al., 2009). However, these types of approaches are generally required to elaborate the practical facilities and proficiency to analyse many complicated analytical data and also understand the results to be achieved (Holubar et al., 2002).

The applications of artificial neural networks (ANNs), alternatively, are recently one of the utmost prevailing computational modelling techniques. It has also many distinctive advantages compared to other nonlinear estimation techniques. ANNs are avoiding complicated mass balance data analysis processes and do not require any previous knowledge about the mathematical relationships that exists among the important parameters (Ozkaya et al., 2007; Strik et al., 2005; Yetilmezsoy et al., 2013). To date, little information is available on the prediction of the methane fraction in the landfill gas generated from the biological pretreated waste in the anaerobic pilot-scale landfills (Chakma and Mathur, 2012). Therefore, the aim of this study is to fill this gab by developing the artificial neural network model for the prediction of the methane fraction from the biological pretreated waste in the pilot-scale bioreactor landfills as a function of the leachate characteristics.

2 Materials and Methods

2.1 Collection of MSW Sample

The MSW samples were obtained in the study from Okhla open dumping site in New Delhi, India and transported to the Indian Institute of Technology Delhi (IIT Delhi) for further examination as per recommendation of the USEPA (USEPA 2002). Figure 2 shows the physical compositions of the raw solid waste sample collected (by wet weight) in this study. The portion of collected raw solid waste sample was biologically pretreated in fully aerobic composting for a period of five weeks.

2.2 Set-up of Simulated Reactor

Two pilot scale anaerobic bioreactor landfills were fabricated from 18 mm thick acrylic sheets and a schematic configuration is illustrated in Fig. 3. A shower tap sprinkler was kept 3 cm below the top of the reactor cover to supply the external

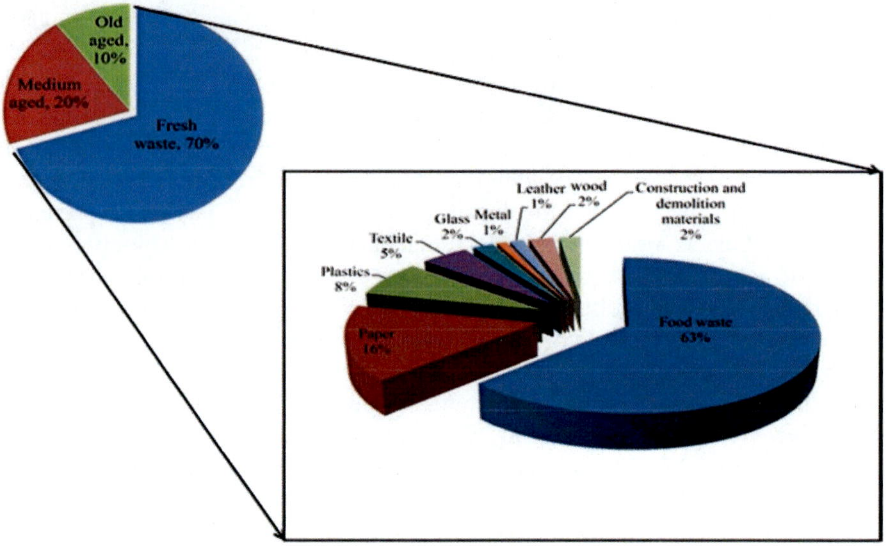

Fig. 2. The percentage compositions of raw solid waste.

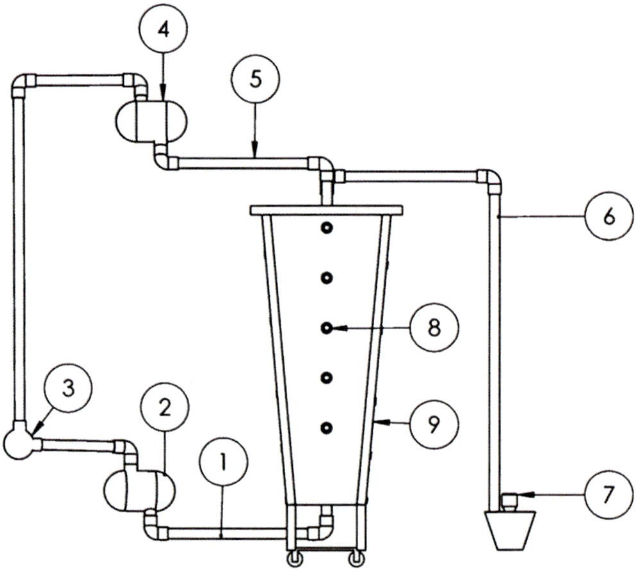

Legend: [1]Leachate Outlet Pipe; [2]Leachate Storage Tank; [3]Peristatic Pump; [4]Elevated leachate Storage Tank; [5]Leachate recirculation line; [6]Gas collection Pipeline; [7]Inverted Cylinder; [8]Temperature Measuring Port; [9]Main body

Fig. 3. Pilot-scale bioreactor landfills.

water. A leachate collection port was also fitted at the centre of the lower part of the reactor for regular assessment of the leachate quality and quantity. Then the designed bioreactors were labelled as reactor B and reactor C. Both of the reactors were filled initially with Old MSW over a 10 cm thick gravel layer and then medium year MSW was added over old MSW as a second layer. Finally, the biologically pre-treated waste was loaded into the reactor. Lastly, a second gravel layer of 5 cm thickness was placed over pretreated waste for evenly distributing external water on the top of landfilled waste.

2.3 Analytical Methods and Instruments

The gas and leachate samples were regularly collected from each of the pilot-scale reactors and the Standard Method (APHA, 2005) was strictly followed to perform the analysis. All the analytical methods and instruments were utilized in this study to determine the global leachate parameters as presented in Table 1. The composition (CO_2 and CH_4) in the landfill gas were performed by gas chromatography with a Porapak Q column (2 m, 80/100 mesh) equipped with a thermal conductivity detector (TCD).

2.4 Neural Network Basic Principles

The schematic representation of a three layers artificial neural network is illustrated in Fig. 4. The neural network structure is established by a sequence of three or more interconnected layers that have different roles namely, input, hidden and output layers (Hornik et al., 1989; Majumdar, 2011).

Table 1. Analytical methods and instruments

Parameters	*Method adopted and instruments used*
pH	Potable pH/Redox-meter (decibel ISO 900 pH/mv meter)
ORP	Potable pH/Redox-meter (decibel ISO 900 pH/mv meter)
TDS	Weight and dried methods at 180 °C APHA 2540 C
Temperature	Digital multi-thermometer
COD	Open reflux method, APHA 5220 B
Chloride	Titration method (Argentometric method) APHA, 4500-Cl^- B
Alkalinity	Titration method, APHA, 2320 B
NH_3-N	Standard method 4500-NH_3 B and 4500-NH_3 C
CH_4	Hewlett-Packard 5850 A Gas Chromatograph

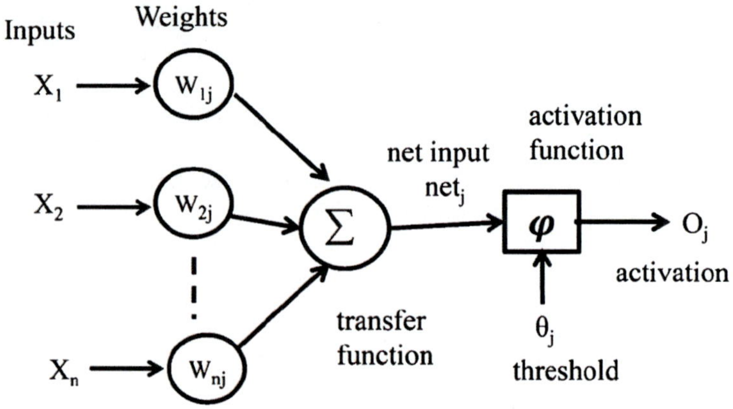

Fig. 4. Schematic representation of a three-layer neural network.

Table 2. The input and output parameters

Input										*Output*
[p_1]	*[p_2]*	*[p_3]*	*[p_4]*	*[p_5]*	*[p_6]*	*[p_7]*	*[p_8]*	*[p_9]*	*[p_{10}]*	*[T_1]*
pH	Alka.	TDS	Chlo.	ORP	NH_3-N	COD	Leach.	Age	Temp.	Methane

Note: Alka = alkalinity; Chlo. = chloride; Leach. = recirculated leachate; Temp. = temperature

2.5 *Modelling Methodology*

2.5.1 Network Architectures

In this study, the ANNs architectural structures had been formulated by splitting the available data sets into two subsets namely input matrix [P] and the target matrix [T] as presented in Table 2. Under this consideration, the 10 principal leachate characteristics were chosen for the input parameters [p]; whereas, the methane fractions in landfill gas were taken as the target parameter [T].

A three-layer feedforward with the tan-sigmoid transfer (tansig) function was used in the hidden layer to introduce the nonlinearity into the network in the present study. The descriptive statistical results of the input parameters used in the present study is given in Table 3. Furthermore, the linear transfer function (purelin) was also used in the output layer neurons due to suitable activation function for continuous valued targets (Yetilmezsoy et al., 2013).

2.5.2 Preparation of Input and Output Parameters

The total observed experimental data obtained from the pilot-scale bioreactor landfills were partitioned into three subsets namely training, validation and test for the application of the artificial neural network structure. Under this consideration, the total experimental data were considered as 100% and then different data subsets

Table 3. Descriptive statistical results for the input parameters

	Reactor B				*Reactor C*			
Input parameters	*Mean*	*Median*	*St. dev.*	*Range*	*Mean*	*Median*	*St. dev.*	*Range*
pH	7.80	7.70	0.66	2.33	8.09	8.15	0.49	2.17
Alkalinity	2823	2650	790	3550	2448.2	2400	417.70	2050
TDS	5397	4346	2056	7396	4647	3842	1601	5736
Chloride	2718.5	2824.10	339.90	1349.6	2459	2474.2	395.80	1524.5
ORP	−133.50	−158.50	82.20	244.00	−148.10	−187	78.50	237.00
NH_3-N	379.60	277.70	219.60	643.90	331.40	265.20	194.90	638.40
COD	19,120	11,340	16,860	54,100	13,541	10,350	10,884	33,860
Leachate	100.86	99.00	42.87	169.00	106.97	113	44.65	180
Age	145.40	145.50	75.50	256.00	145.40	146.5	75.50	256
Temperature	27.71	29.58	8.02	34.58	27.83	29.73	7.94	34.74

Note: St. dev. = Standard deviation

division ratio were tried for the training, validation and testing on the following basis: 50% of the total data subset ratio for training, 25% of the total data for validation and the remaining 25% of the data for testing in the first trial. Then the trials were continued as 60% for training, 20% for validation and 20% for the testing in the second trial. Similarly, it was also about 70% for training, 15% for validation and 15% for the testing in the third trial. The fourth trial was about 70% for training, 15% for validation and 10% for testing. It was fixed as 70% for training, 10% for validation and 15% for testing in the fifth trial. The ratios about 80% for training, 10% for validation and 10% for testing were tried in the sixth trial. Then, each data subsets were being trained in the ANNs model and the processes repeated until all the subdivision was tested. Lastly, the best subdivision simulation result was reported and the best percentage ratio was adopted.

2.5.3 Data Pre-processing and Normalisation

In this study, experimental data such as the leachate characteristics and the methane gas fractions were scaled in the range [−1,1] by using an algorithm function, premnmx in MATLAB environment (Mjalli et al., 2007; Pradhan and Lee, 2010). The data normalisation has also been computed mathematically:

$$X_N = 2 * \frac{X - X_{\min}}{X_{\max} - X_{\min}} - 1 \tag{1}$$

where X_n is the normalised value of certain parameter, X is the measured value of this parameter; $X_{\min}$ and $X_{\max}$ are the minimum and maximum values in the data subset of the parameter.

2.6 ANN Learning

The five commonly used training functions in the literature were tried in this study such as Batch Gradient Descent (traingd), Batch Gradient Descent with Momentum (traingdm), Variable Learning Rate (traingda), Resilient Backpropagation (trainrp) and Levenberg-Marquardt (trainlm). The schematic flowchart in Fig. 5 illustrates the methods that were followed to develop the optimal neural network in this study. The ANN learning rate was being performed by varying the learning rate value from 0.01 to 0.9. The momentum coefficient was being tried in the range of 0.1–0.9 to permit a neural network to respond to the local gradient and also to latest trends in the error surface.

The neurons number in the input layer ($N_I = 10$) and in the output layer ($N_o = 1$) was fixed based on the available input and output parameters to the neural network. In this proposed neural model, the number of neurons were determined by hit and trail methods. It was being tried by varying the numbers of the hidden layer nodes from 5 to 16 and then the result of prediction error compared to find out the optimum number of hidden neurons. Early stopping criterions were set for back-propagated algorithm used in this study by adjusting the number of epochs to 1000 and MSE as 0.00001.

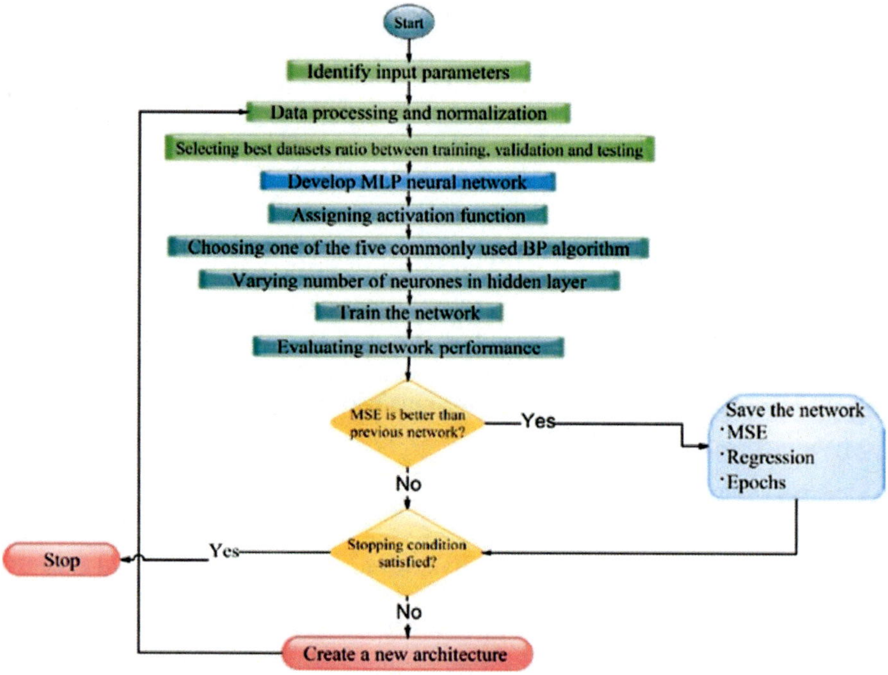

Fig. 5. A simple flow chart of the proposed methodology used in the neural network.

2.7 Performance Evaluation

The performance of the training and test data subsets were computed using the standard statistical performance evaluation criteria such as correlation coefficient (R) and mean square error (MSE) (Coman et al., 2008; Strik et al., 2005). The correlation coefficient R, MAPE and MSE values have been expressed by Eqs. (2)–(4).

$$R = \frac{\sum_{i=1}^{n} e_i p_i - \frac{\sum_{i=1}^{n} e_i \sum_{i=1}^{n} p_i}{n}}{\sqrt{\sum_{i=1}^{n} e_i^2 - \frac{\left(\sum_{i=1}^{n} e_i\right)^2}{n}} \sqrt{\sum_{i=1}^{n} p_i^2 - \frac{\left(\sum_{i=1}^{n} p_i\right)^2}{n}}} \tag{2}$$

$$MAPE = \frac{1}{n} \sum_{i=1}^{n} \left(\frac{|e_1 - p_i|}{e_i} \right) \times 100 \tag{3}$$

$$MSE = \frac{1}{n} \sum_{i=1}^{n} (e_i - p_i)^2 \tag{4}$$

where e_i is the experimental values, p_i is the ith predicated values and n is the number of test. or training samples.

3 Results

The optimal ANN model results are described and discussed herein in detail.

3.1 Selection of the Best Parameters for ANN Topology

The comparative results of various data subset ratios for the prediction of the ANN model with the five widely used training algorithms is presented in Table 4. In this regard, the best data subset was found for reactor B where the data subsets were split about 60% for training, 20% for validation and 20% for testing. On the other hand, the best subset data ratio was found for reactor C where the data subsets were split about 70% for training, 15% for validation and 15% for test as shown in Table 4.

In a similar manner, the five frequently used learning algorithms were trained individually again until the optimum values were obtained. It was undertaken using the chosen subsets as 60% of the data for training, 20% of the data for validation and

Table 4. The comparative results for the data subset ratio in neural network

	Reactor B							Reactor C						
Input ratio	*Training*	*Validate*	*Testing*	*All*	*MSE*	*Neurons*	*Iteration*	*Training*	*Validate*	*Testing*	*All*	*MSE*	*Neurons*	*Iteration*
50/25/25	0.97	0.89	0.87	0.88	0.055	8	9	1	0.82	0.59	0.84	0.098	10	10
60/20/20	0.9	0.78	0.85	0.87	0.052	10	5	0.81	0.89	0.81	0.8	0.122	8	9
70/15/15	0.84	0.98	0.93	0.84	0.065	7	7	0.99	0.97	0.89	0.92	0.07	15	7
75/15/10	0.97	0.81	0.92	0.83	0.087	16	5	0.78	0.97	0.99	0.823	0.089	13	8
75/10/15	0.98	0.92	0.76	0.94	0.028	15	6	0.96	0.87	0.55	0.88	0.059	13	9

Table 5. Performance comparisons of different back-propagated algorithms

	Reactor B				*Reactor C*			
BPA	*Neuron*	*R value*	*MSE*	*Iteration*	*Neuron*	*All*	*MSE*	*Iteration*
Trainlm	10	0.87	0.052	5	13	0.823	0.089	8
Traingd	8	0.86	0.061	107	16	0.78	0.123	78
Traingdm	10	0.82	0.087	22	16	0.75	0.146	17
Traingda	9	0.66	0.348	15	15	0.86	0.076	148
Trainrp	8	0.84	0.066	10	13	0.85	0.073	16

Note: Trainlm = Levenberg-Marquardt; traingd = batch gradient descent; traingdm = batch gradient descent with momentum; traingda = variable learning rate; trainrp = resilient backpropagation

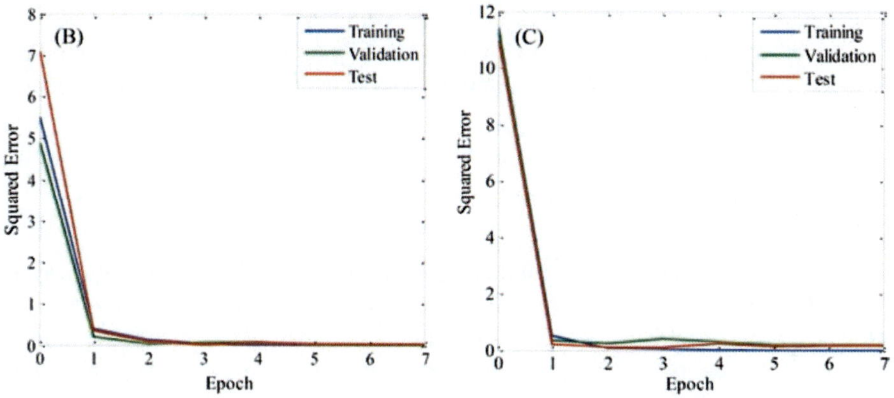

Fig. 6. Square mean errors of training, validation and test for the Levenberge-Marqardt algorithm (a) for reactor B and (b) for the reactor C.

20% of the data for testing for reactor B; whereas, the 70% of the data for training, 15% of the data for validation and 15% of the data for testing for the reactor C.

The results of the five BP algorithms to predict the methane composition in the landfill gas from pretreated waste is presented in Table 5.

The Levenberge-Marquardt (trainlm) algorithm (Table 5) was obtained as the best out of the five BP training algorithm to predict the methane fraction in the landfill gas from both the pilot-scale reactors. It was yielded a minimum MSE of 0.052 with a higher *R* of 0.82 for the reactor B; whereas, it was yielded the MSE value of 0.089 with a higher *R* of 0.87 for the reactor C.

Figure 6 demonstrates the mean square error for training, validation and test for the Levenberge-Marquardt algorithm for the prediction of the methane fraction in landfill gas. The training was ended after No iteration in both reactor B and reactor C because the validation error began to increase. These results were found realistic due to the test subset error and the validation subset errors have showed an identical characteristic. In addition to this, it does not seem that any significant change in fitting was occurred. The gradient descent training algorithm was found as too slow for the prediction and also takes higher impeach compare to the other training algorithms as shown in Table 5. Beside to this, comparatively the variable learning rate

Table 6. The optimum values of the internal parameter for the network architecture

Internal parameters	Optimum values
Learning rate	0.2
Error goal	10^{-5}
Iteration (epoch)	1000
Transfer function in hidden layer	tansig
Transfer function in output layer	purelin
Number of input node	10
Number of hidden layers	1
Number of neurons in output layer	1
Training algorithm	Leven-Marquardt

and resilient back-propagation algorithm were also found to take higher iteration time and similar results were also found in ANN network algorithm development.

3.2 *The Best ANN Architecture*

The optimum values of the internal parameters for the developed neural networks in the study are presented in Table 6. The optimum ANN architecture with a minimum MSE and a higher R value was found during the number of neurons in the hidden layer as 10 for the reactor B and 13 for the reactor C. However, increasing neuron numbers above 10 and 13 created unrealistic results for B and C landfill reactors respectively. It is most likely that the squared mean error began to increase for the training set with increasing neuron numbers more than 10 for the reactor B and 13 for the reactor C. This happened due to the test set and validation set errors which have different features and a significant change over fitting occurred. It is therefore, the optimal number of neurons for the algorithm are 10 for reactor B and 13 for reactor C which created minimum squared mean error with higher R value.

In the hidden layer, the number of neurons were kept as 10 for the reactor B and 13 for the reactor C and then varied the value of epoch from 100 to 5000 to obtain the optimum number of epochs. The optimum training epoch was obtained as 1000 and the performance goal was set as 10^{-5}. In a similar manner, the other internal parameter finding was continued. The value of learning rate was continued by varying its value in the ranges of 0.01 and 0.9. The higher learning rate could lead to the greater step and thus the algorithm may become unstable. In this study, the optimum algorithm was obtained with the higher value of R and a minimum MSE where the lr value was kept 0.2 and a momentum rate was 0.9 for both pilot-scale reactors. Therefore, the best simulation ANN architecture was found as 10–10–1 for the reactor B and 10–13–1 for the reactor C (i.e. the number of neurons in the input–hidden–output layer) as shown in Fig. 7.

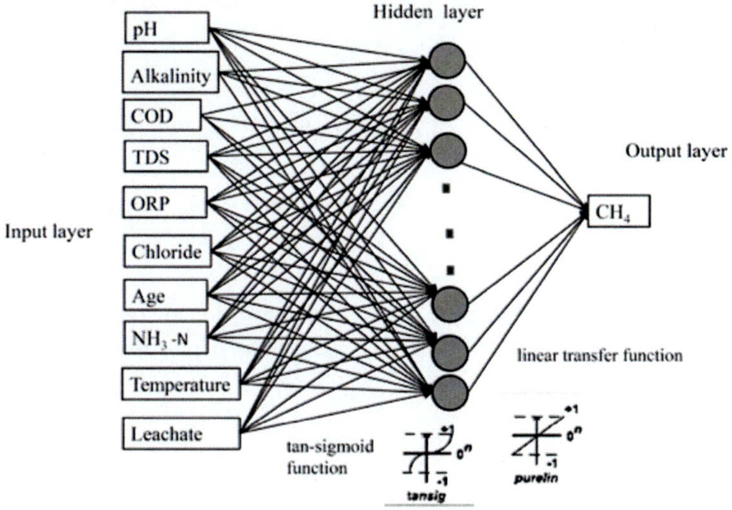

Fig. 7. The developed neural network structure for the prediction of the methane fraction in the landfill gas.

3.3 *ANN Performance Analysis*

The linear regression result between the ANN model outputs versus the corresponding target is illustrated in Fig. 8. A minimum MSE and higher R are commonly used as the measures of closeness of the model to the actual system. The data correlation is good if the slope of the correlation becomes near to one.

In this particular study, a correlation coefficient about 0.87 with a minimum MSE of 0.052 was found for the reactor B; whereas, a correlation coefficient of 0.82 with a minimum MSE of 0.089 was found for the reactor C.

Correspondingly, Fig. 9 presents the methane composition predicted by ANN model and the methane composition data obtained from the experimental studies with time. ANN model shows a reasonable trend prediction for the methane composition from both the reactors. The results were found to be very reasonable for both the reactors, since the trend of the ANN model output and observed data were in good agreement.

4 Conclusions

The study presents a multi-layer feed forward back-propagated neural network to model and predict the methane fraction from the biologically pretreated waste in bioreactor landfills. The Levenberge-Marquardt (trainlm) algorithm was the best training algorithm compared with the others learning algorithm to predict the

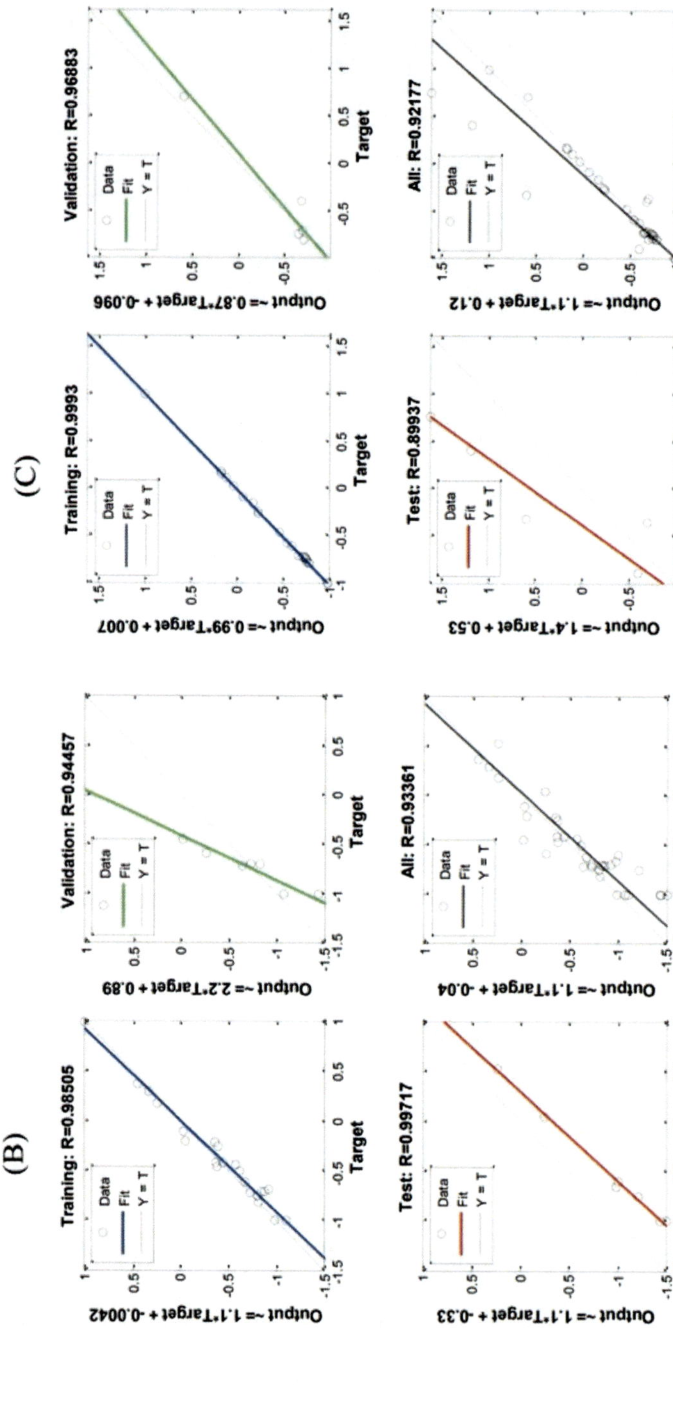

Fig. 8. The linear regression result between the neural network outputs and the targets for Levenberge-Margardt algorithm (a) for the reactor B and (b) for the reactor C.

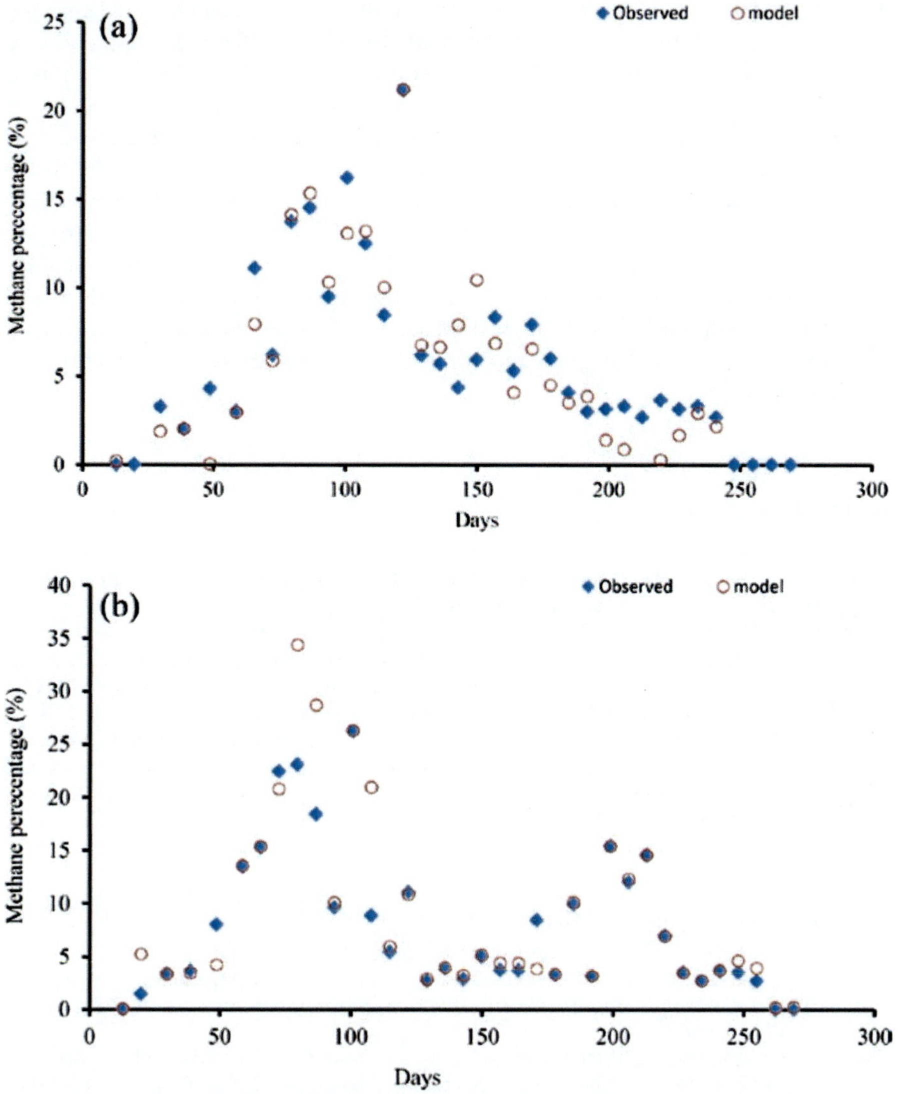

Fig. 9. Observed and predicted the methane fraction in landfill gas from pretreated waste (a) for the reactor B and (b) for the reactor C.

methane fraction in both the reactors. The rigorous study demonstrated that the gradient descent training algorithm was found too slow for the prediction and also takes higher impeach compared to the other training algorithms. In the optimised neural network architecture, a tansig activation function in the hidden layer; whereas, a linear pureline activation function in the output layer were applied. It was found that the optimal numbers of neurons were in the hidden layer as 10 for the reactor B and 13 for the reactor C. The statistical performance evaluation indices indicated that the minimum MSE in the reactor B was 0.052 with a regression coefficient of 0.82; whereas, the minimum MSE was 0.089 with a regression coefficient of 0.87 in the reactor C. The simulation results show that the ANN model is a promising approach which can be used accurately for prediction of the methane fraction in the landfill gas generation from the pretreated waste as functions of leachate characteristics.

References

Amini, H.R., Reinhart, D.R. and Mackie, K.R. (2012). Determination of first-order landfill gas modeling parameters and uncertainties. *Waste Manage*, **32(2):** 305–316. https://doi.org/10.1016/j.wasman.2011.09.021.

Amini, H.R., Reinhart, D.R. and Niskanen, A. (2013). Comparison of first-order-decay modeled and actual field measured municipal solid waste landfill methane data. *Waste Manage*, **33(12):** 2720–2728. https://doi.org/10.1016/j.wasman.2013.07.025.

APHA (2005). Standard Methods for the Examination of Water and Wastewater (21st ed.). American Public Health Association, American Water Works Association, and Water Environmental Federation, Washington, DC.

Bayard, R., de Araujo Morais, J., Rouez, M., Fifi, U., Achour, F. and Ducom, G. (2009). Effect of biological pretreatment of coarse MSW on landfill behaviour: laboratory study. *Water Sci Technol*, **58(7):** 1361–1369. https://doi.org/10.2166/wst.2008.512.

Çelekli, A., Bozkurt, H. and Geyik, F. (2013). Use of artificial neural networks and genetic algorithms for prediction of sorption of an azo-metal complex dye onto lentil straw. *Bioresour Technol*, **129:** 390–401. https://doi.org/10.1016/j.biortech.2012.11.085.

Chakma, S. and Mathur, S. (2011). Estimation of primary and mechanical compression in MSW landfills. *J Hazard Mater*, **16(4):** 298–303. https://doi.org/10.1061/(ASCE)HZ.2153-5515.0000117.

Chakma, S. and Mathur, S. (2012). Postclosure long-term settlement for MSW landfills. *J Hazard Mater*, **17(2):** 81–88. https://doi.org/10.1061/(ASCE)HZ.2153-5515.0000164.

Chakma, S. and Mathur, S. (2016). Modelling gas generation for landfill. *Environ Technol*, 1–9. https://doi.org/10.1080/09593330.2016.1231226.

Coman, A., Ionescu, A. and Candau, Y. (2008). Hourly ozone prediction for a 24-h horizon using neural networks. *Environ Modell Softw*, **23(12):** 1407–1421. https://doi.org/10.1016/j.envsoft.2008.04.004.

De Gioannis, G., Muntoni, A., Cappai, G. and Milia, S. (2009). Landfill gas generation after mechanical biological treatment of municipal solid waste. Estimation of gas generation rate constants. *Waste Manage*, **29(3):** 1026–1034. https://doi.org/10.1016/j.wasman.2008.08.016.

Di Maria, F., Sordi, A. and Micale, C. (2013). Experimental and life cycle assessment analysis of gas emission from mechanically–biologically pretreated waste in a landfill with energy recovery. *Waste Manage*, **33(11):** 2557–2567. https://doi.org/10.1016/j.wasman.2013.07.011.

EC (1999). European Council Directive 1999/31/EC of 26 April 1999 on the Landfill of Waste, Brussels, Belgium. *Off J Eur Commun.*

Holubar, P., Zani, L., Hager, M., Fröschl, W., Radak, Z. and Braun, R. (2002). Advanced controlling of anaerobic digestion by means of hierarchical neural networks. *Water Res*, **36(10):** 2582–2588. https://doi.org/10.1016/S0043-1354(01)00487-0.

Hornik, K., Stinchcombe, M. and White, H. (1989). Multilayer feed forward networks as universal approximators. *Neural Netw*, **2:** 356–359. https://doi.org/10.1016/0893-6080(89)90020-8.

Jahandideh, S., Jahandideh, S., Asadabadi, E.B., Askarian, M., Movahedi, M.M., Hosseini, S., et al. (2009). The use of artificial neural networks and multiple linear regression to predict rate of medical waste generation. *Waste Manage*, **29(11):** 2874–2879. https://doi.org/10.1016/j.wasman.2009.06.027.

Karanjekar, R.V., Bhatt, A., Altouqui, S., Jangikhatoonabad, N., Durai, V., Sattler, M.L., et al. (2015). Estimating methane emissions from landfills based on rainfall, ambient temperature, and waste composition: The CLEEN model. *Waste Manage*, **46:** 389–398. https://doi.org/10.1016/j.wasman.2015.07.030.

Lou, X.F. and Nair, J. (2009). The impact of landfilling and composting on greenhouse gas emissions—A review. *Bioresour Technol*, **100(16):** 3792–3798. https://doi.org/10.1016/j.biortech.2008.12.006.

Majumdar, A. (2011). Soft computing in fibrous materials engineering. *J Text*, **43(1):** 1–95. https://doi.org/10.1080/00405167.2010.527705.

Mjalli, F.S., Al-Asheh, S. and Alfadala, H.E. (2007). Use of artificial neural network black-box modeling for the prediction of wastewater treatment plants performance. *J Environ Manage*, **83(3):** 329–338. https://doi.org/10.1016/j.jenvman.2006.03.004.

Ozkaya, B., Demir, A. and Bilgili, M.S. (2007). Neural network prediction model for the methane fraction in biogas from field-scale landfill bioreactors. *Environ Modell Softw*, **22(6):** 815–822. https://doi.org/10.1016/j.envsoft.2006.03.004.

Pradhan, B. and Lee, S. (2010) Landslide susceptibility assessment and factor effect analysis: Backpropagation artificial neural networks and their comparison with frequency ratio and bivariate logistic regression modelling. *Environ Modell Softw*, **25(6):** 747–759. https://doi.org/10.1016/j.envsoft.2009.10.016.

Siddiqui, A.A., Richards, D.J. and Powrie, W. (2012). Investigations into the landfill behaviour of pretreated wastes. *Waste Manage*, **32(7):** 1420–1426. https://doi.org/10.1016/j.wasman.2012.03.016.

Strik, D.P., Domnanovich, A.M., Zani, L., Braun, R., Holubar, P. (2005). Prediction of trace compounds in biogas from anaerobic digestion using the MATLAB Neural Network Toolbox. *Environ Modell Softw*, **20(6):** 803–810. https://doi.org/10.1016/j.envsoft.2004.09.006.

Thompson, S., Sawyer, J., Bonam, R. and Valdivia, J.E. (2009). Building a better methane generation model: Validating models with methane recovery rates from 35 Canadian landfills. *Waste Manage*, **29(7):** 2085–2091. https://doi.org/10.1016/j.wasman.2009.02.004.

USEPA (2002). State of the practice for bioreactor landfills. United States Environmental Protection Agency. EPA/625/R-01/012. EPA Publication, Washington, DC.

Yetilmezsoy, K., Turkdogan, F.I., Temizel, I. and Gunay, A. (2013). Development of ANN-based models to predict biogas and methane productions in anaerobic treatment of molasses wastewater. *Int J Green Energy*, **10(9):** 885–907. https://doi.org/10.1080/15435075.2012.727116.

Zhang, Y., Yue, D., Liu, J., He, L. and Nie, Y. (2012). Effect of organic compositions of aerobically pretreated municipal solid waste on non-methane organic compound emissions during anaerobic degradation. *Waste Manage*, **32(6):** 1116–1121. https://doi.org/10.1016/j.wasman.2012.01.005.

Chapter 5
Effect of Gasification Zone Length on the Downdraft Gasifier Performance for High Ash Biomass

Haseen Siddiqui, Jayendran Shridharan, Ankush Jain, and Sanjay M. Mahajani

1 Introduction

The fossil fuels have been playing an active role in fulfilling the global energy needs for centuries. The regular use of fossil fuels has raised the major concerns of high depletion rate and the devastation of the earth's environment owing to greenhouse effect and climate change (Martínez, 2011). The underlined concerns have given a push to contemplate for an alternative for energy source. Gasification, in recent decades, has risen up as a boon for the world that has potential to meet the energy needs with added advantage of being environment friendly and carbon neutral (Guo, 2014). The process is categorised as a thermochemical process that involves the conversion of biomass materials into mixture of gases preferably CO, H_2 and trace CH_4. The process is an autothermal process that is feasible at high temperature and atmospheric pressure. The mixture of gases, referred as producer gas, has numerous applications such as electricity generation, transportation fuels, chemical and fertilisers production. Producer gas is a combustible mixture having lower heating value of 4–6 MJ/N m^3 for air being used as gasifying agent whereas it can be enhanced up to 15–20 MJ/N m^3 by using steam as a gasifying agent (Parthasarathy, 2014). Because of having appreciable lower heating value, the producer gas can be used for direct thermal application of cooking. Therefore, in recent decades, people have come up with an idea that led to development of gasifier cook stoves. The application of gasification technology for cooking has been appreciated either by designing compact gasifier cook stoves or by central production with syngas supply network (Bhattacharya, 2005). A typical design of a gasifier cook stove, as shown in

H. Siddiqui (✉) · S. M. Mahajani
Department of Chemical Engineering, Indian Institute of Technology Bombay, Mumbai, Maharashtra, India

J. Shridharan · A. Jain
Department of Chemical Engineering, Thadomal Shahani Engineering College, Mumbai, Maharashtra, India

A. S. Kalamdhad (ed.), *Integrated Approaches Towards Solid Waste Management*, https://doi.org/10.1007/978-3-030-70463-6_5

Fig. 1(a), includes a cylindrical vessel 10–15 mm in diameter and 150–200 mm in height (Bhattacharya, 2005). Air is blown from the bottom required for combustion of a part of feed and for gasification. The gas produced is being burnt at the top of gasifier stove with the help of atmospheric air or supplied secondary air. In the approach of central production with producer gas supply network, a large-scale gasifier is installed centrally and the gas produced is supplied to a tank after gas conditioning. Thereafter, producer gas from storage tank is supplied to different households through piping. The present study presented a third approach that includes pilot scale gasifier with a burner facility connected using about 1 m long duct at the exit of the gasifier as shown in Fig. 1(b). The third approach was mainly developed keeping in mind the community level cooking viz. hostel messes, rural government schools and various functions. The other advantage of this approach is that we can work on the burner design to make the burning of producer gas more efficient.

In cooking application, the volume percentage of the combustible components (CO, H_2 and CH_4) in producer gas is of paramount importance. Various design and operating parameters are identified as controlling parameters that have substantial influence on the composition of combustible components of the producer gas. Siddiqui et al. (2018) carried out a study that talked about how the equivalence ratio, grate rotation period and air split ratio effectuate the lower heating value, the quantity of clinker and the composition of producer gas for high ash garden waste biomass downdraft gasification process. To facilitate the use of the high ash garden waste biomass as a potential feedstock, they came up with a revamped design of downdraft gasifier and a set of desired operating conditions. The quantity of clinker dropped down to 0.08% of feed from 4.13% of feed. Guo et al. (2014) varied the equivalence ratio and biomass feed rate to observe their effect on composition of producer gas and the quantity of tar. The increasing equivalence ratio enhanced producer gas quality and lowered down the tar amount. Zainal et al. (2002) studied

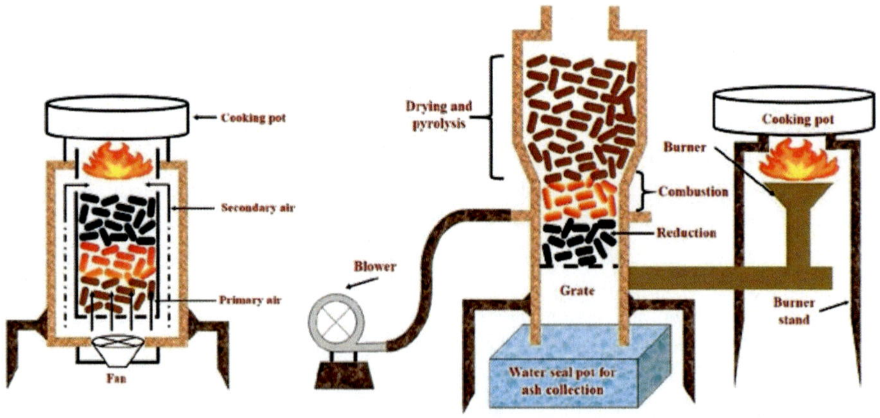

Fig. 1. (a) Gasifier cook stove; (b) gasifier with burner facility for community cooking.

the downdraft gasification process for furniture wood and wood chips as feedstocks to quantify the variation in gas composition, lower heating value and gas production rate for equivalence ratio used as parameter. The flow rate of air used as gasifying agent and feed stocks' moisture content have substantially affected the producer gas quality as investigated by Sheth and Babu (2009) in a downdraft gasifier.

Apart from operating parameters, design parameters also affect the producer gas composition. People have proposed different variations in gasifier design such as two-staged gasifier, three-staged gasifier, stratified gasifiers and throated gasifiers. Guo et al. (2014) reported a higher and more uniform temperature distribution inside the gasifier for three-staged gasifier that helped in enhancing tar cracking through thermochemical pathway. Martínez et al. (2011) reported that 80% of the air supplied through second stage air inlet improved the gas yield, cold gas efficiency and tar cracking. Siddiqui et al. (2018) proposed that complete supply of air from drying zone improved the temperature uniformity inside the gasifier and also caused a potential drop in the clinker formation.

The producer gas composition is also profoundly dependent on the extent of various heterogeneous and homogeneous reactions occurring in different zones during the gasification process. For downdraft gasifier, the zones are formed vertically as shown in Fig. 1(b). The extent of these reactions majorly influenced by the residence time (Alarcon-Gaete, 2016). The quality of producer gas can be improved by increasing the residence time (Hernández, 2010). In our recent parametric study for high ash biomass (garden-waste) in a downdraft gasifier, it was revealed that the residence time was governed by the grate movement time and a decline was reported in quality of producer gas with decreasing residence time (Siddiqui, 2018).

The present study proposed a modification in design that involved a mechanism to move the grate up and down resulting into the change in gasification zone length. The gasification zone length affects the residence time of gases in side gasifier and hence has substantial influence on the producer gas quality. Therefore, present study investigated the effect of gasification zone length on producer gas quality, temperature profile inside gasifier, lower heating value of producer gas and clinker formed. For cooking applications, the flame appearance time and flame stability time also matter a lot and therefore, these parameters are also investigated as a part of this study.

2 Materials and Methods

2.1 Feed-stock and Its Characterization

The feedstock, i.e. garden waste, collected was very low-density material and heavily contaminated with dust, small stones and other residues. Therefore, garden-waste was pre-processed and converted into pellets before using for the gasification process. The pre-processing of the feedstock involved the threshing, shredding and pelletisation (Pradhan, 2018). In rainy season, the dryer is also used to dry the garden waste before sending it to thresher. The pelletisation of garden waste helped in increasing the pellet density to about 1970 kg/m^3. The cylindrical shaped pellets with dimensions 8 × 45 mm were produced using the pelletiser. The calorific value of the pellets was estimated using bomb calorimeter and was found to be 17.35 MJ/kg. The proximate and ultimate analyses of the pellet are reported in Table 1.

2.2 Experimental Setup

The revamped downdraft gasifier was used for performing all the experiments. The detailed description of revamped downdraft gasifier can be found in my earlier work (Siddiqui, 2018). The revamped downdraft gasifier has a provision of supplying air from drying zone and also has a mechanism of moving grate up and down along the vertical axis of the gasifier. The schematic of the revamped downdraft gasifier is shown in Fig. 2. The temperature along the height of the gasifier was recorded using 11 K-type thermocouples as shown in Fig. 2. An ash collection box

Table 1. Characterisation of garden-waste pellets

Proximate analysis, wt% wet basis			
Moisture content (MC)	*Volatile content (VC)*	*Fixed carbon (FC)*	*Ash*
10.12	65.87	15.85	8.16
Ultimate analysis, wt% dry basis			
C	H	N	O
45.21	6.57	1.91	38.15

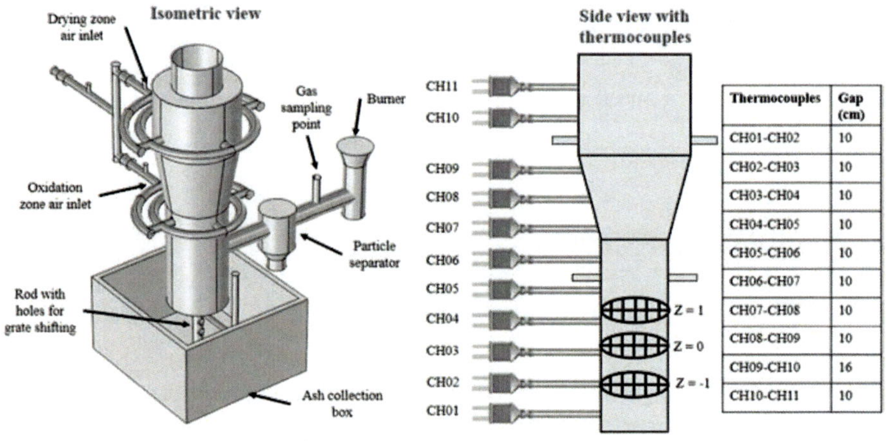

Fig. 2. Revamped gasifier design and its schematic with K-type thermocouples.

filled with water is placed at the bottom of the gasifier to collect the residue and to prevent the gas leakage from the bottom. The particle separator was also introduced to collect the particulate matters before the burner. Gas sampling point is provided in the gas outlet line to collect the gas samples for further analysis using gas chromatography. Three different vertical locations were selected for present study, referred as top one, middle one and bottom one, are being located at 4.5 inch, 9.0 inch and 13.5 inch down the combustion zone air inlet nozzle, respectively. These positions are represented by $Z = 1$, $Z = 0$ and $Z = -1$, respectively in the present study.

2.3 Operating Procedure and Plan of Experiments

All the runs were performed using revamped downdraft gasifier. The gasifier was operated in a semi-continuous mode in which garden waste pellets were fed initially before the start of the run and a continuous flow of producer gas was obtained through the exit pipe of the gasifier. The ash collection box was filled after the feeding of the garden waste pellets so as to avoid gas leakage from the bottom of the gasifier. A 1 hp blower was employed to blow the air, used as a gasifying agent for the present study. Initially, the excess air is supplied through the combustion zone air inlet that helped in ignition of the pellets and establishment of the combustion zone. The air supply at desired flow rate was switched to drying zone air inlet as soon as combustion zone was established. The shifting air supply to upper section caused temperature uniformity and reduced clinker (Siddiqui, 2018). The gas produced was collected at a regular interval of 5 min and was analysed using gas chromatography (GC). All the runs were performed thrice to ensure the repeatability of the results. The clinker formed was collected from the grate and ash collection box and weighted after sun dry for a day.

The experiments were planned for three different vertical locations of the grate, referred as $Z = 1$, $Z = 0$ and $Z = -1$ as shown in Fig. 2. The plan of experiments is tabulated in Table 2. The air flow rate was selected as per the desired conditions obtained for optimised performance (Siddiqui, 2018).

Table 2. Plan of experiments for the present study

Grate vertical position	*Feed amount (kg)*	*Air flow rate (kg/h)*	*Air inlet zone*
Top (Z = 1)	12	11.17	Drying
Middle (Z = 0)	12	11.17	Drying
Bottom (Z = -1)	12	11.17	Drying

3 Results and Discussion

The current section discussed the results obtained for different gasification zone lengths. The parameters viz. temperature variation inside the gasifier, lower heating value, gas composition, flame appearance time and flame stability time are considered to report the performance of the gasifier for the cooking purpose.

3.1 *Temperature Profile*

The temperature variation along the height of the gasifier was recorded by inserting the K-type thermocouples. The temperature variation for all the three grate locations is shown in Fig. 3. The initial point on time axis is the time when ignition of garden waste pellets inside the gasifier was initiated and t = 0 min is the time of flame appearance. The temperature channels CH11, CH10, CH09 were well above the feed level for all cases and therefore, the temperature indicated by them was only

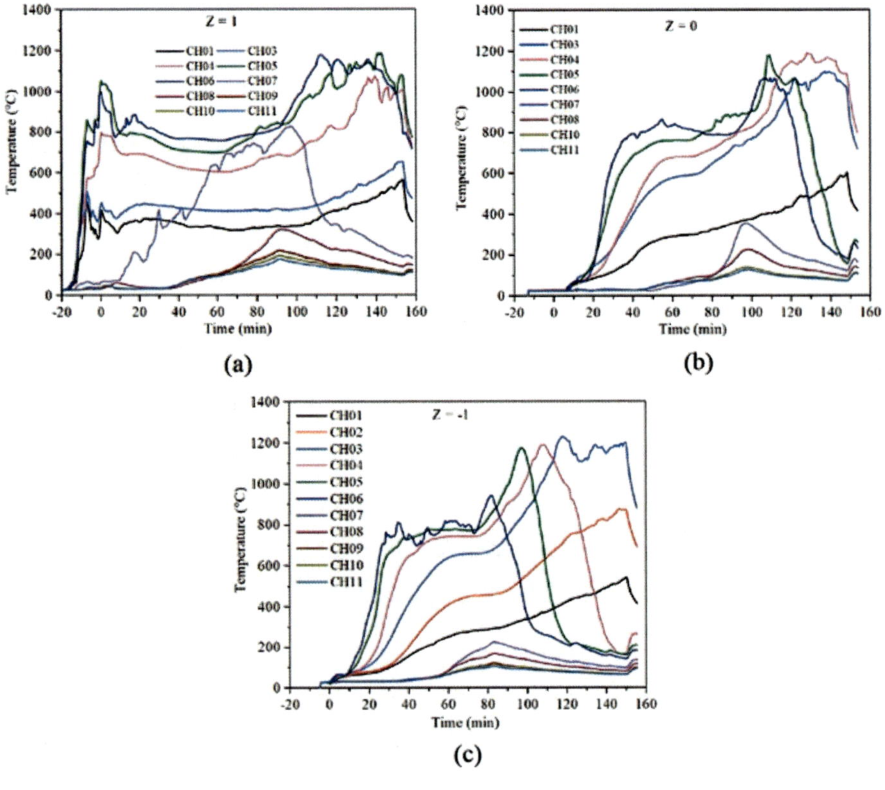

Fig. 3. Temperature variation along gasifier height for (a) Z = 1, (b) Z = 0 and (**c**) Z = −1.

due to hotter gases reaching to these channels. Until time 70 min, the temperature recorded by the channels CH11, CH10 and CH09 was near about 100°C that indicated the cooling of the hot gases coming from combustion zone due to presence of feed bed. After that, the temperature reached up to 300°C indicating less cooling of producer gas due to decreasing feed bed height. For $Z = 1$, more or less uniform temperature was recorded. This is due to the fact that the very low spacing between the grate at the bottom and combustion zone air inlets for $Z = 1$ caused the merging of combustion zone and gasification zone. The channels CH01, CH03 and CH04 are below the grate and therefore these channels recorded the gas outlet temperature. The channels CH05 and CH06 represent the combustion zone for all cases whereas channels between grate and CH05 represent the gasification zone. The average gas outlet temperature as indicated by the channel CH01 was found to be 450°C for all cases. There is an increasing trend in the gas outlet temperature with time due to the evolution of hotter producer gas because of the downward shifting of combustion zone as the feed got consumed. The downward shifting of combustion zone with time can be observed by observing the shifting of temperature peak to the lower channels with time. In case of lowest grate position ($Z = -1$), the average time taken by the temperature peak in shifting from CH06 to CH05 was 11.5 min and from CH05 to CH04 was 9.5 min.

The vertical shifting of grate also caused the change in the vertical position of different zones. For $Z = 1$, when the grate is at top position, the zones were established at higher position as compared to other cases.

3.2 Transient Variation of Gas Composition and Gasification Zone Temperature

The volume percentage of H_2, CO and CO_2 and gasification temperature are plotted for different grate positions with time as shown in Fig. 4. Only those data points are taken that were collected during stable flame. The gasification temperature was found increasing with time in all cases except the case of top grate position ($Z = 1$). The gasification temperature for the case of $Z = 1$ was reported almost constant for the complete experiment. This behaviour can be attributed to the fact of merging of combustion and gasification zone due to very less spacing between combustion zone air inlets and grate. The H_2 vol% goes through a maximum that occurred at temperature of around 800°C for all cases. This behaviour can be explained by the reversible nature of water gas shift (WGS) reaction ($CO + H_2O \rightarrow CO_2 + H_2$; $\Delta H =$ −40.6 MJ/kmol). The WGS reaction is exothermic in nature and backward WGS is favoured at higher temperature. Erlich and Fransson (2011) reported that the equilibrium constant of WGS reaction is equal to 1 at about 800 °C. Above this temperature, WGS reaction proceeds in backward direction resulting into the formation of more CO and H_2O and less H_2. On the other hand, above 800 °C, CO formation was supported by both the reactions *viz.* backward WGS reaction and

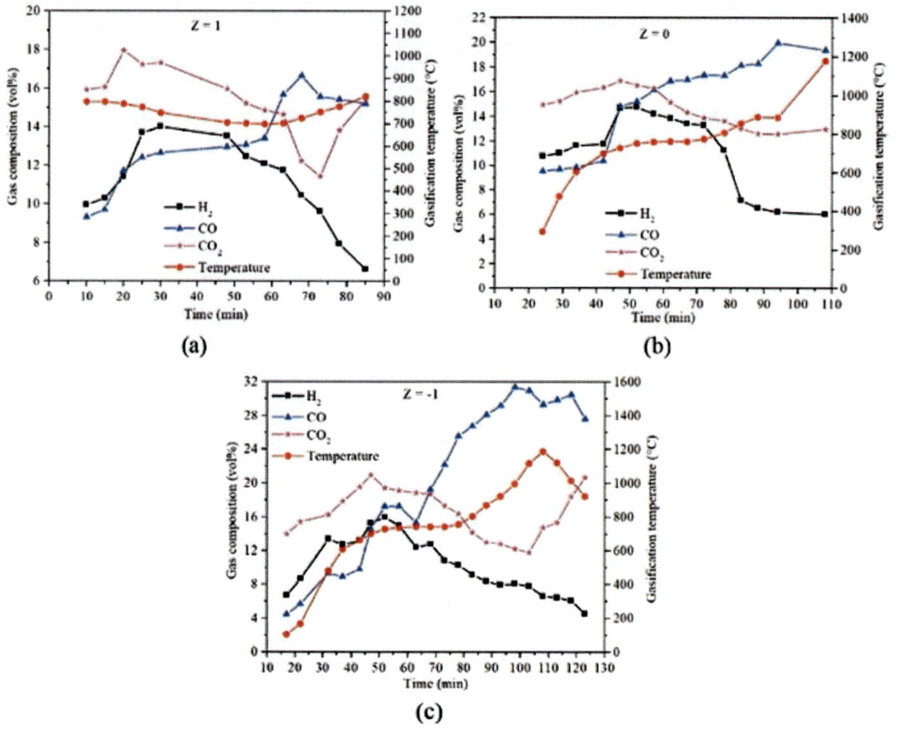

Fig. 4. Gas composition and gasification temperature variation with time (**a**) $Z = 1$, (**b**) $Z = 0$ and (**c**) $Z = -1$.

Boudouard reaction ($C + CO_2 \rightarrow 2CO$; $\Delta H = 172$ MJ/kmol). Therefore, CO concentration showed increasing trend with time and temperature. The vol% of CO_2 gas showed continuous decreasing trend due to major consumption by Boudouard reaction.

3.3 Average Producer Gas Composition and its Lower Heating Value

The gas samples were collected at a time interval of 5 min starting from the flame appearance time to flame goes off time for each run. Simultaneously, gas analysis was performed using Gas Chromatography to get the compositions (vol%) of H_2, N_2, CO, CH_4 and CO_2. The variation of gas composition is shown in Fig. 5(a). There was no significant variation in H_2 vol% with the change of grate positions. The CO volume percentage was found to be decreasing from 20.64% to 13.38% as the grate was shifted from the lowest position to top position. This fall in CO percentage can be explained by the decrease in residence time for the reactions to happen and

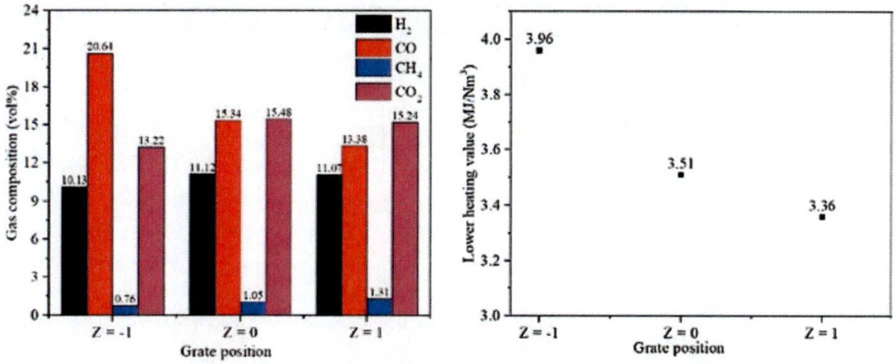

Fig. 5. (a) Average gas composition; (b) lower heating value of producer gas for different grate locations.

comparatively lower gasification temperature in case of $Z = 1$. The CO_2 vol% was found to behave opposite to that of CO vol%.

The lower heating value (LHV) of producer gas depends on its combustible components viz. H_2, CO and CH_4 and calculated as per the relation given in Siddiqui (2018). The variation of lower heating value of producer gas is shown in Fig. 5(b). The LHV of producer gas was found to be decreasing with the shifting of grate from lowest position to top position. It is due to that fact that H_2 and CH_4 volume percentages remained almost constant while CO was found to be decreasing as the grate was shifted to top position. This caused ultimately a decrease in LHV of producer gas.

3.4 *Flame Appearance and Flame Stability Time*

The variation of flame appearance and flame stability time are shown in Fig. 6. The flame appearance time is the time of getting flame at the burner after feedstock ignition and flame stability time is the time for which the flame remains stable. There was not any significant variation in flame appearance time with grate locations. The flame stability time was found to be increasing as the grate was shifted from top position to lowest position. The increase in flame stability time for lowest grate position can be attributed to more residence time that caused the better conversion of feed in to gas. Ash of pellets was obtained along with the coal for lowest position of grate whereas unconverted pellets were also collected with coal in other cases.

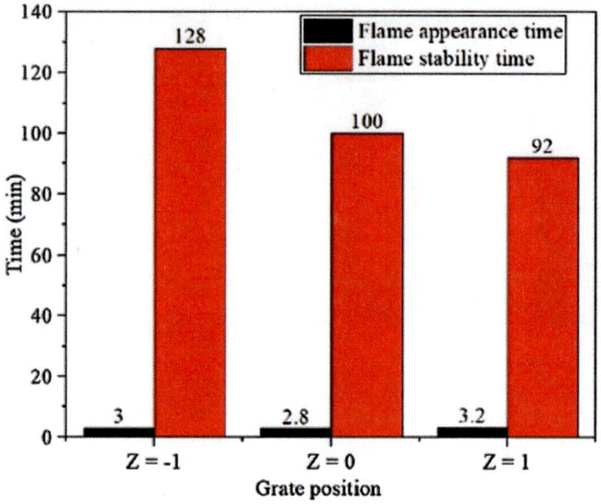

Fig. 6. Flame appearance and stability time for different grate locations.

4 Conclusions

The vertical distance of the grate from the combustion zone air inlet nozzles, decides the length of the gasification zone. Longer the length of gasification zone, higher will be the residence time for the solid and gaseous species. The residence time significantly affect the extent of the reactions occurring in the gasification zone. Therefore, the present study was conducted to obtain an optimal position of grate from combustion zone air inlets and thus, residence time. Comparatively higher gasification temperature was observed in case of lowest grate location ($Z = -1$) that triggered the endothermic reactions. This caused higher production of CO rather than the production of CH_4 and H_2. An appreciable increment in the composition of CO from 13.38% to 20.64% observed on shifting the grate from $Z = 1$ to $Z = -1$. This resulted into significant increment in the lower heating value of the producer gas that is very useful for the cooking point of view and also for other applications. For this position, LHV of gas was reported highest at 3.96 MJ/Nm^3, and flame stable time was maximum at 128 min. The flame appearance time was about 3 min that was not much affected by the grate positions. The amount of clinker formed was not also affected by the vertical positions of the grate.

Future Scope of Work The objective is to come up with the gasifier technology as a better alternative for the community level cooking. The lower heating value of the producer gas is an important parameter for such applications. The technology can be implemented to hostel messes, canteens and ashram schools. Since the technology is capable of incorporating other waste such as garden waste, agricultural waste as a feed, therefore it also enhances the waste management practices. As per the governmental scheme of mid-day meal, food is prepared in bulk in ashram schools and primary schools. They mostly use wood based traditional cook stoves that risk their life. The implementation of this technology would help in reducing the load on wood usage and also give a clean environment to work with.

Acknowledgements The authors would like to acknowledge Prof. Sanjay M. Mahajani, whose guidance and support facilitated the current research work. We would like to extend our thanks to Jayendran Shridharan and Ankush Jain for their valuable support in planning and performing the research work.

Funding This financial aid for the current research work was granted under the project DGDON 422 by the Tata Center for Technology and Design (TCTD), Indian Institute of Technology Bombay, India.

References

Alarcon-Gaete, F.J. and Elicer-Corted, J.C. (2016). Gasification of biomass in an updraft fixed bed reactor: Effect of viscous models. *J Adv Therm Sci Res*, **3(2):** 46–57.

Bhattacharya, S.C. and Leon, M.A. (2005). Prospects for biomass gasifiers for cooking applications in Asia. *In:* World Renewable Energy Regional Conference, Jakarta, Indonesia. pp. 18–21.

Erlich, C. and Fransson, T.H. (2011). Downdraft gasification of pellets made of wood, palm-oil residues respective bagasse: Experimental study. *Appl Energy*, **88:** 899–908.

Guo, F., Dong, Y., Dong, L. and Guo, C. (2014). Effect of design and operating parameters on the gasification process of biomass in a downdraft fixed bed: An experimental study. *Int J Hydrog. Energy*, **39(11):** 5625–5633.

Hernández, J.J., Aranda-Almansa, G. and Bula, A. (2010). Gasification of biomass wastes in an entrained flow gasifier: Effect of the particle size and the residence time." *Fuel Process Technol*, **91(6):** 681–692.

Martínez, J.D., Silva Lora, E.E., Andrade, R.V. and Jaén, R.L. (2011). Experimental study on biomass gasification in a double air stage downdraft reactor. *Biomass Bioenerg*, **35(8):** 3465–3480.

Parthasarathy, P. and Narayanan, K.S. (2014). Hydrogen production from steam gasification of biomass: Influence of process parameters on hydrogen yield – A review. *Renew Energy*, **66:** 570–579.

Pradhan, P., Arora, A. and Mahajani, S.M. (2018). Pilot scale evaluation of fuel pellets production from garden waste biomass. *Energy Sustain Dev*, **43:** 1–14.

Sheth, P.N. and Babu, B.V. (2009). Experimental studies on producer gas generation from wood waste in a downdraft biomass gasifier. *Bioresour Technol*, **100(12):** 3127–3133.

Siddiqui, H., Thengane, S.K., Sharma, S. and Mahajani, S.M. (2018). Revamping downdraft gasifier to minimize clinker formation for high-ash garden waste as feedstock. *Bioresour Technol*, **266:** 220–231.

Zainal, Z.A., Rifau, A., Quadir, G.A. and Seetharamu, K.N. (2002). Experimental investigation of a downdraft biomass gasifier. *Biomass Bioenerg*, **23(4):** 283–289.

Chapter 6
Utilisation of Natural Waste in Freeze-Thaw Affected Soil: A Comparative Analysis

Amit Kumar and Anupam Mital

1 Introduction

Road transport is a major part of the society. This facilitates the trade of goods and serves as a convenient mode of thoroughfare. But, the ease of transport is totally reliant on the condition of roads. In this case, sometimes environmental conditions and soil beneath the pavement may become a hurdle. This situation becomes very severe, when loose soil comes across the way to be considered as subgrade. Additionally, seasonal or frequent worst weather conditions make the situations more complicated. In this scenario, heavy snowfall/rainfall and permafrost are the examples of possible bottlenecks. Soil engineers seek soil stabilisation as a remedy to curb such problems. But, most of the time replacement of soil as a mechanical stabilisation is tedious and extravagant tactic. Then, addition of pertinent supplements is found to be an excellent choice. There are many natural or chemical supplements available in the market to stabilise the soil. In many cases, sort of fibres are also used additionally. But a cheap methodology is always appreciated. In this direction, the present research paper formulated an effective and inexpensive method over classic lime and cement stabilisation, which are popular among the geotechnical engineers.

Addition of cement along some hardening accelerators improved the unconfined compressive strength (UCS) of 7 days cured specimens. Cement, when used with flyash and fibres proved its efficacy to improve frost susceptibility by lessening the heave rate upto $\leq$4 mm/day. Cement stabilisation has been proved superior over lime stabilisation (Liu et al., 2010; Zhang et al., 2016; Chai et al., 2017). Aldaood et al. (2014, 2016) found a significant loss in UCS after repeated freeze-thaw (FT) cycles on lime and gypsum treated soil. While lime when used with natural pozzolana produced appreciable results with clayey soil of Syria (Al-Swaidani et al.,

A. Kumar (✉) · A. Mital
National Institute of Technology, Kurukshetra, Haryana, India

A. S. Kalamdhad (ed.), *Integrated Approaches Towards Solid Waste Management*, https://doi.org/10.1007/978-3-030-70463-6_6

2016). Under FT condition, lime and polymeric solution (CBR^+) provide excellent stability (Jigheh and Azarnia, 2017). Although cement and lime produced prominent results but the engineers found many drawbacks of these additives such as shrinkage or crack propagation with time. To handle this situation, some natural and synthetic fibres have come in light and are being used from a long time. In this direction, use of polypropylene fibre (PPF) in fine grained soil has been found very effective even though in worst environmental conditions too. In a study of natural versus steel fibres; jute fibres were found effective in terms of peak strength over steel fibres. 1.5% addition of natural coir fibres also found compatible with lime stabilised bentonite (Chaduvula et al., 2014; Gullu and Khudir, 2014; Tilak et al., 2015). Another category of fibres called basalt found less pronounced against FT effects than glass fibres and PPF (Orakoglu and Liu, 2017; Boz and Sezer, 2018; Kravchenko et al., 2018).

Calcium and sodium chloride compounds improved the engineering properties of clayey and expensive soils upto some extent under normal and weathered conditions. The improvements were the resultants of cation exchange between alkaline media and soil particles (Bhoi et al., 2013; Zumrawi and Eltayeb, 2016; Yunus et al. 2017; Liu et al., 2018). Due to the industrialisation, various by-products are menacing nuisance so, many researchers tried those globally unaccepted materials as additives for soil stabilisation. This step may spread relief to the society for pleasant and effective way of waste management.

Ashes from industries and natural wastes like flyash, groundnut shell ash, petrowaste, jaggery waste, gallnut powder, waste rubber tyre (chipped and fibered) etc. are the primary wastes; those were used by researchers and produced notable results (Shon et al., 2010; Roustaei and Ghazavi, 2011; Jafari and Ashari, 2012; Wei et al., 2015; Sujatha et al., 2016; James et al., 2018; Khazaei and Moayedi, 2019). Eggshell waste is produced by hatcheries, bakeries and fast food joists etc. and neglected by the dumpsters usually. This natural waste cannot be easily incinerated and although causes pungent environment over the time. So, a wise solution is needed in this direction. As the chemical composition of eggshell powder (ESP) proved it as a rich source of lime (Adogla et al., 2016). So, this waste material can be used as a soil stabiliser effectively. ESP proved its efficacy in soil stabilisation to treat most kind of soils. The ameliorating effects of ESP were found amazing in the past studies (Barazesh et al., 2012; Elizabeth and Betsypaul, 2016; Haruna et al., 2017; Oluwatuyi et al., 2018). This research paper focuses on the combined effect of ESP, NaCl and PPF on fine grained soil under normal and FT conditions. The results were compared with an existing literature to make some comparative analyses regarding usage of cement and other additives in combination.

2 Materials and Methods

Soil was grabbed from Jammu & Kashmir (India). Before processing the soil for lab purposes, soil was pulverised and cleansed from impurities. Laboratory reported soil properties are given in Table 1. ESP, PPF and NaCl were used to treat the soil. The dosage levels are shown in Table 2. Milled and 425 μm sieved ESP was used throughout the testing and its typical chemical composition is CaO 99.83%; Cl 0.009%; Al_2O_3 0.001% and SiO_2 0.001% (Adogla et al., 2016). Crystal pure NaCl (Density at 0 °C 2.17 g/cm^3) was procured from local chemical seller and used in the experiments. PPF (Triangular-12 mm) was the product of Reliance industries and used in arbitrarily distributed manner. Sample matrix was according to Taguchi's L9 orthogonal array as shown in Table 3. Indian Standards for soil testing were used for the determination of soil properties. Liquid limit of the soil specimens was determined by the Cone penetrometer and calculated by using one-point method. For both cured and conditioned (F-T), UCS samples were sized to 38 mm (diameter) × 76 mm (height). Samples were kept under −25 °C and 25 °C for freezing and thawing respectively, while performing the freeze and thaw test. For the determination of mass loss samples were kept for 24 hours in the freezing-thawing chamber while for strength determination samples were kept for 6 hours in each phase since after 6 hours samples produce no significant changes (Ghazavi and Roustaei, 2010; Zaimoglu et al., 2016). During mass loss (ML) determination, soft bristle brush was used to scratch the soil instead of steel wire brush. This precaution avoids the rupturing of samples due to sticking of the fibres in the steel wires. Although the samples were tested according to DOE for all cases viz. UCS and UCS-ML (F-T), but the present paper focuses on the results obtained from optimised samples only with respect to parent soil specimen (Kumar and Soni, 2019a, b).

Table 1. Engineering properties of soil

Property		*Value*
Grain size distribution data	Gravel (%)	0
	Sand (%)	0
	Clay + silt (%)	100
Specific gravity		2.6
Liquid limit (%)		34
Plastic limit (%)		29
Plasticity index (%)		05
Indian standard classification		ML
OMC* (%)		19.33
MDD*(%)		1.72
OMC$^{\#}$ (%)		14.40
MDD$^{\#}$ (g/cc)		1.84

*Obtained by standard proctor test
Obtained by modified proctor test

Table 2. Factors and their levels

Factors			
Level	*ESP (%)*	*PPF (%)*	*NaCl (%)*
1	3	0.05	2
2	6	0.10	4
3	9	0.15	6

Table 3. Experimental programme (orthogonal array)

	Factors		
Trial	*ESP*	*PPF*	*NaCl*
R1	3	0.05	2
R2	3	0.10	4
R3	3	0.15	6
R4	6	0.05	4
R5	6	0.10	6
R6	6	0.15	2
R7	9	0.05	6
R8	9	0.10	2
R9	9	0.15	4

3 Results and Discussion

All the laboratory test results have been presented by bar graphs. Figure 1 shows the graphical presentation of UCS of 7, 14 and 21 days cured optimised samples in normal conditions with respect to parent soil and the variation is also incorporated in the graph.

Figure 1 explains a significant variation in the increment of the compressive strength as the age of the sample increases. This variation can be seen from 6–95% when samples aged about three weeks. Probable reason for the improvement in the strength characteristics was the commencement of the pozzolanic reactions which are the essential property of the calcareous materials and are responsible for the development of binding properties in the soil-additive matrix. These reactions are the reactive resultant of present silica and alumina in the soil and; calcium and chloride in the additives. Some of the possible pozzolanic reactions are as follows and further explanation can be read from Kumar and Soni (2019a).

$$CaO + H_2O \longrightarrow Ca(OH)_2 \tag{1}$$

$$Ca(OH)_2 \longrightarrow Ca^{++} + 2(OH^-) \tag{2}$$

$$\underset{}{Ca^{++}} + 2(OH^-) + \underset{(Silica)}{SiO_2} \longrightarrow \underset{(Gel)}{CSH} \tag{3}$$

$$Ca^{++} + 2(OH^-) + \underset{(Alumina)}{Al_2O_3} \longrightarrow \underset{(Gel)}{CAH} \tag{4}$$

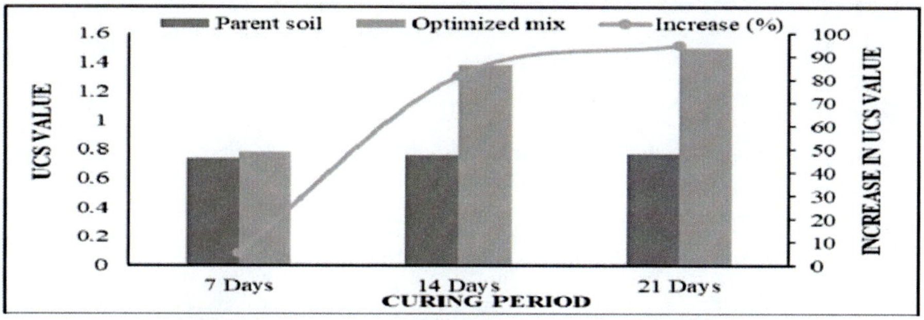

Fig. 1. Variation of UCS values at different curing periods.

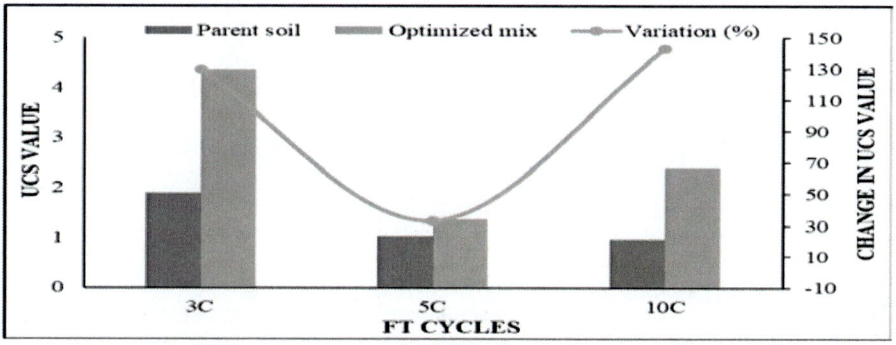

Fig. 2. Variation of UCS values at diffrerent freeze-thaw cycles.

Figure 2 shows the variation of the laboratory reported UCS of the three weeks aged and FT cycles (3, 5 and 10 cycles) affected soil samples. After blow down by the repeated FT cycles, soil experiences some internal chemical reactions that changes the whole orientation of the soil particles. This causes a drastic variation in the results of the strength properties of the subjected soil as shown in Fig. 2. After five FT cycles there was a huge consumption of the calcareous material for the pozzolanic reactions those affected the soil specimens inversely and strength was decreased drastically as shown in Fig. 2. But after the 10th FT cycle soil started to gain strength due to regeneration of hydrous products such as calcium silicate hydrous (CSH) gel.

Figure 3 depicts the durability of the subjected soil and optimised mixes under repetitive FT cycles. A high resistance can also be seen in Fig. 3. Lower mass loss and higher resistance shows the feasibility of the additives to curb the negative impacts of frequently occurred FT cycles. Apart from the formation of CSH gel, possibly fibres were also the reason for the resistance against mass loss because the

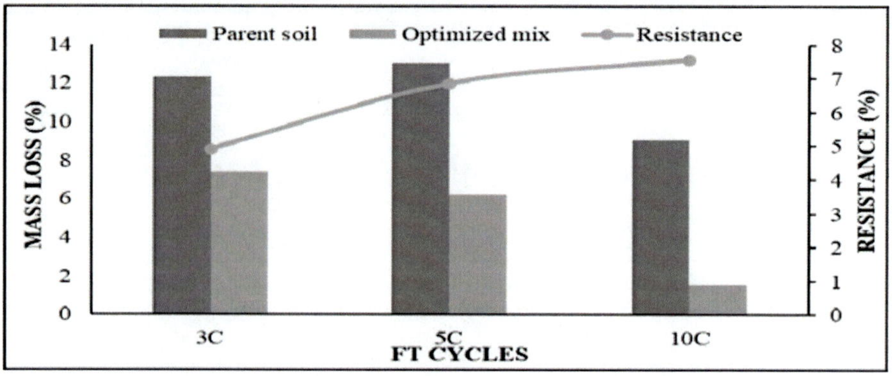

Fig. 3. Variation of mass loss at diffrerent freeze-thaw cycles.

fibres can bind the soil particles due to their sticky nature. More explanation for the results can be found in Kumar and Soni (2019b). The value of the resistance was found 33% better as found by Zaimoglu (2015a, b).

4 Conclusions

The present paper explains the consequences of the incorporation of calcium enriched eggshell powder and sodium chloride in polypropylene fibre reinforced Himalayan soil. The soil strength and durability properties were tested under normal and frozen conditions. After examining the soil samples following conclusions can be drawn.

- Calcareous materials are responsible for the pozzolanic reactions and these reactions are time dependent.
- Presented silica in the soil aid to produce CSH gel but due to scarcity or absence of another pozzolanic material; available gel may get deteriorated.
- Fibres stick to the soil particles and help to bind them together strongly to withstand against mass loss.
- Use of the waste materials like eggshells in soil stabilisation will be effective to solve environmental issues.
- Utilisation of waste materials and low-cost additives will be helpful to produce better end products at low cost, consequently will be economical too.

Authors suggested the researchers to carry out further research using other type of soils and additives for more clarity in results. Higher orthogonal array matrix may also be used to find out the more optimised results.

Acknowledgements Authors wish to thank CRRI, New Delhi for providing lab facility for testing and CSIR, New Delhi for providing the funds for the research. Authors heartily thank Dr. Hari Singh, Professor, Department of Mechanical Engineering, NIT Kurukshetra for their computational assistance regarding MINITAB software.

References

Adogla, F., Yalley, P.P.K. and Arkoh, M. (2016). Improving compressed laterite bricks using powdered eggshells. *Int J Eng Sci (IJES)*, **5(4):** 65–70.

Aldaood, A., Bouasker, M. and Al-Mukhtar, M. (2014). Impact of freeze-thaw cycles on mechanical behaviour of lime stabilized gypseous soils. *Cold Regions Sci Technol*, **99:** 38–45. https://doi.org/10.1016/j.coldregions.2013.12.003.

Aldaood, A., Bouasker, M. and Al-Mukhtar, M. (2016). Effect of water during freeze–thaw cycles on the performance and durability of lime treated gypseous soil. *Cold Regions Sci Technol*, **123:** 155–163. https://doi.org/10.1016/j.coldregions.2015.12.008.

Al-Swaidani, A., Hammoud, I. and Meziab, A. (2016). Effect of adding natural pozzolana on geotechnical properties of lime-stabilized clayey soil. *J Rock MechGeotech Eng*, **8:** 714–725. https://doi.org/10.1016/j.jrmge.2016.04.002.

Barazesh, A., Saba, H., Gharib, M. and Rad, M.Y. (2012). Laboratory investigation of the effect of eggshell powder on plasticity index in clay and expansive soil. *Eur J Exp Biol*, **2(6):** 2378–2384.

Bhoi, A.K., Momin, L.W. and Hajong, A.K. (2013). Alternation of silty clay soil parameter by using chloride compounds. *Int J Conceptions Mech Civil Eng*, **1(1):** 34-37. ISSN: 2357–2760.

Boz, A. and Sezer, A. (2018). Influence of fiber type and content on freeze-thaw resistance of fiber reinforced lime stabilized clay. *Cold Regions Sci Technol*. https://doi.org/10.1016/j.coldregions.2018.03.026.

Chaduvula, U., Desai, A.K. and Solanki, C.H. (2014). Application of triangular polypropylene fibres on soil subjected to freeze-thaw cycles. *Indian Geotech J*, **44(3):** 351-356. https://doi.org/10.1007/s40098-013-0088-9.

Chai, M., Zhang, H., Zhang, J. and Zhang, Z. (2017). Effect of cement additives on unconfined compressive strength of warm and ice-rich frozen soil. *Constr Build Mater*, **149:** 861–868. https://doi.org/10.1016/j.conbuildmat.2017.05.202.

Elizabeth, K.S. and Betsypaul, K. (2016). Improvement of lateritic clay using eggshell powder & marble dust and its comparison with lime stabilized clay. *Int J Scientific Eng Res (IJSER)*, **4(3):** 8–10.

Ghazavi, M. and Roustaie, M. (2010). The influence of freeze–thaw cycles on the unconfined compressive strength of fiber-reinforced clay. *Cold Regions Sci Technol*, **61:** 125–131. https://doi.org/10.1016/j.coldregions.2009.12.005.

Gullu, H. and Khudir, A. (2014). Effect of freeze-thaw cycles on unconfined compressive strength of fine-grained soil treated with jute fiber, steel fiber and lime. *Cold Regions Sci Technol*, **106(107):** 55–65. https://doi.org/10.1016/j.coldregions.2014.06.008.

Haruna, M., Kundiri, A.M. and Yero, S.A. (2017). Effect of compactive effort on strength characteristics of black cotton soil admixed with eggshell powder-gum arabic. *Int Res J Eng Technol*, **4(7):** 316–322.

Jafari, M. and Esna-ashari, M. (2012). Effect of waste tire cord reinforcement on unconfined compressive strength of lime stabilized clayey soil under freeze-thaw condition. *Cold Regions Science and Technology*, **82:** 21–29. https://doi.org/10.1016/j.coldregions.2012.05.012.

James, J., Karthickeyan, S., Chidambaram, S., Dayanandan, B. and Karthick, K. (2018). Effect of curing conditions and freeze-thaw cycles on the strength of an expansive soil stabilized with a combination of lime, jaggery, and gallnut powder. *Advances in Civil Engineering*, 1–9. https://doi.org/10.1155/2018/1813563

Jigheh, H.S. and Azarnia, A. (2017). Effect of liquid polymeric and lime additives on the behavior of fine-grained soil at unfrozen and freeze–thaw conditions. *Indian Geotech J*, **47(4):** 529–536. https://doi.org/10.1007/s40098-017-0274-2.

Khazaei, J. and Moayedi, H. (2019). Soft expansive soil improvement by eco-friendly waste and quick lime. *Arab J Sci Eng*, **44(10):** 8337–8346. https://doi.org/10.1007/s13369-017-2590-3.

Kravchenko, E., Liu, J., Niu, W. and Zhang, S. (2018). Performance of clay soil reinforced with fibers subjected to freeze-thaw cycles. *Cold Regions Sci Technol.* **153:** 18–24. https://doi.org/10.1016/j.coldregions.2018.05.002.

Kumar, A. and Soni, D.K. (2019a). Strength and microstructural characteristics evaluation of a fibre reinforced fine grained soil using Taguchi technique. *Int J Mater Microstruct Propert*, **14(5):** 478–494. https://doi.org/10.1504/IJMMP.2019.102225.

Kumar, A. and Soni, D.K. (2019b). Strength and microstructural characterisation of plastic soil under freeze and thaw cycles. *Indian Geotech J*, 1-13. https://doi.org/10.1007/s40098-019-00372-8.

Liu, J., Wang, T. and Tian, Y. (2010). Experimental study of the dynamic properties of cement- and lime-modified clay soils subjected to freeze–thaw cycles. *Cold Regions Sci Technol*, **61:** 29–33. https://doi.org/10.1016/j.coldregions.2010.01.002.

Liu, L., Li, Z., Liu, X. and Li, Y. (2018). Effect of salt content on freezing temperature and unconfined compression strength of lime-treated subgrade clay. *Appl Clay Sci*, **158:** 65-71. https://doi.org/10.1016/j.clay.2018.03.022.

Oluwatuyi, O.E., Adeola, B.O., Alhassan, E.A., Nnochiri, E.S., Modupe, A.E., Elemile, O.O., Obayanju, T. and Akerele, G. (2018). Ameliorating effect of milled eggshell on cement stabilized lateritic soil for highway construction. *Case Stud Constr Mater*, **e00191:** 1–9. https://doi.org/10.1016/j.cscm.2018.e00191.

Orakoglu, M.E. and Liu, J. (2017). Effect of freeze-thaw cycles on triaxial strength properties of fiber-reinforced clayey soil. *KSCE J Civil Eng*, **21(6):** 2128-2140. https://doi.org/10.1007/s12205-017-0960-8

Roustaei, M. and Ghazavi, M. (2011). Strength characteristics of clay mixtures with waste materials in freeze-thaw cycles. *J Struct Eng Geotech*, **1(2):** 57-62.

Shon, C.S., Donald, S. and Mishra, S.K. (2010). Combined use of calcium chloride and fly ash in road base stabilization. *J Transp Res Board*, **2186:** 120–129. https://doi.org/10.3141/2186-13.

Sujatha, E.R., Dharini, K. and Bharathi, V. (2016). Influence of groundnut shell ash on strength and durability properties of clay. *Geomechanics and Geoengineering: An International Journal*, **11(1):** 20–27. https://doi.org/10.1080/17486025.2015.1006265

Tilak, B.V., Dutta, R.K. and Mohanty, B. (2015). Effect of coir fibres on the compaction and unconfined compressive strength of Bentonite-lime-gypsum mixture. *Slovak J Civil Eng*, **23(2):** 1–8. https://doi.org/10.1515/sjce-2015-0006.

Wei, H., Jiao, Y. and Liu, H. (2015). Effect of freeze-thaw cycles on mechanical property of silty clay modified by fly ash and crumb rubber. *Cold Regions Science and Technology*, **116:** 70–77. https://doi.org/10.1016/j.coldregions.2015.04.004

Yunus, N.Z.M., Wanatowski, D., Marto, A., Jusoh, S.N. (2017). Strength improvement of lime-treated clay with sodium chloride. Geotechnical Research https://doi.org/10.1680/jgere.17.00001

Zaimoglu, A.S., Calik, Y., Akbulut, R.K. and Yetimoglu, T. (2016). A study on freeze-thaw behavior of randomly distributed fiber-reinforced soil. *Periodica Polytechnica Civil Eng*, **60(1):** 3–9. https://doi.org/10.3311/PPci.7533.

Zaimoğlu, A.S. (2015a). The effect of randomly distributed polypropylene fibers borogypsum fly ash and cement on freezing-thawing durability of a fine-grained soil. *Int Schol Sci Res Innov*, **9(8):** 1093–1097. https://scholar.waset.org/1999.3/10003040.

Zaimoğlu, A.S. (2015b). Optimization of unconfined compressive strength of fine-grained soils modified with polypropylene fibers and additive materials. *KSCE J Civil Eng*, **19(3):** 578–582. https://doi.org/10.1007/s12205-015-1425-6.

Zhang, Y., Johnson, A.E. and White, D.J. (2016). Laboratory freeze–thaw assessment of cement, fly ash, and fiber stabilized pavement foundation materials. *Cold Regions Sci Technol*, **122:** 50–57. https://doi.org/10.1016/j.coldregions.2015.11.00523.

Zumrawi, M.M.E. and Eltayeb, K.A. (2016). Laboratory investigation of expansive soil stabilized with calcium chloride. *Int Scholar Sci Res Innov*, **10(2):** 223–227. https://scholar.waset.org/1307-6892/10003855.

Chapter 7
Economic and Ecological Feasibility of Marble Powder in Cement Mortar

Kiran Devi, Babita Saini, and Paratibha Aggarwal

1 Introduction

The generation of industrial wastes is increasing at a faster rate with the growth of the industries; and safe disposal of these wastes has become the main concern of the society. The wastes are directly dumped into open land areas or in watercourses, which pollute the air, land and water. Also, the cost of transportation is an additional burden to the industries. Like other industries, stone industry is an important construction industry and stone is the main building material. India is the third-largest country in stone production. A variety of stones, i.e. marble, limestone, granite, Kota stone etc. are available in India. A large amount of waste is produced during the processing of stones. This waste can be stone pieces, blocks, dust, slurry etc. and is directly dumped into open areas, which causes problems to the environment. This needs safe disposal of stone waste to avoid these environmental problems (Rajamane et al., 2011; Singh et al., 2018; Devi et al., 2018b). In the agreement of the Kyoto Protocol, all the big industries around the world are adopting policies, which encourage reduction in emission of greenhouse gases. The possible way is to reuse the industrial by products in various applications. The construction industry should take a step forward for the sustainable production by utilising various industrial wastes in cement-based materials as a substitution to cement or aggregates. This will conserve the natural raw materials and decrease the nuisance occurring due to disposal problems of waste or by-products (Kabeer and Vyas, 2019).

The utilisation of marble slurry in concrete reduced the cost, energy consumption and carbon footprint (Singh et al., 2017). The incorporation of marble powder with silica fume improved the performance of concrete and reduced harmful impact on the environment (Khodabakhshian et al., 2018). The inclusion of ground granulated blast furnace slag reduced CO_2, which promotes sustainable and green

K. Devi (✉) · B. Saini · P. Aggarwal
Civil Engineering Department, National Institute of Technology Kurukshetra, Kurukshetra, Haryana, India

A. S. Kalamdhad (ed.), *Integrated Approaches Towards Solid Waste Management*, https://doi.org/10.1007/978-3-030-70463-6_7

construction (Kim et al., 2018). The inclusion of sewage sludge ash as cement substitution reduced the global warming potential by 9% (Nakic, 2018). The use of marble powder reduced the strength of mortar (Devi et al., 2018a).

In the present research, marble powder (MP) was used as a partial substitute to cement by 0%, 5%, 7.5% and 10% in cement mortar to study its economic and ecological aspect. The cost of mortar, embodied energy (EE) and embodied carbon dioxide (ECO_2) of mortar containing MP were evaluated and compared with the plain mix. The EE, ECO_2 and cost of mortar per unit strength were also evaluated. The goal of the study was to check the feasibility of marble powder in mortar in terms of cost and sustainability aspects.

2 Materials and Methods

In the present study, ordinary Portland cement of 43 grade conforming to IS: 8112-1989 and coarse sand conforming to IS: 383-2016 were used. The waste (powder) from the marble industry was used as cement replacement by 0%, 5%, 7.5% and 10% (Devi et al., 2018a). The details of various mix proportions have been given in Table 1.

Methodology To assess the sustainability of mortar/concrete, energy consumption (EE) and emission of CO_2 (ECO_2) are the parameters/indicators. Cement is the main ingredient for the emission of CO_2 in the mortar/concrete production followed by aggregates (Long et al., 2017). The values of EE and ECO_2 for each raw material, i.e. cement, sand and water have been taken from literature (Gupta et al., 2019; Siddique et al., 2018) and mentioned in Table 2. These values for MP were considered as nil because it was a by-product and did not require any further processing. These values are used to calculate the value of EE, ECO_2 and cost of concrete by using equation (1). EE, ECO_2 and cost per unit strength was also evaluated (Rajamane et al., 2012). Note that emission and cost related to the transportation of raw materials and production of concrete have not been considered. The price value of materials was taken as per local market rates.

$$EE / ECO_2 / Cost = \sum g_i m_i \tag{1}$$

where $g_i = EE$ per unit mass of materials and m_i indicate the mass of concrete's ingredients i per unit cubic meter.

Table 1. Mix designation of mortar

Mix detail	*MP0*	*MP5*	*MP7.5*	*MP10*
Cement, kg/m^3	575	546.25	531.87	517.5
Sand, kg/m^3	1725	1725	1725	1725
Water, kg/m^3	158.3	158.3	158.3	158.3
Marble powder, kg/m^3	0	28.75	43.13	57.5

Table 2. Factor for EE, ECO_2 and cost of raw materials

Materials	*Cement*	*Sand*	*Water*	*MP*
EE, MJ/kg	4.8	0.081	0.2	–
ECO_2, $kgCO_2/kg$	0.93	0.0051	0.0008	–
Cost, INR/kg	6	1	0.05	–

Table 3. Effect of MP on EE, ECO_2 and cost with or without per unit strength

Mix no.	*EE (MJ/m^3)*	*ECO_2 ($kgCO_2/m^3$)*	*Cost (INR/m^3)*	*EE/28D-S*	*ECO_2/28D-CS*	*Cost/28D-CS*
MP0	2931.38	622.85	5182.91	72.77	15.46	128.67
MP5	2793.38	596.11	5010.41	78.77	16.81	141.29
MP7.5	2724.38	582.74	4924.16	70.91	15.16	128.16
MP10	2655.38	569.38	4837.91	84.75	18.17	154.42
Remarks	Less energy-intensive	Less CO_2 emission	Costlier	Less energy for unit strength	Emit less CO_2 for unit strength	Lower cost for unit strength

Performance evaluation of mortar mixes was also carried out to check the desirability of mixes as per the requirements. The performance indices were evaluated for the individual criteria (El-Dieb and Kanaan, 2018).

3 Results and Discussion

The effect of marble powder as cement substitute in mortar was investigated in terms of economic and ecological aspects. The economic analysis was carried out by calculating the cost of mortar and; ecological analysis was carried out in terms of embodied energy and emission of CO_2 to the environment. In this study, values of EE, ECO_2 and cost of mortar varied from 2655.38 to 2931.38 MJ/m^3, 569.38 to 622.85 $kgCO_2/m^3$ and 4837.9 to 5182.9 INR/m^3, respectively for all the mix proportions and variation in EE, ECO_2 and cost have been given in Table 3.

It has been observed from Table 3 that inclusion of MP reduced the energy consumption and emission of CO_2 along with the reduction in cost of mortar due to reduction in cement content. The reduction of EE, ECO_2 and cost varied from 5–9%, 4–9% and 3–7%, respectively as compared to control mix, i.e. mix without MP. The incorporation of MP reduced the compressive strength of cement mortar due to very high fineness of particles of marble powder (Devi et al., 2018a). On the other hand, the inclusion of MP as cement replacement reduced the energy consumption and emission of carbon dioxide along with a reduction in cost as cement is the main factor responsible for energy consumption and CO_2 emission. The exercise of marble powder in mortar produced less energy-intensive, lower CO_2 emission and economical end products in comparison to control mix, which leads to a small step to green

Table 4. Performance indices for individual criteria

Mix no.	*28 DCS*	*EE*	ECO_2	*Cost*	*EE/28D-CS*	ECO_2*/28D-CS*	*Cost/28D-CS*
MP0	1.00	1.00	1.00	1.00	0.86	0.85	0.83
MP5.0	0.88	0.95	0.96	0.97	0.93	0.93	0.91
MP7.5	0.95	0.93	0.94	0.95	0.84	0.83	0.83
MP10	0.78	0.91	0.91	0.93	1.00	1.00	1.00

construction, i.e. sustainability. The use of MP reduced the energy required for the unit strength, emission of CO_2 for unit strength and cost for the unit strength. The mix MP 7.5 had the lowest values for EE, ECO_2 and cost per unit strength among all the mix proportions because of reduction in cement quantity and strength was also high in comparison to all replacements; while MP10 had the highest one due to reduction in strength.

Performance Evaluation Performance index (PI) tool encourages to find the content of MP to produce the most suitable mixtures conforming with the required performance criteria. The numeric index (R_i) was calculated using equation (2) and the highest numeric index was taken as 1.00 (El-Dieb and Kanaan, 2018). The performance indices for individual criteria of mortar mixes have been given in Table 4.

$$R_i = \frac{\text{Measured performance for each mixture}}{\text{Best measure performance}} \times 1 \tag{2}$$

From the strength criteria, mix with highest value of numeric index is the desired performance; while for cost and ecological aspects lowest value of numeric index is the desired performer. When strength is the main required property, then mix MP0 can be used to achieve the desired properties as observed from Table 4. MP10 can be used to produce more economical and ecofriendly end products as compared to the control mix. If lower energy consumption, CO_2 emission and cost for the unit strength is the main criteria, then mix proportion of MP7.5 can be used to fulfil the desired objective. Results showed that MP at 7.5% can be utilised as cement substitution in mortar with desired performance criteria.

4 Conclusions

In the present study, industrial waste, i.e. marble powder from the stone industry was used as cement replacement in cement mortar to study its economic and ecological aspects. To examine the economic aspect of waste in mortar, cost analysis of mortar mixtures was carried out. The ecological analysis of marble powder in mortar was studied by calculating the emission of carbon dioxide and energy consumption due to inclusion of marble powder. The incorporation of marble powder decreased the energy consumption, emission of CO_2 and cost of mortar production. Although the exercise of marble powder reduced the strength in comparison to control mix, but replacement at 7.5% had better strength than other replacement content. The performance indices of individual criteria were also evaluated and MP7.5 was found to be better in terms of less emission, less energy-intensive and low cost per unit strength. Therefore, marble powder as a partial substitution to cement can be used in the mortar construction where high strength is not the main concern. The utilisation of marble powder not only reduces the cement content but also solves the safe disposal of marble waste consequently encouraging the sustainable construction.

Acknowledgements This work was supported by the Ministry of Human Resources and Development, Govt. of India [Reg. No. 2K15-NITK-Ph. D-6150017].

References

Devi, K., Saini, B. and Aggarwal, P. (2018a). Effect of accelerators with waste material on the properties of cement paste and mortar. *Comput Concrete*, **22:** 153–159.

Devi, K., Acharya, K.G. and Saini B. (2018b). Significance of stone slurry powder in normal and high strength concrete. *In:* Singh, H., Garg, P. and Kaur, I. (Eds.), Proceedings of the 1st International Conference on Sustainable Waste Management through Design. ICSWMD 2018. *Lect Notes Civil Eng*, **21:** 484–492.

El-Dieb Amr, S. and Kanaan, D.M. (2018). Ceramic waste powder an alternative cement replacement – Characterization and evaluation. *Sustainable Mater Technol*, **17**.

Gupta, T., Kothari, S., Siddique, S., Sharma, R.K. and Chaudhary, S. (2019). Influence of stone processing dust on mechanical, durability and sustainability of concrete. *Construct Build Mater*, **223:** 918–927.

IS 383 (2016). Indian Standard Specification for Coarse and Fine Aggregates from Natural Sources for Concrete. Bureau of Indian Standard, New Delhi, India.

IS 8112 (1989). Specification for 43-grade Ordinary Portland Cement. Bureau of Indian Standard, New Delhi, India.

Kabeer, K.I.S.A. and Vyas, A.K. (2019). Experimental investigation on utilization of dried marble slurry as fine aggregate in lean masonry mortars. *J Build Eng*, **23:** 185–192.

Khodabakhshian, A., de Brito, J., Ghalehnovi, M. and Shamsabadi, E.A. (2018). Mechanical, environmental and economic performance of structural concrete containing silica fume and marble industry waste powder. *Construct Build Mater*, **169:** 237–251.

Kim, Y., Hanif, A., Usman, M., Munir, M.J., Kazmi, S.M.S. and Kim, S. (2018). Slag waste incorporation in high early strength concrete as cement replacement: Environmental impact and influence on hydration & durability attributes. *J Cleaner Prod*, **172:** 3056–3065.

Long, W., Gu, Y., Liao, J. and Xing, F. (2017). Sustainable design and ecological evaluation of low binder self-compacting concrete. *J Cleaner Prod*, **167:** 317–325.

Nakic, D. (2018). Environmental evaluation of concrete with sewage sludge ash based on LCA. *Sustain Prod Consum*, **16:** 193–201.

Rajamane, N.P., Nataraja, M.C. and Lakshmanan, N. (2011). An introduction to geopolymer concrete. **November:** 1–4.

Rajamane, N.P., Nataraja, M.C., Dattatreya, J.K., Lakshmanan, N. and Sabitha, D. (2012). Sulphate resistance and eco-friendliness of geopolymer concretes, **January:** 13–22.

Siddique, S., Chaudhary, S., Shrivastava, S. and Gupta, T. (2018). Sustainable utilisation of ceramic waste in concrete: Exposure to adverse conditions. *J Cleaner Prod*, **210:** 246–255.

Singh, M., Choudhary, K., Srivastava, A., Sangwan, K.S. and Bhunia, D. (2017). A study on environmental and economic impacts of using waste marble powder in concrete. *J Build Eng*, **13:** 87–95.

Singh, H., Garg, P. and Kaur, I. (2018). Proceedings of the 1st International Conference on Sustainable Waste Management through Design. Springer, Nature Switzerland AG.

Chapter 8
Use of Gas Turbine Operated by Municipal Solid Waste to Obtain Power and Cooling Assisted by Vapour Absorption Refrigeration System

Kamaljyoti Talukdar

1 Introduction

At present, waste such as MSW (municipal solid waste) is increasing at a large scale. With rapid modernisation and improved standard of living MSW generation is increasing at a greater magnitude. If MSW is utilised constructively, it will be a great advantage. The present paper is based on the utilisation of MSW by a gas turbine to operate the turbine to produce electrical power and cooling by vapour absorption refrigeration system in May and January. Many works on cooling by gas turbines have been done. Masri (1986) identified cooling technologies from the high-temperature air coming out from gas turbines such as air-transpiration, open-loop and closed-loop water cooling. Masri (1987) used the second law model to show stage-by-stage turbine cooling flow and loss analysis calculations. Bayley and Turner (1970) described a program of experimental and analytical research for designing to evaluate the aerodynamic and thermodynamic performance of transpiration-cooled porous surfaces in the high-temperature gas turbine. Torbidoni and Horlock (2005) developed a more sophisticated method from the earlier work (i.e. a computational method of estimating the cooling flow requirements of blade rows in a high-temperature gas turbine, for convective cooling alone and convective plus film cooling) and used it to calculate the cooling flow required for a nozzle guide vane (the first blade row) of a high-temperature gas turbine, with given inlet gas temperature and coolant inlet temperature. Zhang et al. (2014) demonstrated the use of pressure-sensitive paint (PSP) to measure film cooling effectiveness on a turbine nozzle surface in a high-speed wind tunnel. Bhargava et al. (2010) discussed the gas turbine performance enhancement approach in which modified Brayton cycles were used with humidification or water/steam injection or with fuel cells, obtained for hybrid cycles.

K. Talukdar (✉)
Department of Mechanical Engineering, Bineswar Brahma Engineering College, Kokrajhar, Assam, India

A. S. Kalamdhad (ed.), *Integrated Approaches Towards Solid Waste Management*, https://doi.org/10.1007/978-3-030-70463-6_8

2 Material and Methods

Figure 1 shows a brief layout of the gas turbine plant. Air at ambient temperature (Kolkata city) (Tiwari, 2004) enters the compressor at 1. It is compressed and sent to the combustion chamber at 2 where MSW (fuel) is burnt to heat compressed air to 1000 K and sent to the turbine at 3. The composition of MSW (fuel) is obtained from Talukdar (2019) and calorific value from Kaushik and Singh (2013). In the turbine, 1 MW electrical power is obtained and remaining exhaust air after producing power from the turbine is sent to the generator of VARS maintained at 80 °C to produce cooling.

The description of the *T–S* diagram in Fig. 2 is similar as mentioned in Talukdar (2019) for process 1–2–3–4. The detailed calculations for calculating temperatures at 1, 2′, 2, 3, 4, 4′ are obtained from Talukdar (2019). The mass flow rate of air (m_a) to compressor and fuel (MSW) (m_f) to the combustion chamber are obtained from Talukdar (2019) after detailed calculations. After 4, the combusted air is sent to the VARS generator till the exhaust air reaches 5′ where temperature attains 80 °C (VARS generator temperature) and released to ambient air. For the next cycle, again fresh air is taken at 1.

The working of VARS shown in Fig. 3 is not described here since it is available in various literature. The assumed values of temperatures for evaporator (T_e) is 10 °C, absorber (T_a) and condenser (T_c) is the ambient temperature of Kolkata city for May and January (Tiwari, 2004), and generator temperature (T_g) is 80 °C. Heat absorbed by evaporator and generator is given by Q_E and Q_G respectively. Heat rejected by absorber and condenser is given by Q_A and Q_C respectively. For calculating actual COP (coefficient of performance), (COP_{actual}), Carnot COP (COP_{Carnot}) is to be known based on temperatures of evaporator, absorber, generator and condenser. Based on Carnot COP, actual COP is assumed to some degree less than Carnot COP.

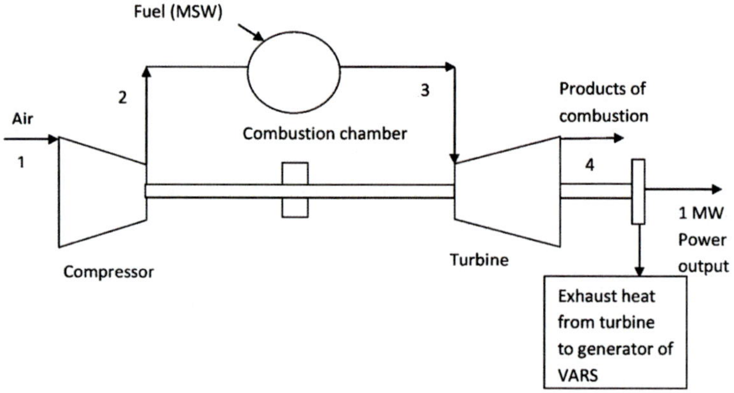

Fig. 1. Schematic view of gas turbine plant.

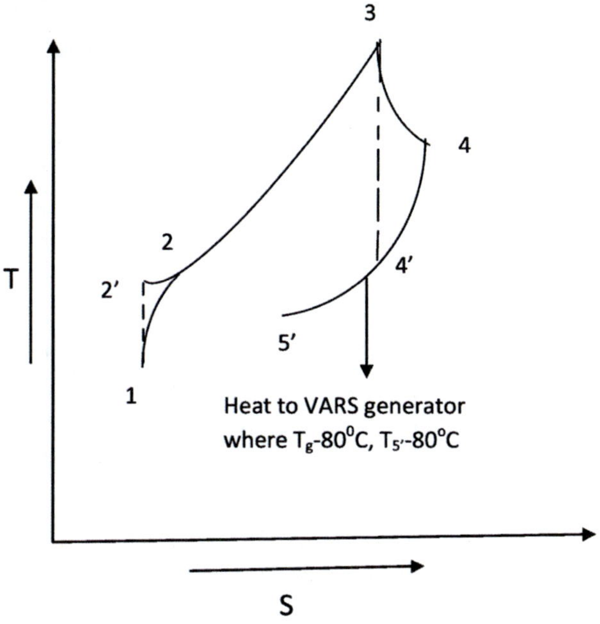

Fig. 2. Temperature–entropy (*T–S*) diagram of gas turbine layout shown in Fig. 1.

$$\mathrm{COP}_{\mathrm{Carnot}} = \left(\frac{T_g - T_a}{T_g}\right) \times \left(\frac{T_e}{T_c - T_e}\right) \tag{1}$$

The amount of heat available to the VARS generator (Q_G) from exhaust gas from the turbine is given by equation (2).

$$Q_G = (m_a + m_f) \times C_{pg} \times (T_4 - Tg) \tag{2}$$

where, m_a is mass flow rate of air to the compressor, m_f is the mass flow rate of fuel (MSW) to the combustion chamber, C_{pg} is the specific heat of combustion gas coming out from combustion chamber and entering turbine (1.147 kJ/kg K), T_4 is the exhaust temperature of combustion gas from the turbine to the generator of VARS, T_g the VARS generator temperature (80 °C).

After $\mathrm{COP}_{\mathrm{actual}}$ and Q_G are known, the desired cooling load (Q_E) can be obtained from equation (3).

$$Q_E = \mathrm{COP}_{\mathrm{actual}} \times (Q_G + W_{sp}) \tag{3}$$

where W_{sp} is solution pump power (can be neglected) as it is very small.

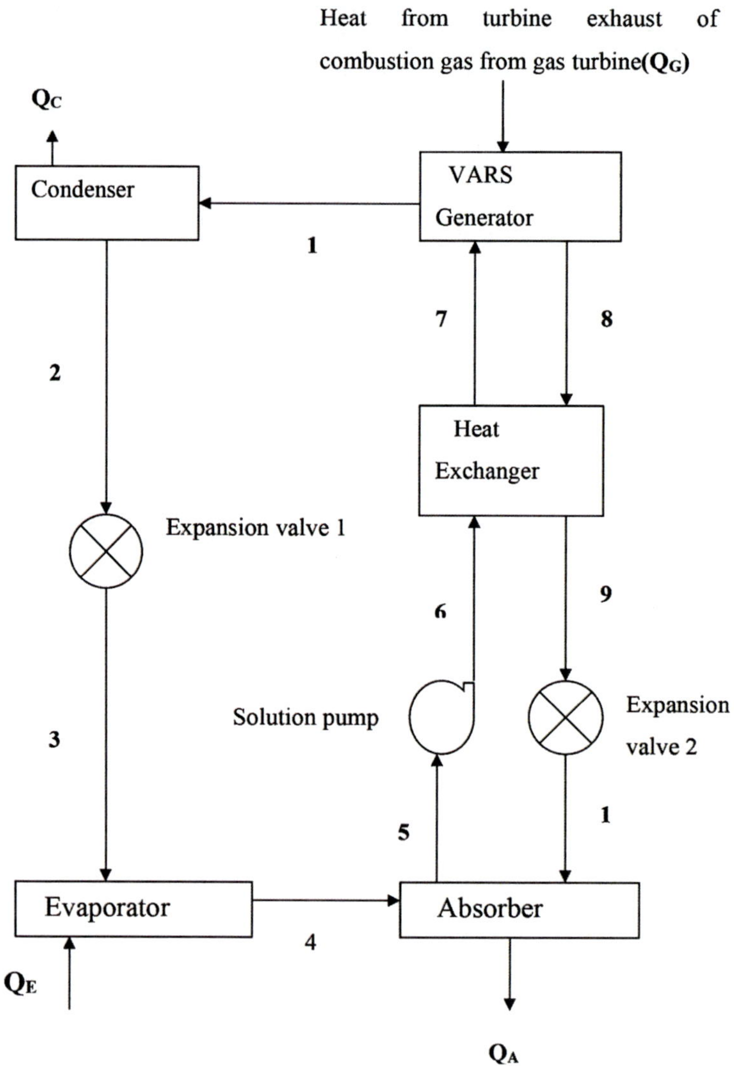

Fig. 3. Schematic view of water–LiBr VARS operated by exhaust heat from gas turbine.

3 Results

Figure 4 shows the variation of Carnot COP for January and May calculated by equation (1). It is seen that minimum Carnot COP occurs at 15:00 hours for January (0.395) and May (0.214). Based on that, actual COP is considered less than above mentioned Carnot COP throughout the day, i.e. 0.32 for January and 0.16 for May.

It is seen that Carnot COP increases from 1:00 hour to 5:00 hours because T_a and T_c decrease from 1:00 hour to 5:00 hours as they are dependent on the ambient

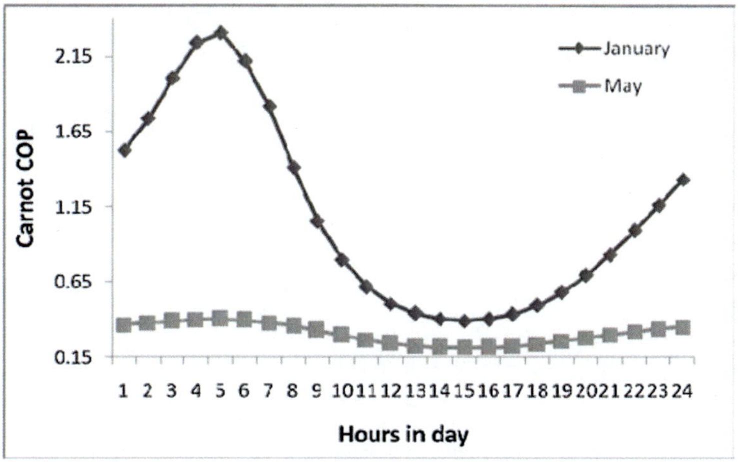

Fig. 4. Variation of Carnot COP for the month of January and May.

temperature. Hence, the term ($T_g - T_a$) in numerator increases and ($T_c - T_e$) in denominator decreases. So, the increase in the numerator part and decrease in the denominator part leads to an increase in Carnot COP from 1:00 hour to 5:00 hours. Next Carnot COP decreases from 6:00 hours to 15:00 hours as T_a and T_c increase from 6:00 hours to 15:00 hours. As a result, ($T_g - T_a$) decreases and ($T_c - T_e$) increases. So, the decrease in the numerator part and an increase in the denominator part leads to a decrease in Carnot COP from 6:00 hours to 15:00 hours. At last Carnot COP increases from 16:00 hours to 24:00 hours because of T_a and T_c decrease to 24:00 hours. Hence, the term ($T_g - T_a$) in numerator increases and ($T_c - T_e$) in denominator decreases and Carnot COP increases from 16:00 hours to 24:00 hours.

It is seen that Carnot COP for January is more than May due to lesser ambient temperature in January than May. Hence, ($T_g - T_a$) is more and ($T_c - T_e$) is less in equation (1). Therefore, greater value in the numerator ($T_g - T_a$) and lesser value in the denominator ($T_c - T_e$) in January lead to greater Carnot COP.

Figure 5 shows the variation of generator heating load for January and May. The variation that is seen is because the summation of the mass flow rate of air to compressor (m_a) and fuel (MSW), i.e. ($m_a + m_f$) decreases from 1:00 hour to 5:00 hours, increases from 6:00 hours to 15:00 hours (maximum) and again decreases from 16:00 hours to 24:00 hours, as generator heating load is dependent on ($m_a + m_f$) according to equation (2). Remaining terms, i.e. C_{pg} and ($T_4 - T_g$) remain constant. Also, the generator heating load is found to be more in May than January due to greater ($m_a + m_f$).

Figure 6 shows the variation of cooling load for January and May. It is seen that the cooling load follows the same trend as that of generator heating load (Q_G) and COP_{actual} remains constant. However, the cooling load is more for January due to greater COP_{actual} (0.32 for January) than May (0.16).

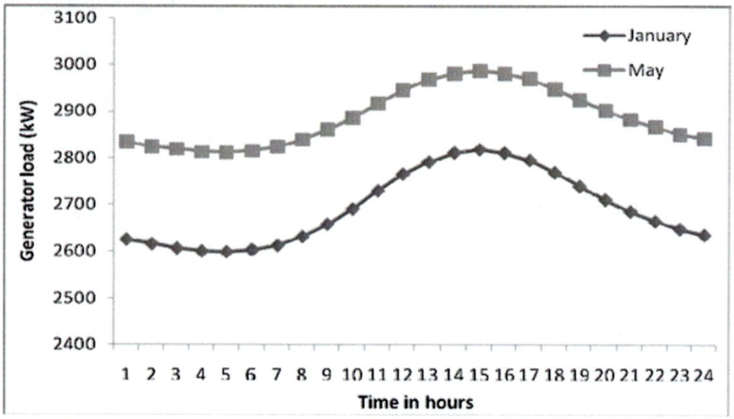

Fig. 5. Variation of generator heating load for the month of January and May.

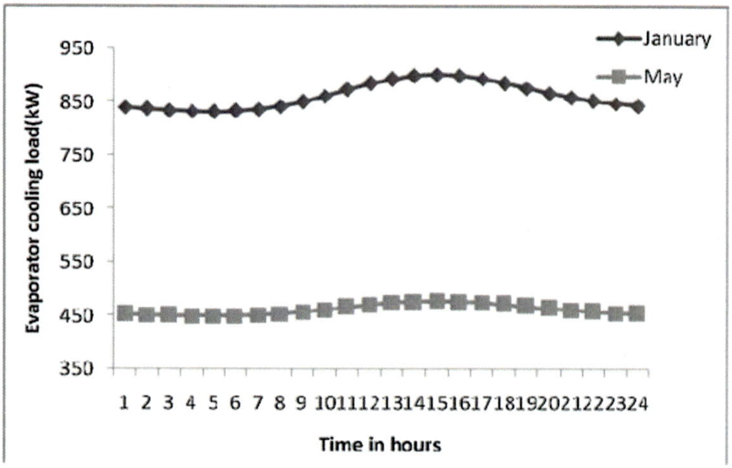

Fig. 6. Variation of evaporator cooling load for the month of January and May.

4 Discussion

In Fig. 4, it is seen that Carnot COP increases from 1:00 hour to 5:00 hours, decreases from 6:00 hours to 15:00 hours where minimum Carnot COP occurs at 15:00 hours and accordingly actual COP is considered for analysis throughout the day for January (0.32) and May (0.16) since it will not operate at 15:00 hours if actual COP is considered greater than COP based on Carnot COP at 15:00 hours.

The generator heating load is dependent and directly proportional to the summation of airflow rate (m_a) and fuel (MSW), i.e. ($m_a + m_f$) according to equation (2) and the variation seen in Fig. 5 is due to variation of ($m_a + m_f$).

The cooling load variation seen similar to the generator heating load and is directly proportional to the product of COP actual and ($Q_G + W_{sp}$) as in equation (3). However, W_{sp} is very small and is neglected.

5 Conclusions

Finally, it can be seen that the generator heating load and evaporator cooling load is not constant and varies on an hourly basis with a maximum value at 15:00 hours. If greater values of COP are taken, the cooling load will increase. Also, if output power from the gas turbine increases, the cooling load will increase due to a greater mass flow rate of air to compressor and fuel (MSW) to the combustion chamber of the gas turbine, thereby increasing generator heating load (Q_G). Also, for the study two months are taken, i.e. January and May. If the system works well in January (the coldest month) and May (the hottest month) the system will work well throughout the year.

References

Bayley, F.J. and Turner, A.B. (1970). The transpiration-cooled gas turbine. *J Eng Power*, **92(4):** 351–358. doi: https://doi.org/10.1115/1.3445364.

Bhargava, R.K., Bianchi, M., Campanari, S., Pascale, A.D., di Montenegro, G.N. and Peretto, A. (2010). A parametric thermodynamic evaluation of high performance gas turbine based power cycles. *J Eng Gas Turbines Power*, doi: https://doi.org/10.1115/1.3155782.

Kaushik, S.C. and Singh, O.K. (2013). Estimation of chemical exergy of solid, liquid and gaseous fuels used in thermal power plants. *J. Thermal Anal. Calorim.*, **115:** 903–908.

Masri, M.A. (1986). On thermodynamics of gas-turbine cycles: Part 3—Thermodynamic potential and limitations of cooled reheat-gas-turbine combined cycles. *J Eng Gas Turbines Power*, **108(1):** 160–168. doi: https://doi.org/10.1115/1.3239864.

Masri, M.A. (1987). Exergy analysis of combined cycles: Part 1—Air-cooled Brayton-cycle gas turbines. *J Eng Gas Turbines Power*, **109(2):** 228–236. doi: https://doi.org/10.1115/1.3240029.

Talukdar, K. (2019). Modeling of gas turbine operated by municipal solid waste. *IOSR J Mech Civil Eng*, **16(2):** 14–19. doi: https://doi.org/10.9790/1684-1602011419.

Tiwari, G.N. (2004). Solar Energy—Fundamentals, Design, Modeling and Applications. Narosa Publishing House, New Delhi.

Torbidoni, L. and Horlock, J.H. (2005). A new method to calculate the coolant requirements of a high-temperature gas turbine blade. *J Turbomach*, **127(1):** 191–199. doi: https://doi.org/10.1115/1.1811100.

Zhang, L., Baltz, M., Pudupatty, R. and Fox, M. (2014). Turbine nozzle film cooling study using the pressure sensitive paint (PSP) technique. doi: https://doi.org/10.1115/99-GT-196.

Chapter 9
Experimental Study on Recycled Polyethylene Terephthalate (PET) Bottle Fibre Reinforced Concrete

Rony Kutum, Piyush Singh, and Anirban Saha

1 Introduction

Plastic waste is one of the major environmental concerns. Eight million to 12.7 million tonnes are dumped into the sea (https://www.sas.org.uk/our-work/plastic-pollution/plastic-pollution-facts-figures/). If plastic wastes are thrown in the land, it makes the soil less fertile and it also pollutes water and harms sea life. Plain concrete has a limited ductility and resistance to cracking (https://theconstructor.org/concrete/). The addition of fibres can improve the properties of concrete. These fibres are uniformly dispersed into the concrete mix which increases its structural integrity. This type of concrete is known as fibre reinforced concrete. This can be natural, animal fibres or polymers. Incorporating plastic into concrete is one possible environmentally friendly approach for their safe disposal. Polyethylene terephthalate is one of the most important and extensively used plastics in the world. PET (polyethylene terephthalate) is a form of polyester that is extracted or moulded into plastic bottles and containers (https://www.worldbank.org/en/news/immersive-story/2018/09/20/what-a-waste-an-updated-look-into-the-future-of-solid-waste-management). PET as fibre reinforcement in structural concrete can provide crack control and ductility enhancement for concrete (Kim et al., 2010). In recent years, a lot of experimental studies were carried out on using waste plastic bottles. The use of PET bottle fibres has a positive effect on the properties of the concrete (Subramani and Fizoor Rahman, 2017). PET-fibres as the partial replacement of aggregate in the Portland cement aggregate concrete. It shows an increase in compression and tensile strength (Ramadevi and Manju, 2012). Akcaözoglu et al. (2010) shredded waste PET bottles granules and used it as a lightweight aggregate in mortar. It was concluded that the use of shredded waste PET granules as aggregate resulted in the production of structural lightweight concrete. The PET-fibres increase the durability

R. Kutum (✉) · P. Singh · A. Saha
Department of Mechanical Engineering, Assam Engineering College, Guwahati, Assam, India

A. S. Kalamdhad (ed.), *Integrated Approaches Towards Solid Waste Management*, https://doi.org/10.1007/978-3-030-70463-6_9

of concrete and it can be used to improve the ductility of the concrete (Foti, 2011). Long term durability performance of recycled PET-fibre reinforced cement composite can be achieved (Won et al., 2010). In this study, the effect of different sizes of PET fibres on concrete was evaluated. These PET fibres are produced from PET bottles. Also, the study of strength and failure mode of PET-fibres reinforced concrete is performed.

2 Materials and Methods

2.1 Materials

The following materials have been used in the present study.

2.1.1 Cement

Various types of cement are available in the market grades (M20, M25, M30 etc.), out of this IS 1489-1, Portland Pozzolana Cement is used for the preparation of specimens, which is most known and available everywhere. The pozzolanic materials commonly used are:

- Volcanic ash
- Calcined clay
- Fly ash
- Silica fumes

The initial setting time of PPC is 30 min (minimum) and the final setting time of PPC is 600 min (maximum). Fineness should not be less than 300 m^2/kg.

2.1.2 Aggregate

Aggregates are one of the important materials for concrete mix. The shape of aggregate, composition and size all have significant impact on the durability, workability and mixing of the concrete. Fine aggregate is essentially any natural sand particles obtained from the land through the mining process. Fine aggregates of natural sand or any crushed stone particles that are 1/4 inch or smaller (Subramani and Fizoor Rahman, 2017). The fine aggregate of size $\leq$0.5 mm was used for making the specimen. Coarse aggregates are obtained from the rock quarried from ground deposits. This can be river gravel, crushed stone from rock etc. Coarse aggregates can be characterised as either round or smooth (river gravel) or angular (crushed stone). Angular have greater interparticle fraction and also required more cement paste as compared to rounded one. Coarse aggregate of size 10–12.5 mm was used for preparing the specimen.

2.1.3 Water

Water plays an important role in the hardened concrete. The impurities in the water may affect the setting of the cement and can affect the strength of the concrete. Fresh clean water without any impurities and oil has to be used for casting and curing of the specimen. The water–cement ratio of 0.5 is taken.

2.1.4 Fibre Reinforcement

Different types of plastic fibres are available like HDPE, LDPE etc. for this project; PET fibre is selected. PET fibres are obtained from waste plastic bottles from households and industries. Polyethylene terephthalate has the resin identification code 1. PET fibres have strength and rigidity for many applications. It is often used for carbonated beverages, water and food products because it is strong and light. PET fibres have good heat resistance and low creep at high temperatures. PET fibres have good stability because of low water absorption.

Basic physical properties of PET materials are:

- Tensile strength: 2.5 N/mm
- Impact strength: 1.5–3.5 kJ/m
- Density: 1.37 g/cm

For this experiment, the PET fibres are cut into the length of 50 mm, 100 mm and 150 mm and a width of 20 mm is used in concrete blocks. The bottles were cut after removing the top and bottom parts.

2.2 Methods

This experimental scheme comprises of the tests like compressive strength test and water absorption test.

2.2.1 Mixed Design

In this study, the mix design selected was M20 grade (IS 10262:2009) having a mix proportion of 1:1.5:3 with a water–cement ratio of 0.5.

2.2.2 Compressive Strength Test and Water Absorption Test

A total of 12 concrete blocks are prepared with different sizes of PET-fibres (50 mm, 100 mm and 150 mm) and cured under normal conditions for 28 days. A compressive strength test and water absorption test were done for different sizes of fibre blocks after curing. The compressive strength of the specimens is calculated by the following formula:

$$\text{Compressive strength} = \frac{\text{Compressive load}}{\text{Cross}-\text{sectional area of the specimen}} \tag{1}$$

And the water absorption test of the specimens is calculated by the following formula:

$$\%\text{ of water absorption} = \frac{(W_2 - W_1)}{W_1} \tag{2}$$

where W_1 = weight of the dry block and W_2 = weight of the wet block.

An experimental investigation was conducted for the strength of PET-fibres and normal concrete blocks. The PET-fibre size of 100 mm was taken. Total six specimens of size 150 × 150 × 120 mm were used for the experiment (with and without fibres). The concrete blocks are cast with a design mix of 1:1.5:3 with a water–cement ratio of 0.5 (Tables 1 and 2). The mixture was eventually distributed and filled into the moulds and PET-fibres are added in layers. The specimens are then cured for 7 days inside the water tank (Figs. 1 and 2).

Another, experimental investigation was conducted for the pet fibre of size 50 mm, 100 mm and 150 mm on concrete. PET fibres are collected from the waste

Table 1. Experimental design for concrete specimens (with and without fibres)

Serial number	*Water–cement ratio*	*Sample*		*Size (mm)*	*Cement (vol%)*	*Fine aggregate (vol%)*	*Coarse aggregate (vol%)*	*Fibre content (vol%)*	*Fibre size (mm)*	*Fibre types*
1	50%	S1	a	150 × 150 × 120	20%	30%	50%	0%	0	
			b	150 × 150 × 120	20%	30%	50%	0%	0	Without
			c	150 × 150 × 120	20%	30%	50%	0%	0	fibre
2	50%	S2	a	150 × 150 × 120	20%	30%	40%	10%	100	
			b	150 × 150 × 120	20%	30%	40%	10%	100	PET-bottle
			c	150 × 150 × 120	20%	30%	40%	10%	100	fibre

Table 2. Experimental design for concrete specimens

Serial number	*Water–cement ratio*	*Sample*		*Block size (mm)*	*Cement (vol%)*	*Fine aggregate (vol%)*	*Coarse aggregate (vol%)*	*Fibre content (vol%)*	*Fibre size (mm)*	*Fibre types*
1	50%	S1	a	150 × 150 × 120	20%	30%	40%	10%	50	PET-bottle fibre
			b	150 × 150 × 120	20%	30%	40%	10%	50	
			c	150 × 150 × 120	20%	30%	40%	10%	50	
2	50%	S2	a	150 × 150 × 120	20%	30%	40%	10%	100	PET-bottle fibre
			b	150 × 150 × 120	20%	30%	40%	10%	100	
			c	150 × 150 × 120	20%	30%	40%	10%	100	
3	50%	S3	a	150 × 150 × 120	20%	30%	40%	10%	150	PET-bottle fibre
			b	150 × 150 × 120	20%	30%	40%	10%	150	
			c	150 × 150 × 120	20%	30%	40%	10%	150	

Fig. 1. PET-fibre reinforced concrete after 7 days.

plastic bottles from the households and industries. These pet bottles are then cut into the uniform desired length of 50 mm, 100 mm and 150 mm as shown in Fig. 3.

A total of nine concrete specimens of size 150 × 150 × 120 mm were used for the experiment with different sizes of PET-fibres. The concrete blocks, cast with a

Fig. 2. Plain concrete after 7 days.

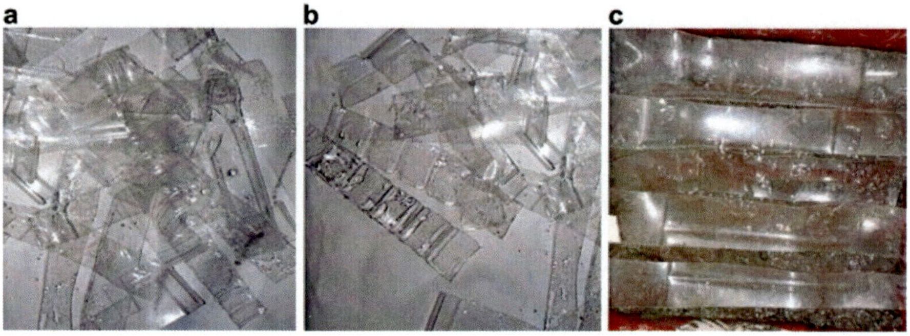

Fig. 3. Different size of PET-bottle fibres. (a) Fibre size: 50 mm (b) Fibre size: 100 mm (c) Fibre size: 150 mm.

design mix of 1:1.5:3 and 0.5 of the water–cement ratio, were taken for the mix. After mixing the process slump test was conducted to check the workability of the concrete mixes concrete. The zero slump was obtained as shown in Fig. 4. The mix is then eventually distributed into the mould.

The PET fibres are then placed over the mix Fig. 5. Three layers of PET fibres are placed into mould.

The mix and then rest of the mould portion was filled with the concrete mix. A total of 12 specimens were prepared with PET fibre of size 50 mm, 100 mm and 150 mm as shown in Fig. 6.

After this, the concrete blocks are placed in a tanker filled with water for the curing process (28 days) and then tested.

3 Results and Discussion

The concrete block with PET-fibre mesh shows higher compressive strength as compared to the control concrete block after curing for 7 days as shown in Table 3. The crack propagation rate was less in PET-fibre reinforced concrete. The concrete with a 150 mm fibre size shows a higher compressive strength of 33.03 N/mm^2. The

Fig. 4. Zero slump.

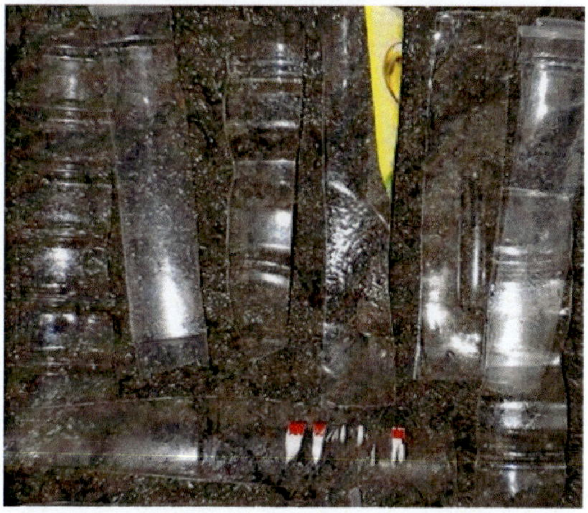

Fig. 5. Layering of PET-bottle fibres in the mould.

concrete with 100 mm fibre size shows the compressive strength of 29.62 N/mm^2 and the concrete with 50 mm size fibre shows the compressive strength of 26.95 N/mm^2 after curing for 28 days as shown in Table 4. Zero slumps were obtained during the workability test. The PET-fibres hold the cement aggregate united during the cracking process as shown in Fig. 7. The compressive strength test of concrete blocks is shown in Fig. 8 (Table 5).

This method is used to determine the rate of absorption of water by cement concrete by measuring the increase in the mass of the specimen resulting from absorption of water (Figs. 9 and 10).

Fig. 6. PET-fibres reinforced concrete blocks.

Table 3. Compressive test for cube after 7 days

Serial number	*Percentage of fibre (vol%)*	*Technical replicates*	*Load (kN)*	*Average load (kN)*	*Compressive strength (N/mm²)*	*Average compressive strength (N/mm²)*	*Fibre types*
1	0	T1	360		16		Without fibre
				376.66		16.736	
		T2	300		13.33		
		T3	470		20.88		
2	10	T1	600		26.66		PET-bottle fibre
				600		26.51	
		T2	590		26.22		
		T3	600		26.66		

4 Conclusion

In this paper, the comparative study of different sizes of PET-fibres on concrete blocks has been investigated. The following conclusions were made:

1. The use of PET-fibre reinforcement improved the compressive strength of concrete.
2. The compressive strength was found to increase with an increase in fibre size. The highest compressive strength of 33.03 N/mm^2 was observed on PET-fibre with a size 150 mm.
3. It has been found that PET-fibres reinforced concrete have low water absorption (2.88%).

Table 4. Compressive test for cubes after 28 days of curing

Serial number	*Percentage of fibres*	*Technical replicates*	*Load (kN)*	*Average load (kN)*	*Compressive strength (N/mm²)*	*Average strength (N/mm²)*	*Fibre sizes (mm)*
1	10	T1	600	606.67	26.66	26.95	50
		T2	620		27.55		
		T3	600		26.66		
2	10	T1	650	667	28.88	29.62	100
		T2	670		30.22		
		T3	680		33.33		
3	10	T1	750	743.33	33.33	33.03	150
		T2	750		33.3		
		T3	730		32.44		

Fig. 7. Cracking of PET-fibre reinforced concrete block.

Fig. 8. Compressive strength test on concrete blocks after 7 days.

Table 5. Water absorption test for concrete blocks 28 days

Sample	*Weight of dry blocks (kg)*	*Average weight W_1 (kg)*	*Weight of wet blocks W_2 (kg)*	*Average weight W_2(kg)*	*% of water absorption*
S1	6.4	6.38	6.6	6.58	3.12
S2	6.35		6.57		3.46
S3	6.4		6.59		2.88

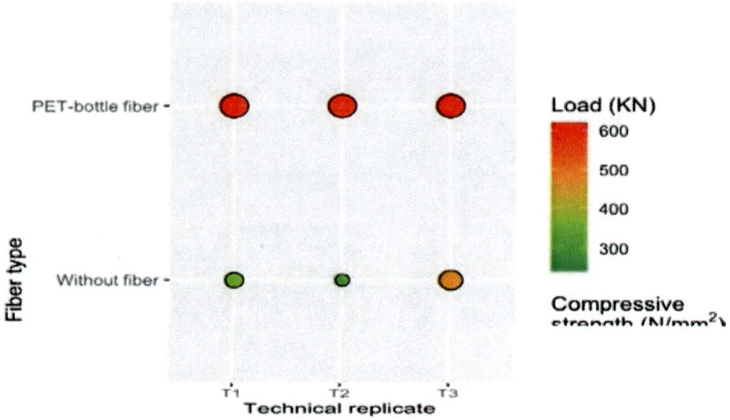

Fig. 9. Bubble heat-map showing load and compressive strength in concrete blocks (7 days).

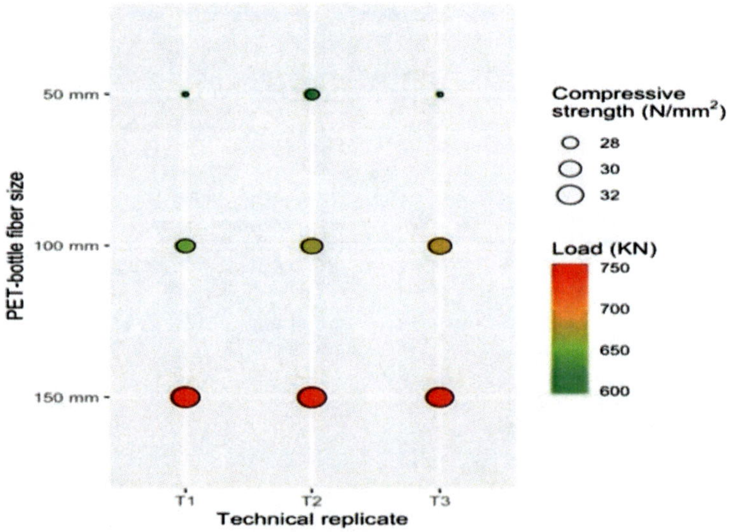

Fig. 10. Bubble heat-map showing load and compressive strength in different fibre size (28 days).

Acknowledgements The authors would like to express our profound gratitude and sincere thanks to the Department of Mechanical Engineering and Department of Civil Engineering (Strength of Material LAB), Assam Engineering College, Guwahati for their valuable supervision, guidance and constructive suggestion throughout this work

References

Akçaözoğlu, S., Atiş, C.D. and Akçaözoğlu, K. (2010). An investigation on the use of shredded waste PET bottles as aggregate in lightweight concrete. *Waste Manage*, **30(2):** 285–290. doi: https://doi.org/10.1016/j.wasman.2009.09.033.

An Updated Look into the Future of Solid Waste Management (2018). https://www.worldbank.org/en/news/immersive-story/2018/09/20/what-a-waste-an-updated-look-into-the-future-of-solid-waste-management (10/12/2019).

Concrete: Concrete Technology. https://theconstructor.org/concrete/ (8/12/2019).

Foti, D. (2011). Preliminary analysis of concrete reinforced with waste bottles PET fibers. *Construct Build Mater*, **25(4):** 1906–1915. doi: https://doi.org/10.1016/j.conbuildmat.2010.11.066.

Kim, S.B., Yi, N.H., Kim, H.Y., Jay Kim, J.-H. and Song, Y.-C. (2010). Material and structural performance evaluation of recycled PET-fiber reinforced concrete. *Cem Concrete Composites*, **32.3:** 232–240. doi: https://doi.org/10.1016/j.cemconcomp.2009.11.002.

Plastic pollution how much plastic pollution is there? https://www.sas.org.uk/our- work/plastic-pollution/plastic-pollution-facts-figures/ (8/12/2019).

Ramadevi, K. and Manju, R. (2012). Experimental investigation on the properties of concrete with plastic PET (bottle) fibres as fine aggregates. *Int J Emerg Technol Adv Eng*, **2(6):** 42–46.

Subramani, T. and Fizoor Rahman, A. (2017). An experimental study on the properties of PET-fiber reinforced concrete. *Int J Applic Innov Eng Manage (IJAIEM)*, **6(3):** 058–066.

Won, J.-P., Jang, C.-I., Lee, S.-W., Lee, S.-J. and Kim, H.-Y. (2010). Long-term performance of recycled PET fibre-reinforced cements composites. *Construct Build Mater*, **24(5):** 660–665. doi: https://doi.org/10.1016/j.conbuildmat.2009.11.003.

Chapter 10
Estimation of Demolition Waste Using Waste Generation Rate and Logistic Flows in Surat, India

D. J. Patel, Prabhat Shrivastava, and D. A. Patel

1 Introduction

The Central Pollution Control Board (CPCB, 2016) defines the construction and demolition waste (CDW) as 'waste generated from the construction, repairing, remodelling or demolition of building etc. which include waste materials, soil, debris etc.'. The CPCB has estimated solid waste generation in India to be around 48 million tonnes per annum of which the construction industry accounts for approximately 25%, i.e. 12 million tonnes per year (TIFAC 2015). However, the Ministry of Urban Development (MoUD) India estimated that the Indian construction industry generates about 10–15 million tonnes per year. The estimates given by CPCB and MoUD are outdated. The waste generation rate (WGR) is considered as a significant factor in decision making and waste management (Lu et al., 2011). Ram and Kalidindi (2017) demonstrated WGR method to estimate the CDW in India under limited data availability. The usages of the WGR method is to predict the quantity of demolition waste (DW) produced from the demolition work, predicting fees and number of trucks required for disposal (Yuan and Shen, 2011).

Various case studies were used to determine the factors considered for the estimation of CDW at project level and regional level by different researchers. Total floor area is a significant influencing factor in estimating the CDW at regional level (Song et al., 2017; Zheng et al., 2017; Asgari et al., 2017). Bernardo et al. (2016) established relationship to estimate demolition waste using density, ageing index of the building, building density and land use. Case studies for project level estimation of waste consist of building internal and external factors. Internal factors of the building such as geographical location, use of the building, and public–private nature significantly affect the demolition waste generation measured by weight (Chen and Lu, 2017). Ram and Kalidindi (2017) used total floor area, type of

D. J. Patel (✉) · P. Shrivastava · D. A. Patel
Civil Engineering Department, S.V. National Institute of Technology, Surat, Gujarat, India

A. S. Kalamdhad (ed.), *Integrated Approaches Towards Solid Waste Management*, https://doi.org/10.1007/978-3-030-70463-6_10

structure, the material used in the calculation of waste generation by volume in Chennai, India. Ding and Xiao (2014) considered structure type, building design and structure codes in the estimation of regional CDW.

Most of the earlier estimation models for CDW or DW are from the developed country, because their demolition sector records are easily available. However, in developing countries like India, the records related to demolition sector are very limited. Swachh Bharat Abhiyan launched on 2 October 2014, stated that the 100% recycling of solid waste produced in India by 2 October 2019 also includes CDW. The demolition activities generate 300–500 kg m^{-2} DW (CPCB, 2016). However, in Surat, it is 500 kg m^{-2}. It shows significant difference between sources. Reliable and updated estimation is not available for generation of DW at regional level and project level. In addition, these old standards are used in making national policies, rules and regulation for managing DW. As a result, inadequate policies, the absence of regulations and underestimated recycling facilities lead to unauthorised dumping of DW in landfills. In Surat, data related to demolition waste generation is unreliable and unaccountable. Therefore, there is a need to validate the WGR for demolition activity and quantify the amount of demolition waste produced by the demolition of a building unit. This will act as a benchmark to estimate the demolition waste produced within the boundary of Surat municipal corporation (SMC). This will also help in formulating the sustainable solution for managing the building-related waste. Objectives of this study are: (1) to investigate a scenario of demolition waste (DW) flow after demolition and (2) to develop a demolition waste estimation model based on the mean value of demolition waste generation rate (DWGR).

2 Research Methodology

In order to achieve the set of objectives, literature review and knowledge acquisition approach were adopted. This study adopted site visit (qualitative), questionnaire survey (quantitative) and interviews to gather information's from various stakeholders such as SMC construction and demolition representatives; builders/building and demolition contractors; entrepreneurs/private contractors involved in demolition waste handling/processing; construction raw material manufacturers; and construction and demolition waste recycling industry. Mainly close ended questions were formed with enough space for respondent to give additional information related to subject matters. A few open-ended questions were addressed to capture the domain knowledge. Forty-one questionnaires were sent to gather the information of the demolition waste flow and generation of demolition waste from various stakeholders of demolition work. Site visit is the key source to interact and to collect information from the demolition workers involved. Forty-nine site visits were conducted for data collection. Microsoft Office Excel (2007) tool was used to conduct the data analysis. The data were converted into MS Excel file. Three-way ANOVA was demonstrated to find out the correlation between the factors and value of demolition waste generation rate (DWGR) to estimate the demolition waste produced from the building. The IBM SPSS statistics v25 software was used for ANOVA analysis.

Gujarat's demolition sector is unorganised so it is quite difficult to get permission to visit demolition sites. So, this study was limited to cover and visit demolition sites from seven zones of Surat city located in western part (Gujarat state) of India as it was convenient to approach those sites and interact with persons.

3 Data Collection

The different methods of data collection used by various researchers are divided into three categories: (1) the data from the existing research literature and bill of quantities (Guzman et al., 2009); (2) the flow of waste materials after building demolition such as government building permit records (Ram and Kalidindi, 2017), government annual reports (Song et al., 2017), truckload records (Ghosh et al., 2016) and (3) the actual measurements before the building construction and demolition (Cha et al., 2017), questionnaire survey (Asgari et al., 2017) and data collection from the active construction companies (Kern et al., 2015). More reliable data can be achieved using the third category, estimation of CDW as compared to other categories (Chen and Lu, 2017).

Six different assumptions need to be satisfied to produce a valid result for a three-way ANOVA (Statistics.laerd.com, 2018). They are: (1) the dependent variable should be measured at the continuous level (i.e., it is an interval or ratio variable); (2) each of the three independent variables should consist of two or more categorical, independent groups; (3) there should be an independence of observations which means that there is no relationship between the observations in each group or between the groups themselves; (4) there should be no significant outliers; (5) the dependent variable should be approximately distributed for each combination of the groups of the three independent variables and (6) there should be the homogeneity of variances for each combination of the groups of the three independent variables.

The data collection was divided into two parts i.e. primary data and secondary data. The primary data was collected from two questionnaire survey, site visits and telephonic interview and the secondary data was collected from government's guidelines, norms and SMC draft guidelines for managing DW. The first survey was carried out to find out the common practices among demolition contractor for estimating DW generated from the building. Methods and factors used to estimate the waste generated from building demolition are also listed out. A good amount of sample data was obtained from the demolition contractors based on their previous work and past experience. The data used for validation of the study were collected from the actual measurements during/before the building demolition work. The factors considered for the study were the age of the building, use of the building, type of structure and the gross floor area of the building. These factors were selected from the literature review and from the interview with the demolition contractor.

4 Construction and Demolition Sector In Surat

Surat Green Precast private limited (SGPPL) is the recycling industry for recycling of CDW produced within the city limit. According to SGPPL database approximately 1500 metric tonnes (MT) per day of total solid waste is generated in Surat. The CDW is the 20–22% of the total solid waste generated, i.e. 320 MT per day. However, municipal waste statistics on the website of SMC reveals that the total solid waste generated in Surat city is 2000 MT per day in 2016. Out of which 25–30% of the waste is from construction and demolition works (CDW), i.e. approx. 500 MT (minimum)–600 MT (maximum) per day (as per guideline of CPCB 2016). SGPPL is designed for recycling capacity of 300 MT per day. Tender is given for 20 years under PPP (public-private partnership) BOOT (build, own, operate and transfer) model for waste collection, recycling and disposal of construction and demolition waste (CDW). Figure 1 indicates that the growth of CDW (MT/day) and MSW (MT/day) has been significantly increasing yearly. In Surat, private companies and contractors play a significant role in managing DW. Contractual arrangements require that DW is disposed by the contractor at his cost. Private waste disposal contractors are paid on the bases of the number of trucks used and distance from the disposal site. Private waste disposal contractors dispose waste into municipal waste collection site or illegal disposal to landfill.

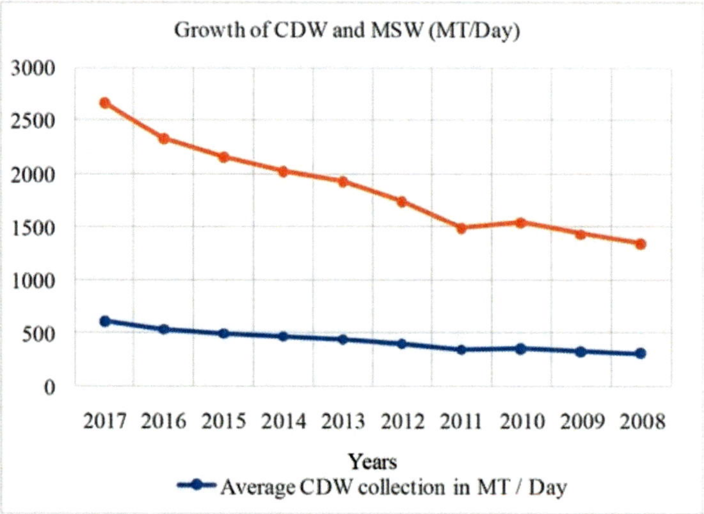

Fig. 1. Growth of CDW (MT/day) and MSW (MT/day) yearly.

5 Establishing Waste Generation Rate

During the demolition of the building, materials which have economic value are sold in the secondary market for reuse. Major portion of the debris consists of concrete and brick masonry. Therefore, this study emphasises on the prediction of concrete and masonry debris, which are not appropriately handled in the city. Stratified sampling is considered for selection of buildings to conduct the study. The classification of the collected data according to use of the building, i.e. residential and non-residential building (mixed-use, public use buildings), the age of building, i.e. old and new building and type of structure, i.e. load bearing and framed structure within Surat municipal area.

The gross floor area (GFA) of the buildings is calculated by the equation,

$$\text{GFA} = \text{Floor area} \times \text{Number of floors} \tag{1}$$

where; GFA is the gross floor area of the building (m^2), floor area is the built-up area of one floor (m^2) and the number of floors is the total number of the floors in the building.

The total amount of waste is calculated by the equation,

$$\text{Total amount of waste} = \text{Numbers of tipper used} \times \text{average carrying capacity} \tag{2}$$

where the total amount of waste from demolition of building (kg), the number of Tipper used for disposal of DW and the average carrying capacity is the average amount of waste carry in one trip.

Generally, Tippers are used for disposal services in Surat. The average amount of waste carried by single axel tipper 9.25 tonnes per trip and 14.5 tonnes for double axel tipper. Lastly, the DWGR is determined by the following equation,

$$\text{DWGR} = \text{Total amount of waste} / \text{Gross floor area}\left(m^2\right) \tag{3}$$

where DWGR is demolition waste generation rate (kg/m^2), total amount of waste is the total quantity of waste coming out from the building (kg) during demolition and gross floor area of the building.

6 Estimating the Proportion of Various Types of Buildings

In Surat city, the major buildings are categorised in two types of building structure, i.e. load bearing structure and reinforced concrete (RC) framed structure. The old buildings mostly consist of two to three floors load-bearing structure and old low-rise RC framed structure. Load bearing structures (LBS) are generally old building located in the old part of the city, i.e. central zone and Athwa area. Some of the old LBS buildings in the old part of Surat are nearly 100 years old. In this study, total

49 numbers of buildings were surveyed. Forty-one buildings were taken for the ANOVA analysis and eight on-going demolishing buildings were closely observed for the validation of the study. The proportion of LB structure was about 44% and RC framed was about 56% of the total number of demolished buildings. The old buildings were considered having an age above 30 years and the new buildings having age below 30 years. The proportion of old buildings were about 54% and the new building was 46% of the total number of the surveyed buildings. 53% of the buildings were residential and 47% of the buildings were non-residential.

7 Research Question

The research questions used for this study are:

(H0) *Null Hypothesis*: There is no significant difference in the mean value of the demolition waste generation rate of the building and it does not depend upon building characteristics such as the use of the building, the age of building and type of structure.

(H1) *Alternative Hypothesis*: There is a significant difference in the mean value of the demolition waste generation rate of the building and it depends upon building characteristics such as the use of the building, the age of building and type of structure.

8 Testing for Assumptions

Assumption 1: In this study, the dependent variable was the waste generation rate (WGR) which was measured as a continuous variable (measured as scale in SPSS), therefore, the assumption one was validated.

Assumption 2: In this study, there were three independent variables, i.e. age of the building, type of structure and use of the building. Each of the three independent variables has two levels. Therefore, assumption two was validated.

Assumption 3, 4, 5: Testing for normality and outliers following steps were used for the test in SPSS. As shown in Table 1, the significant values under the Shapiro–Wilk test are greater than 0.05. Therefore, it is concluded that 'WGR' for this particular subset of individuals is normally distributed.

Assumption 6: In this study, Levene's test for homogeneity of variances was used for checking the homogeneity of variances for each combination of the groups of the three independent variables. The result of the Levene's test for homogeneity of variances shows that the significant level (Sig.) is greater than 0.05, $F(7, 33) = 0.327$, $p = 0.936 > 0.05$, i.e. non-significant, the variance is not significantly different, so the null hypothesis is accepted and equal variance are assumed. Furthermore, the data are fulfilling all the six assumptions.

Table 1. Shapiro–Wilk test for testing normality for each category

WGR for each category	*Sig.*	*Remarks*
Type of structure = load bearing, use of structure = residential, age of structure = old	0.967	Significant
Type of structure = load bearing, use of structure = residential, age of structure = new	0.967	Significant
Type of structure = load bearing, use of structure = non-residential, age of structure = old	0.850	Significant
Type of structure = framed structure, use of structure = residential, age of structure = old	0.876	Significant
Type of structure = framed structure, use of structure = residential, age of structure = new	0.421	Significant
Type of structure = framed structure, use of structure = non-residential, age of structure = old	0.783	Significant
Type of structure = framed structure, use of structure = non-residential, age of structure = old	0.783	Significant
Type of structure = framed structure, use of structure = non-residential, age of structure = new	0.967	Significant

9 Result and Discussion

There are significant effects of type, use and age of building on DWGR value, which needs to be checked by testing of Between-Subjects effects of the ANOVA. Result shows F statistics = 59582.525, p-value $\leq$ 0.0001; F statistics = 2825.966, p-value $\leq$ 0.0001; F statistics = 586.575, p-value $\leq$ 0.0001; F statistics = 176.948, p-value = 0.0001; F statistics = 4.266, p-value = 0.047; F statistics = 18.746, p-value $\leq$ 0.0001; F statistics = 25.345, p-value $\leq$ 0.0001 for type of structure, use of structure, age of structure, type of structure * use of structure, type of structure * age of structure, use of structure * age of structure and type of structure * use of structure * age of structure respectively. The result shows that p-value is less than 0.05 for all the above cases with level of significance 5%. So, type of structure, use of structure and age of structure as well as their interactions are needed to determine waste generation rate of the building. Therefore, the null hypothesis is rejected and the alternative hypothesis is accepted that the waste generation rate is significantly affected by the use of building, structure type and age of the structure.

The mean value of DWGR value is higher for non-residential building, i.e. 541.50 kg/m^2 as compared to residential building, i.e. 512.50 kg/m^2. The mean value of DWGR for load-bearing building produces 593.70 kg/m^2 as compared to the DWGR of the framed structure, i.e. 460 kg/m^2. The old load bearing non-residential produces 621.00 kg/m^2 the largest amount of waste as compared with other building categories. On the other hand, new framed residential building produces 442.00 kg/m^2 least amount of demolition waste in the group.

10 DWGR Estimation Model Performance Prediction

Table 2 shows the testing of the predicting performance of the three-way ANOVA analysis. The data from eight on-going demolition works were taken to validate the result of this study. The final amount of DW produced from eight buildings was calculated by using equation (2). Building demolition waste was calculated on the basis of the result of the three-way ANOVA. Equation (3) was used for predicting the DW produced from eight buildings. The result confirms that the difference between results is between +10% and −10%.

11 Recommendations for Government Agencies

On the basis of the above study, the various recommendations are as follows: (1) detailed building documents need to be collected before issuing permit. This data set will act as primary tools for stakeholders in effective and sustainable management of DW; (2) there is a need for proper regulation and monitoring towards the collection and disposal of DW in Surat. This will help in reducing illegal dumping as well as the smooth collection of DW by recycling plant; (3) DW disposal site needs proper demarcation of boundary and better road accessibility; (4) SMC should provide a market environment for recycled DW materials along with suitable incentives and publicity to motivate private recycling sectors; (5) the central body should introduce standards and testing norms for the use of recycled DW products; (6) SMC and Road and Building (R&B) Department should encourage the use of recycled DW construction materials in public works with an innovative certification program.

Table 2. Validation of waste generation rate of the building

Gross floor area	*Type of structure*	*Age of building*	*Use of building*	*Estimate of waste generated*		*Percentage error*
				On-site	*Using result*	
780	L.B.	O.B.	R	496782	451620	9.09%
1040	L.B.	N.B.	R	559187.2	594880	−6.38%
1250	L.B.	O.B.	N.R.	836104.5	775625	7.23%
1540	L.B.	N.B.	N.R.	713575	763525.25	−7.00%
1840	F.S.	O.B.	R	907295.9	840088.8	7.41%
1450	F.S.	N.B.	R	672640.5	640610	4.76%
2820	F.S.	O.B.	N.R.	1428837.6	1347960	5.66%
2480	F.S.	N.B.	N.R.	1093184	1150720	−5.26%

Note: L.B. = load bearing building, F.S. = framed structure, O.B. = old building, N.B. = new building, R = residential and N.R. = non-residential building.

12 Conclusion

This study shows the present scenario of the DW flow and its management in the Surat city. In Surat, decadal population growth rate was 83% in 2011 compared to 2001 census. The DW scenario is going to worsen if not managed properly. The DW generation has grown from 330 tonnes per day (TPD) in 2012 to 450 tonnes per day in 2016. Around 30% (for old building) to 70% (for a new building) of DW can be reused and recycled as a secondary construction material.

In this study, a methodology is demonstrated to produce the reliable estimate of DW at the project level. The waste generation rate is very significant information for accurate prediction and successful management of the demolition waste. Total 49 low rise residential and non-residential buildings were surveyed after and during the building demolition process to acquire the data for building characteristics and quantity of waste output for each building. The DWGR results derived through the ANOVA analysis showed the factors influencing the demolition waste generation and enlist the detailed DWGR values. The partial eta squared value shows that the 43.4% of the variation on the DWGR is explained by the variation in the type of structure, use of the structure and the age of the structure. The result also shows that old, non-residential, load-bearing structure generates 621 kg/m^2, which are the largest amount of DW as compared to the other eight categories. The new, framed residential building produces 442 kg/m^2 least amount of demolition waste among eight categories. The result of the study will greatly contribute to the DW department of SMC and help the private contractor in the prior estimation of the waste output from the building, fees for disposal service and the number of trucks required for disposal of waste.

The results of this study help policymaker and urban planner for better planning and sustainable management of DW. Furthermore, the data collection method helps researchers from developing countries to overcome the problem of limited availability of data.

Acknowledgements The authors thank the Surat municipal corporation staff, private sector construction and demolition contractors and engineers from the recycling industry (SGPPL) for their valuable support in conducting this study.

Declaration of Conflicting Interests The authors declared no potential conflicts of interest with respect to the research, authorship, and/or publication of this article.

Funding Sources The authors received no financial support for the research, authorship, and/or publication of this article.

References

Asgari, A., Ghorbanian, T., Yousefi, N., Dadashzadeh, D., Khalili, F., Bagheri, A., Raei, M. and Mahvi, A. (2017). Quality and quantity of construction and demolition waste in Tehran. *J Environ Health Sci Eng*, **15:** 1.

Bernardo, M., Gomes, M. and de Brito, J. (2016). Demolition waste generation for development of a regional management chain model. *Waste Manage*, **49:** 156–169.

Cha, G., Kim, Y., Moon, H. and Hong, W. (2017). New approach for forecasting demolition waste generation using chi-squared automatic interaction detection (CHAID) method. *J Clean Prod*, **168:** 375–385.

Chen, X. and Lu, W. (2017). Identifying factors influencing demolition waste generation in Hong Kong. *J Clean Prod*, **141:** 799–811.

CPCB (Central Pollution Control Board) (2016). Guidelines on environmental management of C & D Wastes. Ministry of Environment, Forests & Climate Change, New Delhi, India

Ding, T. and Xiao, J. (2014). Estimation of building-related construction and demolition waste in Shanghai. *Waste Manage*, **34(11):** 2327–2334.

Ghosh, S., Haldar, H., Chatterjee, S. and Ghosh, P. (2016). An optimization model on construction and demolition waste quantification from building. *Proc Environ Sci*, **35:** 279–288.

Guzman, J.S., Marrero, M., Montes-Delgado, M. and Ramirez-de-Arellano, A. (2009). A Spanish model for quantification and management of construction waste. *Waste Manage*, **29(9):** 2542–2548.

Kern, A.P., Dias, M.F., PivaKulakowski, M. and Gomes, L.P. (2015). Waste generated in high-rise buildings construction: A quantification model based on statistical multiple regression. *Waste Manage*, **39:** 35–44.

Lu, W., Yuan, H., Li, J., Hao, J., Mi, X. and Ding, Z. (2011). An empirical investigation of construction and demolition waste generation rates in Shenzhen city, South China. *Waste Manage*, **31(4):** 680–687.

Ram, V. and Kalidindi, S. (2017). Estimation of construction and demolition waste using waste generation rates in Chennai, India. *Waste Manage Res*, **35(6):** 610–617.

Sgppl.co.in. (2018). SGPPL. http://sgppl.co.in. Accessed 4 April 2018.

Song, Y., Wang, Y., Liu, F. and Zhang, Y. (2017). Development of a hybrid model to predict construction and demolition waste: China as a case study. *Waste Manage*, **59:** 350–361.

Statistics.laerd.com (2018). How to carry out and report a three-way ANOVA using State Laerd Statistics. https://statistics.laerd.com/stata-tutorials/three-way-anova-using-stata.php. Accessed 2 July 2018.

TIFAC (Technology Information, Forecasting and Assessment Council) (2015). Utilization of waste from construction industry. http://tifac.org.in. Accessed 2 July 2018.

Yuan, H. and Shen, L. (2011). Trend of the research on construction and demolition waste management. *Waste Manage*, **31(4):** 670–679.

Zheng, L., Wu, H., Zhang, H., Duan, H., Wang, J., Jiang, W., Dong, B., Liu, G., Zuo, J. and Song, Q. (2017). Characterizing the generation and flows of construction and demolition waste in China. *Constr Build Mater*, **136:** 405–413.

Chapter 12
Sustainability of Natural Aggregates by Utilising CDW in Concrete

B. Jeyanth, R. Sudharsanan, M. Thillai Backiam, and M. Vijay

1 Introduction

Concrete is a composite material with cement, fine and coarse aggregate which is well known for its strength and durability. The rapid growth in urbanisation and industrialisation increases its consumption to a huge quantity. Since natural aggregate occupies more than 70% of volume, any rise in use of concrete will intensify its quantity consumed. All over the world every year about 40 billion tons of aggregate is produced in which India contributes 13% (Tam et al., 2010). Further, the increase in concrete consumption makes the concrete into a most non-sustainable material due to maximum consumption of natural resources. The generation of source of natural aggregate requires thousands of years. The rate of generation of natural aggregate is very much smaller compared to its present rate consumption which leads to depletion of natural resource. Many research works are already in progress to identify and use a suitable alternative construction material, which is economical as well as feasible both in terms of strength and durability characteristics. As a result, researches have been increased to recover the recycled aggregates from construction and demolition waste and various methods for the utilisation of derived materials (Tam et al., 2018). The demolition of a structure may be done either by conventional method or by selective demolition which results in high economic viability (Silva et al., 2011).

Rapid growth in urbanisation leads to construction of new structures by demolishing the old one which results in generation of huge quantity of construction and demolition waste (CDW). Every year India generates about 15–23 million tons of CDW and handling this huge quantity creates lot of problems in the solid waste

B. Jeyanth (✉) · R. Sudharsanan · M. Thillai Backiam
Velammal Engineering College, Chennai, Tamilnadu, India

M. Vijay
Kongunadu College of Engineering & Technology, Trichy, Tamilnadu, India

A. S. Kalamdhad (ed.), *Integrated Approaches Towards Solid Waste Management*, https://doi.org/10.1007/978-3-030-70463-6_12

management. Generally, CDW is disposed by means of landfilling technique, which occupies huge area and creates lot of environmental related issues causing water, soil and land pollution which results in poor fertility of soil due to increase in alkalinity (Pathak et al., 2014).

Since CDW is derived from the construction waste it possesses similar properties with that of the natural aggregates with high water absorption characteristics which can be reduced by adding fine granular particles (Mymrin and Correa, 2007). The use of CDW as an alternative aggregate for concrete facilitates reduction in consumption of natural aggregates as well as landfill issues. Based on the composition, CDW has been categorized as recycled concrete aggregate (RCA), recycled ceramic aggregate (CA) or recycled mixed aggregate (RMA). Among those types of CDW, the RCA has been utilized by the researchers a lot to know its viability in concrete (Junaka and Sicakova, 2016; Cantero et al., 2019). The CDW management system is considerably progressed from 1980s in the developing countries, particularly Australia, Western Europe and North America. About 98% of recycled aggregates are used for construction applications in Japan making it a leading country in recycling concrete waste. The largest producers of CDW aggregates in the present decade are Asia/Pacific, Russia and South American regions (Tam et al., 2010). The introduction of Industrialised Building System (IBS), is an improved construction technique in developed countries, has facilitated the increase of recycling rate of Construction waste (Esa et al., 2016). Over last decade, a significant research on RCA and its application in concrete is carried out (Junaka and Sicakova, 2016).

Low slump value of the concrete is caused by the use of RCA and that can be improved by using saturated surface dry coarse aggregate (Shahidana et al., 2017). Replacing 25% of natural coarse aggregate with RCA has no significant adverse effect on its structural performance. When the replacement ratio increased to 50%, reduction in compressive strength ranges from 7% to 13% with a smaller decrease in splitting and elastic modulus. The flexural strength of the concrete made with RCA is slightly lower compared to the concrete with natural coarse aggregates (Kabir et al., 2016). The use of recycled coarse aggregate increases the abrasion resistance of concrete (Bravo et al., 2015; Kabir et al., 2016). Acceptability of recycled material for concrete is hampered due to a poor image of recycling activity, and lack of confidence on the products made with it. The recycling operation is directly affected by the cost of disposal of waste from construction industry to the aggregate preparation plant. Even though, more number of new construction activities are taking place in developing countries, the low dumping costs and space availability for land filling act as a barrier to recycling activities. To divert more volume of waste for recycling, charges must be imposed on sanitary landfill (Rao et al., 2007). The main aim of this research is to check the feasibility of use of CDW in concrete as aggregate, which leads the concrete as sustainable material.

2 Methodology

The schematic representation of the methodology adopted for checking the feasibility of CDW in concrete is shown in Fig. 1.

3 Result and Discussion

3.1 Preparation of Cultivated Aggregates

The concrete waste from the construction and demolition is collected, crushed, sieved and segregated as coarse and fine aggregates. The aggregates passing through the 20 mm and retained in 12.5 mm sieve is used as cultivated coarse aggregates or C-coarse (Fig. 2). The C-coarse used are well graded and lay within the particle size ranges recommended as per IS 383-2016. The aggregates passing through the 4.75 mm sieve and retained on 75 μm sieve is collected and named as cultivated fine aggregate or C-sand (Fig. 3). Both the aggregates are further washed to remove the dust and finer particles from the surface of aggregate.

3.2 Testing of Aggregates

Both natural and cultivated aggregates must be tested to carry out mix design of CC mix and mix proportions of the CDWC mixes. The testing of aggregate starts with the mechanical sieve analysis to conform the corresponding zone of aggregate. Then the physical properties such as specific gravity, bulk density and water absorption are also measured.

Mechanical sieving of C-coarse is carried to measure the weight and percentage retained on each sieve. The cumulative of percentage retained and passing are calculated and presented in Table 1 and Fig. 4. Finally, the aggregate passing 20 mm and retained in 12 mm are used as C-coarse.

Both natural sand and C-sand are sieved separately and percentage of passing through each sieve is measured and presented in Table 2. The result shows that the percentage of natural sand and C-sand passing the sieves are conforming the zone II. The percentage passing of natural sand and C-sand is plotted and shown in Fig. 5. Both fine aggregates follow almost similar pattern of passing through each sieve, which shows that there is no significant variation among them in terms of volume of aggregate of each size. Hence, the C-sand can be used as an alternative for natural fine aggregate.

The specific gravity of natural and cultivated aggregates are obtained separately. The values of physical properties of natural and cultivated aggregates are presented in Table 1.

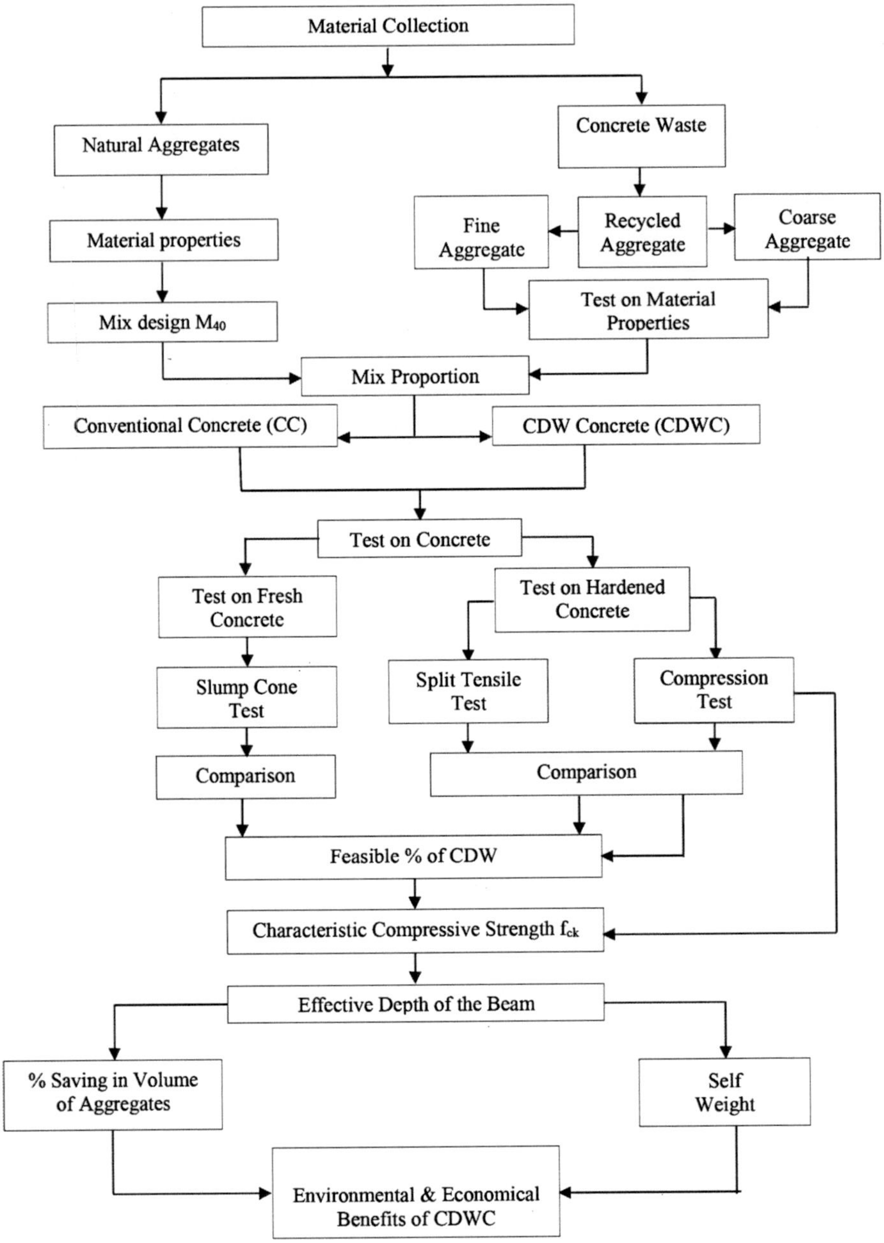

Fig. 1. Research methodology flow chart.

Fig. 2. Cultivated coarse aggregate.

Fig. 3. Cultivated fine aggregate.

3.3 Mix Design and Mix Proportion

The mix design for M40 grade concrete is done as per IS 10262-2019 for the CC mix. The natural aggregate, both coarse and fine, is replaced by cultivated aggregate in 25%, 50%, 75% and 100% of its volume with 0.4 W/C. The mix proportion used for CC and CDWC mixes is shown in Table 2.

3.4 Test for Workability of Concrete

The slump value of concrete defines its workability. For each concrete mix, without and with CDW replacement, the slump value is measured and presented in Fig. 6. The difference in slump value between the CC and CDWC is less than 5 mm which shows that the partial replacement of Cultivated Aggregates does not affect the workability of the concrete. Particularly the 50% replacement results in slump value almost equal to that of CC mix. Therefore, the M3 mix can be taken as suitable mix in terms of workability.

Table 1. Physical property of natural and cultivated aggregates

S. no.	*Property*	*Natural coarse*	*Natural sand*	*C-coarse*	*C-sand*
1	Specific gravity	2.65	2.7	2.48	2.57
2	Bulk density (kg/m^{3}aa	1129	1570	1478	1757
3	Water absorption	1.00	2.20	1.20	2.50

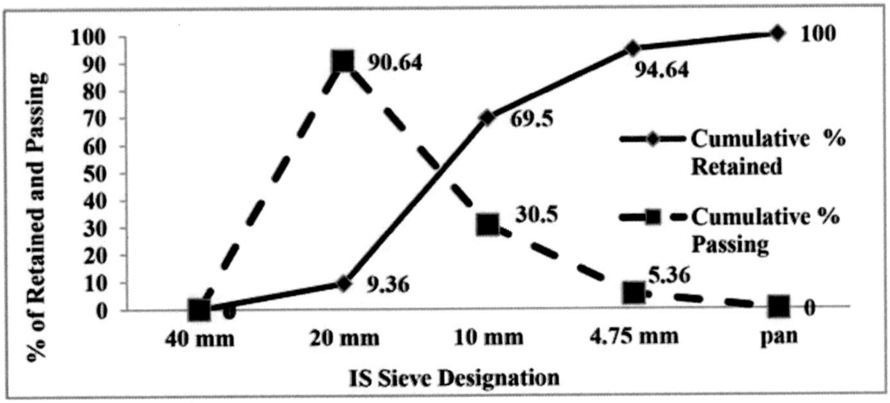

Fig. 4. Sieve analysis for C coarse.

Table 2. Mix proportion for CC and CDWC mixes

				Fine aggregate		*Coarse aggregate*	
Nature of mix	*Type of mix*	*Unit*	*Cement*	*Natural*	*Cultivated*	*Natural*	*Cultivated*
CC mix	M1	kg	450	652	–	1146	–
			1	1.45	–	2.54	–
CDWC mix	M2	kg	450.00	489.04	155.16	859.59	268.15
			1.00	1.09	0.34	1.91	0.60
	M3	kg	450.00	326.03	310.33	573.06	536.30
			1.00	0.72	0.69	1.27	1.19
	M4	kg	450.00	163.01	465.49	286.53	804.45
			1.00	0.36	1.03	0.64	1.79
	M5	kg	450.00	0.00	620.66	0.00	1072.60
			1.00	0.00	1.38	0.00	2.38

3.5 Test on Hardened Concrete

3.5.1 Compressive Strength Test

The progressive increase in compressive strength of hardened concrete is obtained by testing the concrete cubes after proper curing of 7, 14 and 28 days respectively. The CC mix shows an average compressive strength of 26.7 N/mm^2, 35.4 N/mm^2 and 43.4 N/mm^2 for 7, 14 and 28 days of curing respectively. The inclusion of

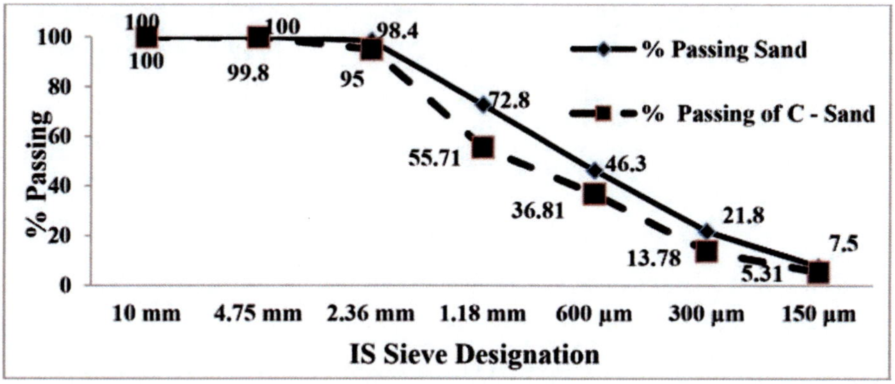

Fig. 5. Comparative analysis of sieve for sand and C-sand.

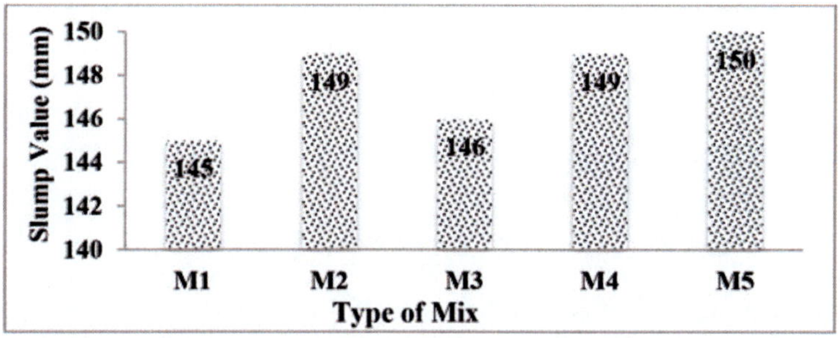

Fig. 6. Slump cone values of various mixes.

cultivated aggregate modifies the cube compressive strength to the significant level. The CDWC mix shows a slight decrease in compressive strength due to the presence of cultivated aggregates. The use of cultivated aggregates causes this decrease in compressive strength. The comparison of average compressive strength yields a better result for M3 mix obtained by replacing 50% of natural aggregates by cultivated aggregates. The benefits gained by reducing the consumption of natural aggregates and use of construction demolition waste is much higher than that of the problems created due to a small decrease in compressive strength. Hence, the M3 mix can be effectively replaced instead of CC in all construction activity.

The average cube compressive strength of each mix at the end of 28 days of curing is taken as a characteristic compressive strength (f_{ck}) for the corresponding mix and shown in Table 3.

Table 3. Characteristic compressive strength of each mix

S. no.	*Type of mix*	*Compressive strength (N/mm^2)*
1	M1	43.44
2	M2	39.10
3	M3	41.80
4	M4	37.24
5	M5	34.71

Table 4. Tensile strength of mixes

S. No.	*Type of mix*	*Tensile strength (N/ mm^2)*
1	M1	4.12
2	M2	3.79
3	M3	3.73
4	M4	3.61
5	M5	3.13

3.5.2 Split Tensile Strength

Table 4 describes the split tensile strength is carried out for CC and CDWC mixes at the end of 28 days of curing. The result shows that the CC mixes possess maximum tensile strength. The increase in percentage of replacement of cultivated aggregates decreases the tensile strength than that of the CC mix. The maximum reduction in tensile strength is about 25% for M4; whereas 25% and 50% replacements (M2 and M3) produce a minimum reduction of about 9%. Hence, M2 and M3 mixes can be considered for construction applications.

3.5.3 Optimum Percentage of Replacement of CDW

The CDWC mix is obtained by using various percentage of replacement of cultivated aggregates. Optimum percentage of replacement has been identified based on the test results of both fresh and hardened concrete. From the workability test, M3 mix yields similar result to that of CC mix. The compressive strength test result also reveals that the M3 mix achieves the characteristic compressive strength similar to M40 grade concrete, and hence, it can replace the CC mix. Whereas both M2 and M3 mixes shows significantly closer tensile strength characteristics, which means both of them can be selected as suitable mixes. By considering all the test results it is concluded that the M3 mix is identified as a suitable mix, and hence, 50% of cultivated aggregate as optimum percentage of replacement.

3.5.4 Calculation of Effective Depth of the Beam

In a structural member effective depth of the beam controls moment resisting capacity, deflection characteristics and construction cost. The depth of the beam influences on consumption of material, self-weight of the beam and overall economy of the structure. The effective depth of the beam mainly depends upon the characteristic compressive strength of concrete. Any variation in the characteristic strength will alter the effective depth of beam. The effective depth of the beam is evaluated for the concrete with different mixes. The depth varies depending on the characteristic compressive strength of the concrete, which is obtained from the experimental results. The required effective depth (d) is obtained by using the following moment equation as per IS 456-2000.

$$\mu = 0.138\, f_{\mathrm{ck}}\, bd^2$$

The width of beam (b) is assumed as 250 mm and span (l) as 4 m subjected to an UDL of 100 kN/m over the entire length. Assuming the beam as simply supported, the maximum bending moment developed is 200 kN m. The effective depth of beam for this bending moment is calculated for each type of mix and presented in Table 5.

The effective depth of beam for the given maximum bending moment is 365 mm for CC mix and 372 mm for the CDWC mix with optimum percentage of replacement. From this result, there is no significant increase in effective depth of the beam with 50% CDW aggregate replacement. Hence, the use of CDW in concrete is feasible in terms of effective depth of beam.

3.5.5 Volume of Aggregate Saved

The use of cultivated aggregate reduces the consumption of natural aggregates, which results in solid waste management. Hence, the calculation of volume of aggregates saved must be calculated accurately. The volume of natural aggregates saved for every 100 m^3 of concrete work by utilising the optimum CDWC mix (M3) is calculated as 12.075 m^3 (about two trucks) of fine aggregate and 21.625 m^3 (about 3.60 trucks) of coarse aggregate.

Table 5. Effective depth of beam for various type of mix

S. no.	*Type of mix*	*Characteristic compressive strength (N/mm²)*	*Effective depth (mm)*
1	M1	43.44	365
2	M2	39.10	385
3	M3	41.80	372
4	M4	37.24	395
5	M5	34.71	409

4 Conclusion

This study reveals the feasibility of using cultivated aggregates for the sustainability of natural aggregates. The physical properties of cultivated aggregates were tested and compared with the conventional aggregates. Both natural sand and C-sand are conforming to the zone II and the sieve analysis results and almost similar pattern of % passing through each sieve. The workability of the CDWC was not affected by the partial replacement of the cultivated aggregate upto 50%.

The strength of the CDWC was evaluated by comparing the conventional concrete mix with different mixes of CDWC. The test result clearly shows that M3 mix which was obtained by using 50% of cultivated aggregates yields a better compressive and tensile strength. Even when there is slight reduction in the strength compared to Conventional Concrete it does not have an adverse effect. Thus, from the study, it is concluded that M3 is a suitable mix for CDWC.

Further, the study has been extended to determine the effective depth of the beam by using the cultivated aggregates and it confirms that there is no significant increase in effective depth of the beam with M3 mix which results in less consumption of materials. About 12.075 m^3 of fine aggregate and 21.625 m^3 of coarse aggregate can be saved for every 100 m^3 of concreting work by utilising CDWC. Hence, the CDW will be a suitable alternative for the conventional natural aggregates. The use of CDWC will reduce the volume of solid waste generated highly and corresponding environmental pollution. Further, the volume of saving of natural aggregate and use of CDW will reduce the cost of construction as well as solid waste management issues.

References

Bravo, M., de Brito, J., Pontes, J. and Evangelista, L. (2015). Mechanical performance of concrete made with aggregates from construction and demolition waste recycling plants. *J Cleaner Prod*, 1–16.

Cantero, B., Saez del Bosque, I.F., Matías, A., Sanchez de Rojas, M.I. and Medina, C. (2019). Inclusion of construction and demolition waste as a coarse aggregate and a cement addition in structural concrete design. *Arch Civil Mech Eng*, 1338–1352.

Esa, M.R., Halog, A. and Rigamonti, L. (2016). Strategies for minimizing construction and demolition wastes in Malaysia. *Resour Conserv Recycl*, doi: http://dx.doi.org/10.1016.

IS 10262:2019 (2019). *Concrete Mix Proportioning – Guidelines* (second revision). Bureau of Indian Standards, New Delhi, India.

IS 383:2016 (2016). *Coarse and Fine Aggregate for Concrete Specification* (third revision). Bureau of Indian Standards, New Delhi, India.

IS 456:2000 (2007). *Plain and Reinforced Concrete Code of Practice* (fourth revision). Bureau of Indian Standards, New Delhi, India.

Junaka, J. and Sicakova, A. (2016). Concrete containing recycled concrete aggregate with modified surface. International High-Performance Built Environment Conference-A Sustainable Built Environment Conference 2016 Series (SBE16), IHBE 2016.

Kabir, S., Al-Shayeb, A. and Khan, I.M. (2016). Recycled construction debris as concrete aggregate for sustainable construction materials. *Int Conf Sustain Des Eng Constr Procedia Eng*, **145:** 1518–1525.

Mymrin, V. and Correa, S.M. (2007). New construction material from concrete production and demolition wastes and lime production waste. *Constr Build Mater*, 578–582. doi: https://doi.org/10.1016/j.conbuildmat.2005.09.007.

Pathak, A., Kumar, S. and Jha, V.K. (2014). Development of building material from geopolymerization of construction and demolition waste (CDW). *Trans India Ceram Soc*, **73(2):** 133–137. doi: https://doi.org/10.1080/0371750X.2014.922429.

Rao, A., Jha, K.N. and Misra, S. (2007). Use of aggregates from recycled construction and demolition waste in concrete. *Resour Conserv Recycl*, **50:** 71–81.

Shahidana, Moh. A., Azmib, M., Kupusamy, K., Sharifah Salwa, M.Z. and Alie, N. (2017). Utilizing construction and demolition (C&D) waste as recycled aggregates (RA) in concrete. *Global Congr Manuf Manage Procedia Eng*, **174:** 1028–1035.

Silva, R.V., de Brito, J. and Dhir, R.K. (2011). Properties and composition of recycled aggregates from construction and demolition waste suitable for concrete production, doi: https://doi.org/10.1016/j.conbuildmat.2014.04.117.

Tam, V.W.Y., Tam, L. and Le, K.N. (2010). Cross-cultural comparison of concrete recycling decision-making and implementation in construction industry. *Waste Manage*, **30:** 291–297. doi: https://doi.org/10.1016/j.wasman.2009.09.044.

Tama, V.Y.W., Soomro, M. and Jorge Evangelista, A.C. (2018). A review of recycled aggregate in concrete applications (2000–2017). *Constr Build Mater*, **172:** 272–292.

Chapter 13
Experimental Study on Incorporation of Plastic Wastes as a Binding Material in Pervious Pavers

M. Bharathi, K. K. Indirajeevan, S. S. Mathivanan, R. Muniyandi, and S. Praveenkumar

1 Introduction

In India, groundwater plays a significant role in meeting the basic needs of the people. It satisfies around 63% of the irrigational requirements and 80% of the domestic water requirements in urban and rural areas. According to the NITI Aayog Composite Water Management Index (CWMI) report, groundwater recharging in most of the Indian states is less than 50%, and also, it states that by the year of 2020 and 2021, majority of the cities in India are going to face severe effects due to the depletion of the groundwater (CWMI, 2019). India is the topmost country in the world in abstracting the groundwater, which consumes 250 km^3 volume of water per year (Pervious Concrete Pavements, 2010). According to the analysis done by the Central Ground Water Board (CGWB), of 14,465 wells, 61% of them have a drastic reduction in the groundwater level in the past 10 years (CGWB, 2016–17). So, the country faces difficulties in recharging the groundwater level, and the decrease in rainfall still makes the situation worse. On the other hand, rainwater is wasted on roads as surface runoff because of the impervious nature of roads. Around 61 lakh litres of rainwater is wasted every year per kilometre of roads, which is calculated as per the guidelines given by the MORTH, Government of India (Circular by Government of India: Ministry of Road Transport & Highways, 2013). It is due to the lack of rainwater harvesting and improper water management techniques. In addition, the climatic change is one of the adverse effects of global warming, which in turn causes the variations in the rainfall and causes extreme precipitation in a shorter period that also leads to the wastage of rainwater as runoff (Min et al., 2011). The runoff water which is polluted on the road causes more than 14% of pollution in rivers (Thives et al., 2018). In India, mostly the rainwater runoff is carried by the combined sewers, which make this runoff water unusable and also the stagnation of

M. Bharathi · K. K. Indirajeevan · S. S. Mathivanan · R. Muniyandi · S. Praveenkumar (✉)
Department of Civil Engineering, PSG College of Technology, Coimbatore, Tamilnadu, India
e-mail: spk.civil@psgtech.ac.in

A. S. Kalamdhad (ed.), *Integrated Approaches Towards Solid Waste Management*, https://doi.org/10.1007/978-3-030-70463-6_13

water on the road weakens its structural integrity. All these problems are mostly due to the impermeable covering of the road surface.

The pavers are one of the impervious coverings over the ground surface. These pavers are the discrete blocks available in different shapes and sizes that are arranged in a certain pattern, which are used in places such as pedestrian footpaths, street roads, parking areas, tennis courts etc. To resolve the problems due to the imperviousness of the pavers, pervious concrete pavers were introduced.

The first pervious concrete was invented in 1800s. The pervious concrete is otherwise called no-fine concrete which has the voids by eliminating the use of fine aggregates. These voids help the surface runoff water to percolate into the ground. It consists of only the coarse aggregates that are bound together by cement.

Nowadays, many environmental issues are increasing day by day because of various developmental activities. In the construction field, cement is the most widely used material which has more advantages. But, the major effect of cement is the carbon emission during manufacturing and depletion of natural resources like limestone etc., which are the raw materials used. Therefore, a lot of researchers are working in this concern to replace the cement to some extent by alternative materials like ground granulated blast-furnace slag (GGBS), fly ash, glass waste, and so on. All these are done to reduce the environmental ill-effects of the cement. The main aim of this project is to completely eliminate the ill-effects of cement by fully replacing it with the plastic wastes which have no methods to dispose and causing various environmental pollutions. The essence of this project is sorting out these kinds of problems by making the pavers pervious using the plastic wastes as the binding material.

2 Materials

In recent years, pervious concrete is emerging as the commonly used pavement material, which leads to sustainable development because of its environmental influence. Concerns have been increasing among public agencies, planners and developers towards reducing the pollutants in water supplies and the environment. Recharging the groundwater supplies, reducing the quantity of storm water generated from developed areas, improving the storm water quality, reducing the discharge of pollutants in water supplies, and minimising the effects of development on watersheds have become the primary focus area while developing a natural land.

2.1 Plastic Wastes

Plastics are synthetic materials, which are the by-products of crude oil. Plastics are basically made up of polymers, and their structure may be either homogeneous or heterogeneous. They become softer at higher temperatures and harder at lower temperatures. This property of plastics makes it usable as a binding material. The different types of plastic wastes were used as binding materials (Fig. 1).

Fig. 1. Types of plastic wastes used.

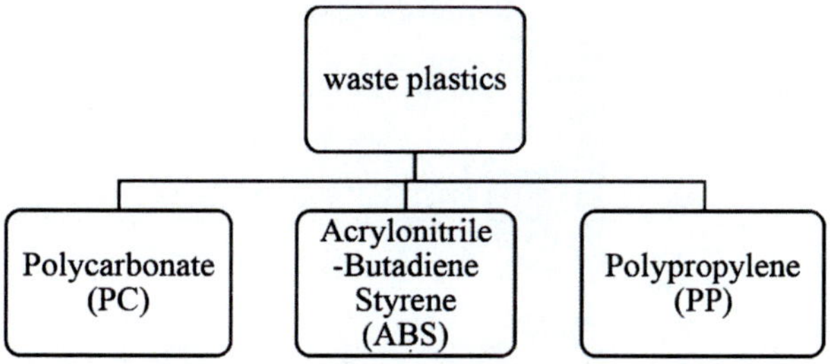

2.2 *Acrylonitrile Butadiene Styrene (ABS) Plastics*

Acrylonitrile butadiene styrene is a copolymer plastic that is formed by the emulsification of three different monomers such as acrylonitrile, butadiene and styrene. The acrylonitrile creates a polar attraction when it is mixed with the other two monomers. This imparts tough and highly durable properties. It has the chemical formula $(C_8H_8)(C_4H_6)(C_3H_3N)_z$. It has a wide range of domestic and commercial applications.

2.3 *Polycarbonate (PC) Plastics*

Polycarbonates are thermoplastics that are formed by the carbonate groups (–O–(C=O)–O) and have the chemical formula $C_{15}H_{16}O_2$. They are amorphous, transparent and tougher in nature. These plastics have high impact resistance, and hence, these are employed in the bullet proofing glasses. Apart from this, it is also used in vehicle headlight covers, medical devices etc.

2.4 *Polypropylene (PP) Plastics*

Polypropylene is a thermoplastic. It is made by the combination of propylene monomers. It belongs to the polyolefin family. It is a rigid, tough and crystalline plastic. Its chemical formula is $(C_3H_6)_n$. It is used in packing industries, consumer products such as bottles etc. (Fig. 2).

2.5 *Aggregates*

2.5.1 Coarse aggregates

The coarse aggregates of various sizes ranging from the aggregates passing through 10 mm IS sieve and retained in 4.75 mm IS sieves were used as filler materials. The properties of the coarse aggregates are shown in Table 1.

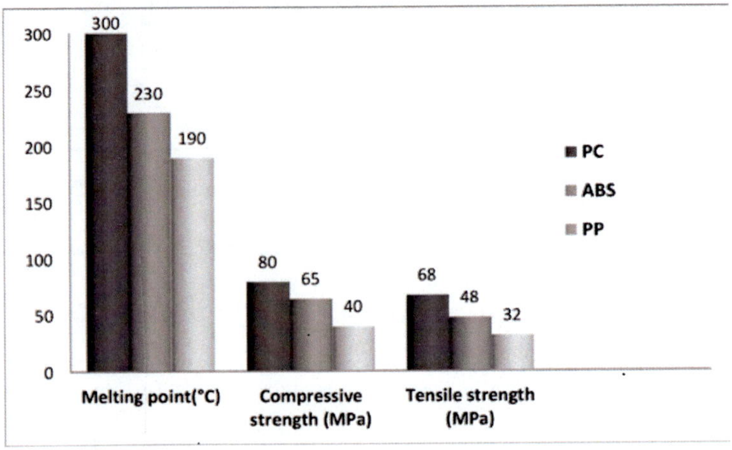

Fig. 2. Properties of the plastics.

Table 1. Properties of coarse aggregates

Properties	*Values*
Specific gravity	2.79
Bulk density (kg/m^3)	1650
Abrasion (%)	20

Table 2. Properties of fine aggregates

Properties	*Values*
Specific gravity	2.6
Bulk density (kg/m^3)	1560

2.5.2 Fine aggregates

M-Sand was used as the fine aggregate in the pavers. The fine aggregates are those which are less than 4.75 mm IS sieve in size. These fine aggregates were used to fill the voids between the coarse aggregates. The properties of the fine aggregates are shown in Table 2.

3 Methodology

Plastic wastes are used as the binding material in the pervious green pavers, which is an alternative to cement that causes various environmental issues. The thermoplastic is a very good binding material; its viscosity reduces when it is heated at a particular temperature. Due to this, it coats over the surface of the aggregates. This wet mix undergoes re-solidification when the temperature falls down and binds the aggregates together which imparts strength. Modelling pervious green pavers has no standard references. Hence, the procedure as shown in Fig. 3 was used.

Fig. 3. Steps involved in sample preparation.

3.1 *Segregation*

The solid wastes consist of various types of plastic wastes in different quantities. So, there is a need for segregation of the plastic wastes based on their types in order to have uniformity. The segregation of plastics was done by the following methods:

1. Burning of plastics
2. Smelling of plastics
3. Immersion in water

Among the plastic wastes, polycarbonates, ABS and polypropylene were selected, which satisfied the requirements such as toughness, compressive strength, tensile strength and impact resistance.

3.2 *Grinding*

Grinding the segregated plastic wastes was done to reduce the size and to ease the mixing process.

3.3 *Mix Proportion*

The proportioning of the plastics was done by the trial and error method to find the optimum mix. Table 3 shows the various mix proportions.

3.4 *Melting and Mixing*

The aggregates were first heated up to 150 °C. Then, the ground plastic wastes were added and further heated up to 250 °C in the case of PC & ABS and heated up to 180 °C in the case of PP until the surface of the aggregates were completely coated with these plastics.

3.5 *Moulding*

Heating was stopped after mixing, and the mix was transferred immediately to the mould within 30 seconds. It was then compressed to make it denser.

Table 3. Mix proportions for the plastic and aggregates

Block no.	*Size of coarse aggregate (mm)*	*Mix proportions*	*Plastic to aggregate ratio (weight batch)*	
Full pervious pavers			**Pervious (P:CA)**	
1	10	PC: 65% & ABS: 35%	1:7	
2	6	PC: 65% & ABS: 35%	1:6	
3	10	PC: 70% & ABS: 30%	1:7	
4	6	PC: 70% & ABS: 30%	1:6	
5	10	PP: 100%	1:7	
6	6	PP: 100%	1:7	
Integrated pavers			**Pervious (P:CA)**	**Impervious (P:FA:CA)**
7	10	PC: 65% & ABS: 35%	1:7	1:1:2
8	6	PC: 65% & ABS: 35%	1:6	1:1:2

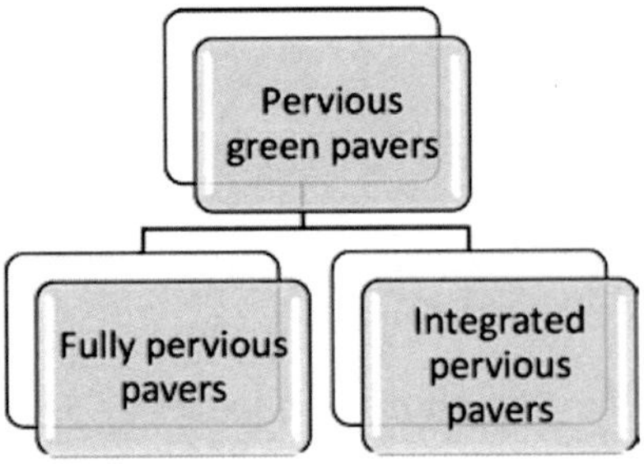

Fig. 4. Types of pervious green pavers.

3.6 *Demoulding*

Demoulding was done after 3 minutes and kept in the room temperature for 2–3 hours. The pervious green pavers were cast into two types as shown in Fig. 4.

3.6.1 Full pervious pavers

The full pervious pavers are those similar to the conventional pervious concrete paver in its design, where the entire surface will be pervious. It consist only the coarse aggregates that were bound together by the plastic wastes which leave larger voids in it.

3.6.2 Integrated pervious pavers

These are the special type of pavers, which are composed of both pervious and impervious segments. The outer impervious ring consists of fine aggregates, coarse aggregates and plastic wastes, which impart strength to the pavers. The inner core of the paver was pervious which consists of only coarse aggregates and plastics that impart the pervious property to the pavers. Thus, these pavers satisfy both strength and pervious parameters.

4 Experimental Programme

4.1 Block Compressive Strength

Compressive strength is the ability of a structure to resist loads acting on it without any deformation or fracture. Compression test is the basic and important test, and hence, this test is considered as the key value for the design of structures. Compressive strength is usually measured using universal testing machine. The test was done as per IS 15658- 2006. The test was performed at two different temperatures. For the first test, the pavers were kept at room temperature, and for the second test, the pavers were kept at the elevated temperature. All the compressive strength results shown for each block are the average values of three blocks of same kind.

4.1.1 Compression Test at Room Temperature

The compression test was done by keeping the pavers at the atmospheric or room temperature similar to conventional compressive strength test for paver.

4.1.2 Compression Test at Elevated Temperature

This type of compression test was done to determine the compressive strength when there is an increase in the ambient temperature. In the real-time application, the intensity of temperature varies accordingly with respect to the location. To account for this variation, this test was conducted by increasing the ambient temperature of the block by an indirect method called steam bath. The specimen was placed in the steam bath and the temperature was elevated to about 100 °C; then, the compressive strength was obtained. The integrated pavers were tested at the elevated temperature.

4.2 Abrasion Test on Blocks

Abrasion test which indicates the amount of scraping, wearing or the depreciation of the surface of the block is the next important test for pavements. The paver blocks which are used as the pavement surface are exposed to constant traffic flow which in turn causes wear and tear, the vulnerable one that leads to a decrease in strength. This also leads to the disintegration of the surface and crushing of the block. The abrasion test was performed as per IS 15658-2006.

4.3 Water Permeability Test on Blocks

The water permeability test is the measure of the rate of flow of water or the ability of water to penetrate through the pores. The study of the permeability property of the pervious pavers is required to calculate their efficiency. This constant head permeability test was approached in two ways. In the first type, the test was conducted directly on the blocks without filling it with sand. In the second type, the blocks were filled with sand, which usually happens when the runoff water runs over the blocks. This permeability test was performed as per IS 2720-2017.

5 Results and Discussions

5.1 Block Compressive Strength

The compressive strength increases with the decrease in the size of the coarse aggregates independent of the type of plastic wastes used. This is because reduction in the size of the aggregates increases the effective contact area between them, which makes it denser. The compressive strength of the integrated pervious pavers showed a higher value than that of the full pervious pavers. This is because, in full pervious pavers, the outer edges will be discontinuous, which reduces the compressive strength, whereas in integrated pervious paver, the outer edges were strengthened by the impervious ring, which increases its compressive strength. As per the strength criteria of the IS code for pavers, the full pervious pavers can be used in non-traffic areas such as pedestrian footpaths etc., and the integrated pervious pavers can be used in the traffic areas such as roads, parking areas, etc. (Fig. 5).

The compressive strength of the integrated pervious pavers was tested at the elevated temperature of 100 °C to check its stability in the field when it is exposed to higher temperatures. The result shows that even at 100 °C, it loses only 14% of its total strength. Among the various mixes, the mix PC:ABS (70%:30%) gives the maximum strength. The compressive strength of the blocks at room temperature and at elevated temperatures are as in Tables 4 and 5.

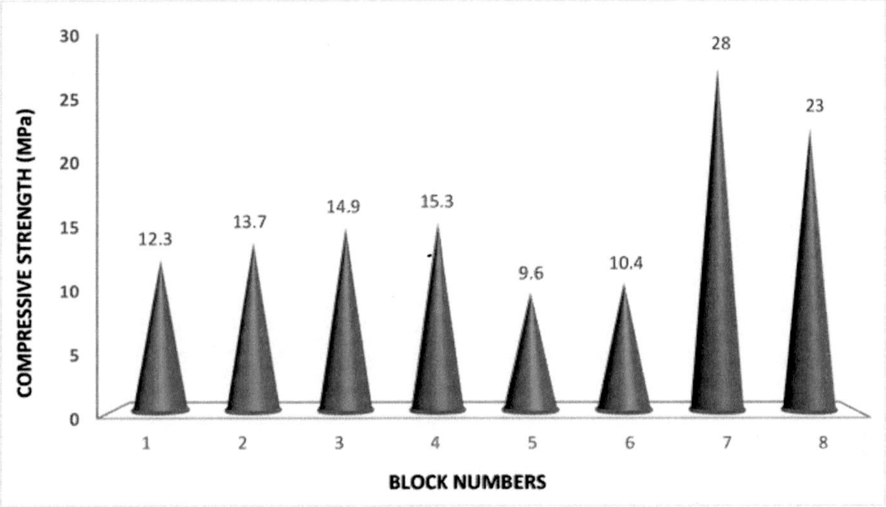

Fig. 5. Compressive strength at atmospheric temperature.

Table 4. Compressive strength results in atmospheric temperature

Block no.	*Size of coarse aggregate (mm)*	*Compressive strength (MPa)*
Full pervious pavers		
1	10	12.3
2	6	13.7
3	10	14.9
4	6	15.3
5	10	9.6
6	6	10.4
Integrated pavers		
7	10	28
8	6	23

Table 5. Compressive strength at elevated temperature

Block no.	*Size of coarse aggregate (mm)*	*Compressive strength (MPa)*
Integrated pavers		
9	6	24
10	10	18.5

Table 6. Abrasive strength of the paver

Trial no. (PC: 70% & ABS: 30%), mix 1:1:2	*Abrasive strength (in terms of % weight loss)*	*Average abrasive strength*
1	0.545	0.537%
2	0.530	
3	0.535	

5.2 *Abrasion Test on Blocks*

The wear on the surface of the pavers is considerably lower (0.54%). Hence, these pavers can be used in the traffic and non-traffic areas positively. The results are as in Table 6.

5.3 *Water Permeability Test on Blocks*

The permeability of the pervious green pavers is higher than that of the pervious cement pavers. This is due to two reasons,

(1) The cement grains act as filler material and fill the voids.
(2) The thickness of the cement slurry coated over the aggregates is high.

These lead to the reduction in the pore size, thereby reducing the permeability. But these problems were not encountered while using the plastic wastes as binder. These pervious green pavers showed permeability 41% more than the previous cement pavers, even when they were clogged with the specks of dust. The permeability of the paver blocks without sand fill and with sand fill are shown in Tables 7a and 7b respectively.

6 Conclusion

From the tests and studies, it was found that the conventional paver blocks can be replaced with the pervious green pavers. The pervious green pavers reduce the emission of carbon dioxide (CO_2), which is released abundantly during the production of cement, which is a major threatening issue in the environmental aspects. In addition, the plastic wastes that are dumped in the environment can be efficiently utilised for the production of these pavers. Hence, it is regarded as ecofriendly. Although plastic wastes were used as binders, it had shown an adequate compressive strength in reference to standard specifications. The full pervious pavers can be used in the pedestrian footpaths. Subsequently, an increase in the strength was achieved in the integrated pavers, which comprises both pervious and impervious segments, and these can be used in the traffic areas such as parking, roadways etc.

Table 7a. Permeability test on paver blocks

Test done with filling the voids without sand				
Block no.	*Trial no.*	*Permeability of water (cm/s)*	*Type of paver*	*Average permeability (cm/s)*
7	1	0.369	Integrated	0.4052
	2	0.4316		
	3	0.3923		
	4	0.428		
4	1	0.147	Full pervious	0.1463
	2	0.143		
	3	0.149		
	4	0.146		

Table 7b. Permeability test on paver blocks

Test done with filling the voids with sand				
Block no.	*Trial no.*	*Permeability of water (cm/s)*	*Type of paver*	*Average permeability (cm/s)*
7	1	0.0339	Integrated	0.03925
	2	0.0428		
	3	0.0357		
	4	0.0446		
4	1	0.496	Full pervious	0.4930
	2	0.499		
	3	0.490		
	4	0.487		

The pervious green pavers show sufficient abrasive strength to be used as the covering material on roads. The permeable property of the paver allows the surface runoff water to penetrate into the ground which in turn helps groundwater recharging. Using the pervious green pavers, merely 1380 litres of water per hour per square meter can be either used in groundwater recharging or in other useful purposes such as irrigation, gardening etc. The implementation of these pavers in the real-time application will show nearly 70% of a rapid increase in permeability than the conventional pervious concretes do. The usage of water is completely ruled out in the green paver blocks, whereas in conventional pavers, large quantity of water is required during the manufacturing and curing process. Also, the time taken for curing was completely reduced, i.e. the maximum strength was attained in one day, whereas it is a time-consuming process in the case of conventional cement pavers. The cost of green paver blocks is approximately the same as that of the conventional pavers, which can be reduced further in mass production using appropriate methods. Thus, the pavers being ecofriendly, efficient and economical, it is suggested for public usage.

Acknowledgements We wish to record our sincere and heartfelt thanks to Dr. K. Prakasan, Principal, PSG College of Technology, for providing us with excellent facilities and accessories to complete the project.

We wish to record our deep sense of gratitude to Dr. G. Sankarasubramaian, Head of the Department, Department of Civil Engineering, PSG College of Technology, for his colossal guidance and support throughout the project.

We express our gratitude to Mrs. V. Vanathi, Assistant professor, Department of Civil engineering, PSG College of Technology, for her guidance and advice.

We express our sincere thanks to all the industries and firms who aided us to complete the project successfully.

References

ACI 522.R-10 (2010). Report on Pervious Concrete.

Government of India, Ministry of Road Transport & Highways, Circular: No. RW/NH-33044/14/2003-S&R(R) (Pt.II), dated on 5 September 2013, Tentative Guidelines for drainage through Rain Water Harvesting and Artificial recharging along National Highways.

IS: 1199 (1959). Methods of sampling and analysis of concrete. Bureau of Indian Standards.

IS: 15658 (2006). Precast concrete blocks for paving. Bureau of Indian Standards.

IS: 4082 (1996) (Reaffirmed 2003). Stacking and storage of construction materials and components at site – Recommendations. Bureau of Indian Standards.

IS: 456 (2000). Code of practice for plain and reinforced concrete. Bureau of Indian Standards.

Min, S.K., Zhang, X., Zwiers, F.W. and Hegerl, G.C. (2011). Human contribution to more-intense precipitation extremes. *Nature*, **470:** 378–381.

Pervious Concrete Pavements (2010). Maintained by National Ready Mixed Concrete Association (NRMCA). http://www.perviouspavement.org.

Report of Central Ground Water Board (CGWB), 2016–17.

Report of NITI Aayog Composite Water Management Index (CWMI) (2019).

Liseane Padilha Thives, Enedir Ghisi, Natália Mattos da Silva. (2018). Potable Water Savings in Multifamily Buildings Using Stormwater Runoff from Impermeable Paved Streets. European Journal of Sustainable Development, 7, 3, 120–130.

Chapter 14
Biogas Development in India: A Sustainable Energy for Future

Nikita Kanaujia, Jiwan Singh, and Ajay S. Kalamdhad

1 Introduction

Rapid urbanisation and subsequent increase in population density accelerate the generation of waste and demand for energy incessantly. In order to manage these wastes and to fulfil the surging need of energy efficiently without affecting the environment is an exhausting task. Today's knee-jerk use of fossil fuel has induced greenhouse gases to the atmosphere with carbon dioxide (CO_2) as a powerful contributor that adversely affect the atmosphere, which is a global concern. CO_2 concentration in the atmosphere is 411 parts per million (ppm) in 2019 in contrast to 283 ppm in 1880 (pre-industrial revolution period) (Abbasi et al., 2012). Therefore, to combat this situation, waste to energy and renewable energy are dynamic concepts, which evolved over the years in accordance with the changing needs and circumstances of our society. Among them, biomass pertaining energy allures the most, which could manage the need of energy and organic waste efficiently. The municipal solid waste (MSW) and agriculture sector waste kick in most in the environmental distress in India. Per capita MSW generation in India is 400 grams/day and based on the lifestyle and size of the city, mostly comprises of degradable organic matter (51–53%) (Bhatt et al., 2018). Due to limited and uncertain fossil fuel resources, moreover it causes environmental pollution, despite people use it perpetually, owing to its low cost and accessibility, so that it is difficult to roll back the use of fossil fuel that acquires normalcy, completely withdraw the use of fossil fuel is a formidable problem, but one has to control its use for the betterment of the environment. Hence, renewable energy appears to be a promising and affordable

N. Kanaujia · J. Singh (✉)
Department of Environmental Science, Babasaheb Bhimra Ambedkar University, Lucknow, India

A. S. Kalamdhad
Department of Civil Engineering, Indian Institute of Technology Guwahati, Guwahati, Assam, India

A. S. Kalamdhad (ed.), *Integrated Approaches Towards Solid Waste Management*, https://doi.org/10.1007/978-3-030-70463-6_14

alternative, since it has innumerable health and environmental benefits. Use of renewable energy is beneficial for the environment in general and human in particular. In order to achieve sustainable waste management and energy recovery from waste, anaerobic digestion is supposed to be proven and vital technology. Anaerobic digestion (AD) is resurrected process of the biomass wherein microorganisms degrade materials which are biodegradable in nature within the absence of oxygen and the end product is CH_4, CO_2 and other trace gases. AD offers numerous advantages such as sustainable management of organic waste, rightly controlled parameter of anaerobic digestion lead to high reduction of organic waste, pathogen removal, production of biogas (Khalid et al., 2011). About 85% of world energy demand is fulfilled by conventional energy sources. India too virtually depends on coal base energy as it shares the gigantic portion of total energy generation. However, it now seems that the renewable energy sources has massive potential to completely displace the conventional source of energy or fossil fuel.

1.1 Bio-energy Scenario in India and Its Impact on the Environment

Clean energy is vital for inclusive growth of modern society (Stougie et al., 2018). Demand for energy disperse unevenly across the world as six major countries (China, USA, Russia, India, Japan and Germany) consumed more than half of the available energy (Quaschning, 1969). India lies in fourth biggest after China, USA and Russia in energy consumption. India's peak energy demand in 2017–2018 was 1,64,066 MW and peak met is 1,60,752 therefore apparently energy deficits of –3314 MW (https://powermin.nic.in/en/content/power-sector-glance-all-india). Society is gradually heading towards waste minimization along with clean, sustainable and affordable fuel (Sims, 2003). The steadily increasing worth and unfortunate environmental impact of fossil fuel have emboldened the production of bioenergy to the unprecedented extent over the last 15 years (Popp et al., 2014). As it has numerous advantages such as zero emission of carbon, highly flexible compared to all other renewable energy sources and versatile as well since it can generate electricity, heat and transport fuel in other words solid, liquid and gaseous fuel derived from different type of biomass by its pretreatment and conversion technologies (Roder and Welfle, 2019). Therefore, it is needed to enhance the bioenergy production for inclusive growth of country, there is a paradigm shift towards the production of bioenergy or there has been more and more focus to harness maximum potential of biomass and enhance its efficiency for bioenergy production (Popp et al., 2014). Figure 1 shows that renewable energy contributed 16% out of total installed capacity, however, bio power contributed 16% of total renewable energy.

Hence, prudent policy reform for solid waste management can help India to become resilient to climate change.

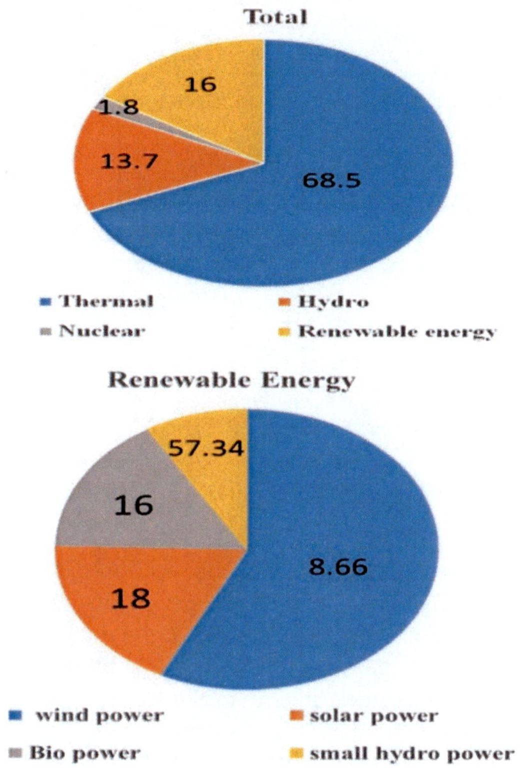

Fig. 1. Source wise energy installed capacity in India (source: https://mnre.gov.in/physical-progress-achievements).

2 Waste Biomass to Bioenergy

An important issue pertaining to the usage of waste biomass as an energy source is the need to process it into a suitable and usable form. The input in these processes could be intently grown energy crop, wood, plants, oilseed or agricultural residue collectively known as biomass and the output could be usable heat in the sort of solid, liquid and gases like charcoal, wood pallets, biogas, bioethanol and biodiesel which are biofuels produced from biomass. Out of these, some happen to be very efficient and reliable technologies while some are still facing challenges in its production. Waste to energy could manage a large quantity of waste generated from the agriculture sector, municipality, industrial and animal waste by different technologies such as incineration, gasification, pyrolysis, combustion, AD. Among them, AD is widely popular in India because the organic component in the waste is approximately 52% and the paper component is 10% (Singh et al., 2015). As shown in Fig. 2, grid interactive power and off-grid captive power from biomass.

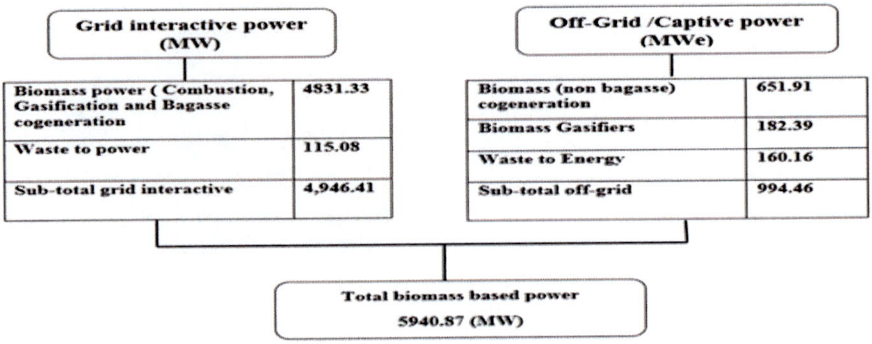

Fig. 2. Grid interactive power and off -grid captive power from biomass (source: https://mnre.gov.in/physical-progress-achievements).

3 Different Generation of Biomass

First generation biomass is obtained from edible things, produced from agriculture sector in which carbohydrate rich starch and sugars feedstocks are involved. Moreover, at present, various forms of biofuel are under development so as its source too, hence second-generation biofuel has come into force. Unlike first generation feedstock, it is produced from non-edible raw materials mainly from lignocellulosic biomass, which comprises three basic elements cellulose, hemicellulose, and lignin, such as agricultural and forestry residue. While third generation biofuels are extracted from microalgae and micro-bacterium, these are considered as excellent feedstocks due to the presence of oil, carbohydrate or protein content. As far as third generation, biomass is concerned it can produce high yield as compared to first- and second-generation biomass and do not require any particular pasture for growing. It could be cultivated in bioreactors with splendid yield (Srirangan et al., 2012; Saladini et al., 2016; Boyle, 1996). Table 1 exhibits generations of biomass as well as its pros and cons.

4 Past, Present, Future Perspective of Anaerobic Digestion and Its Implication on Energy Supply in India

Biogas process are being used from the ancient time, the very first existence of biogas was perceived by Plinius when he noticed a peculiar thing, a mysterious flame coming from subsurface of organic waste. In addition, number of affirmations are there to support that biogas was used in Assyria during 10th century before Christ (BC) to heat the water bath, later in the 17th century Alessandro Volta an Italian physicist and chemist discovered a methane content in the marshes and as well as determine the amount of organic material was used for AD and amount of biogas

Table 1. Generation of biomass resources (Srirangan et al., 2012)

Feedstock	*Categories*	*Example*	*Advantage*	*Limitation*
First-generation	Starch crops Sugar crops Oilseed crops	Cereal grains, corn (maize), wheat, sorghum, cassava, potatoes Sugarcane Rapeseed, soybean, sunflower, peanut, palm, coconut, safflower, linseed hemp	Outstanding energy content Stable and mature technology	Requires tropical arable land Geographic limitations
Second-generation	Organic waste residue Dedicated energy crops	Lignocellulosic feedstock, i.e. corn cobs, corn Stover, wheat straw rice hulls cane bagasse Perennial grasses (e.g. Switch grass, miscanthus) Woody energy crops (e.g. Polars, willows, eucalyptus)	Free from competition with food industries Less controversial High availability	Relatively immature Requires attentive, labourious and costly pre-treatment technologies
Third-generation	Microalgae and microbacterium	Algae	Fast growing Non-arable land Contribute in CO_2 mitigation, since they can grow in high CO_2 concentration Have excessive photosynthetic efficiency, which cause high productivity	Relatively immature

production is interlinked and complement to each other following this, Cruikshank in 1801 manifests, the absence of oxygen molecules in methane and in 1804, Dalton gave the exact formula for methane. Afterward in 19th century, Bechamp in 1868 demonstrated that microbiological process is a foundation of AD and microbes play a critical role in this process. Subsequently, with gradual progress in the AD process, major advancement was formulated by Sohngen in 1910 is the establishment of the stoichiometric equation of hydrogenotrophic methanogenesis

$$CO_2 + 4H_2 \rightarrow CH_2 + 2H_2O$$

And the theoretical calculation of methane potential was formulated in 1933 by Buswell and Boruff. It is evident that the First biogas plant was established in India in 1859 (Abbasi et al., 2012). In 1890, biogas was used as fuel to light up street lamp in Exeter, UK (Bond and Templeton, 2011). Today's various research have been conducted over the past decade to overcome complication regarding biogas. Various types of substrate has been experimented to find out the suitability for AD as well as optimize optimum condition such as pH, alkalinity, temperature, hydraulic retention time, organic loading rate for better performance. William O. Travis innovated two stage AD process wherein solid and liquid material are separated. In first stage hydrolysis, acidification and solubilisation occurs producing a leachate and soluble organic compound such as volatile fatty acids further in second stage leachate is converted into biogas (Parawira et al., 2008). Currently, in India nearly five million which is acceptable, household biogas digesters of capacity 1–10 N m^3 biogas d^{-1} had been installed in rural areas till 2016, although trial has been made to explore it to urban areas too, for management of MSW, vegetable or fruit waste from fruit market and waste food from restaurant that's been only effective under suitable condition. Approximately 1000 small-scale AD plants with capacity 30–200 Nm^3 daily, had been set up in Indian municipalities till 2015. Today's municipalities in India are extensively investing in AD process for organic waste management under 'Swachh Bharat Mission' launched in 2014 (Breitenmose et al., 2019).

4.1 Ministry of New Renewable Energy (MNRE) Programs for the Development of Biogas in India

Several schemes pertaining to biogas development have been traverse for managing huge quantity of organic waste generated from household, agriculture, industries and other forms of organic waste and also for providing emission free fuel. Figure 3 shows different initiatives taken by the government of India to embolden the AD

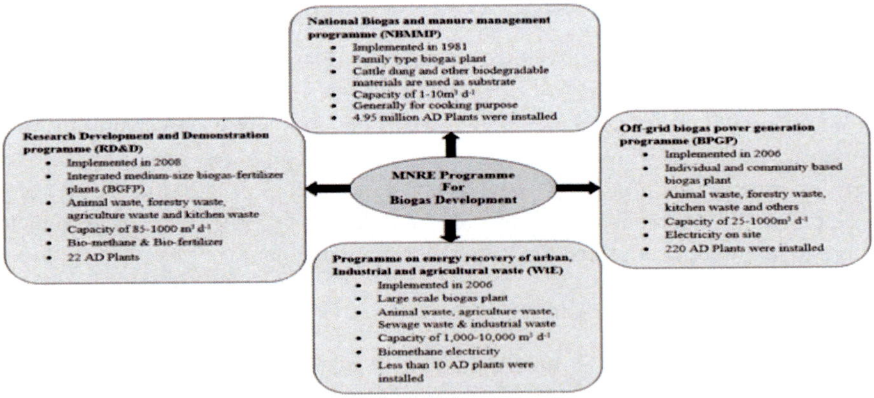

Fig. 3. Several MNRE programmes for development of biogas in a country (Breitenmoser et al., 2019).

technologies for handling the continuously producing organic waste from the various sources.

There is scope for further development in AD technologies, apparently new technologies regarding AD has been explored as scientists all over the world are trying to develop an idea for improving biogas production to get pollution free low-cost reliable energy. AD is effortless technique to recapture energy content from biomass in the form of biogas although different techniques relevant to improvement in AD method have been developed over the years. These techniques including the pretreatment of biomass, development of two-stage AD, membrane bioreactor (MBRs), improved reactor design, advanced monitoring and control, improved microbiology, latest purification technology of biogas, removal of inhibitors which aid to increase the biogas production (Maurya et al., 2019).

4.2 Technologies Used for Improved Biogas Production: An Effect of Pretreatment Method

There is determined urge for the enhancement of biogas so it can be accomplished either by pretreatment method or by adding additives, it has been observed that these are the critical factors contributing to the enhancement of biogas production. Pretreatment method should be selected on the basis of composition and characteristics of biomass and it is mainly required in lignocellulosic biomass (Karthikeyan et al., 2018). The intricate structure of the cell wall in lignocellulosic biomass is primarily characteristic to lignin content, cellulose crystallinity and accessible surface area. Hence, the reason behind pretreatment is to alter lignin from lignocellulosic biomass, to reduce crystallinity of cellulose and to improve the surface properties by enlarging the surface area. Conventional anaerobic digestion is unable to enhance biogas production by mere conversion of biomass into biogas. Pretreatment is required for most of solid organic waste specifically those materials, which have high lignocellulosic content i.e. cellulose, hemicellulose and lignin for altering its composition in order to acquire better methane production. Generally, the pretreatment method can be categorised in three forms namely, physical, chemical and biological method. Physical pretreatment is applied in most of the cases prior to any other method. In physical pretreatment mechanical operations are applied (Amin et al., 2017). Chemical and microbes are not used during the pretreatment. Chemical pretreatment is mainly accomplished by chemicals like acid, bases and ionic liquid to alter the recalcitrant material lignin, particularly from lignocellulosic biomass and improve the digestion of such biomass (Zheng et al., 2014). Although thermochemical methods are popular for biomass pretreatment, however, it is noted there are certain environmental and economic issues regarding these approaches in this order. Biological pretreatment could be a promising option to the traditional methods as it uses bacteria, fungi and enzymes under mild conditions with low energy input and low disposal cost (Zabeda et al., 2019). Biological pretreatment is a potent method, which conserves the hemicellulose fractions, and does not produce any toxic material that could inhibit the growth of fermentative bacteria (Wagner et al., 2018). The main objectives of biological pretreatment is delignification of lignocellulosic biomass leading to reduction in the degree of polymerization of cellulose and partial hydrolysis of hemicellulose. Delignification of hemicellulose improves the accessibility of cellulose and makes it more susceptible to the degradation process. Break down of lignin structure in the cell wall using microbes as a catalyst biological pretreatment mainly focused on fungal and bacterial strain. The benefit of biological pretreatment over non-biological pretreatment is that there is low energy requirement, and no chemical requirement (Amin et al., 2017). Various bacteria have been used in biological method among them Actinomycetes bacteria were observed to be effective on grasses in contrast to fungi and are widely used in degradation of plant-cell. However, the performance of biological pretreatment is dependent upon microbial growth and utilization of accessible sugars by the microorganism (Wagner et al., 2018).

4.3 *By Using Additives*

Various endeavours have been made to enhance biogas production by stimulating microbes with the aid of different types of chemical and biological additives under different operating condition in an enclosed space. Biological additives include different types of plant biomass, numerous types of weeds, forest and crop residues accompanied with microbial strains, inorganic additives although it has not used or less popular in terms of habitat (Satyanarayan and Murkute, 2008), but it could enhance biogas production significantly to create a favourable condition for the microbes, however the suitability of additives certainly depend upon type of feedstocks used (Pan et al., 2016).

4.3.1 Addition of Biochar

Biochar is a carbon-based material prepared through pyrolysis of biomass at a temperature range of 180–950 °C and has a potential to improve biogas production. (Sunyoto et al., 2016; Zaidi et al., 2019) Recent studies found that biochar has immense potential to increase the biogas production by act as an adsorbent for inhibitors during digestion such as heavy metals and pesticides. Biochar produced at high temperature can maintain the optimum pH in the reactor, reduce acid inhibition and promote the methanogenic activity (Zaidi et al., 2019). The highly porous biochar improves the microbial activity by biofilm formation. Luo et al. (2015) propounded that fruitwood biochar (at 800 °C) of 10 g/L were added in pulp sewage and glucose for AD, can rise the methane production up to 70.6%. Further, Pan et al.

(2016) have found that wood chips biochar (at 550 °C) of 5% dose improve the methane production by 68.97%

5 Conclusion

As far as management of waste generation and fulfilment of surging demand of energy is concerned, anaerobic digestion is supposed to be a promising technology for heat, electricity and fuel considering low cost technology as compared to other waste to energy technologies like gasification, pyrolysis and incineration. Generation of energy from waste is an attractive and affordable alternative in developing countries in order to achieve sustainable development. Waste to energy gives assistance to multiple sectors for the efficient management of waste such as agriculture waste, municipal solid waste, sewage waste, industrial waste. On the basis of previous research, it is found that the effectiveness of different types of pretreatment methods such as physical, chemical and biological is dependent upon the attribute of feedstock, although digestion of livestock manure gives the competent result as compared to lignocellulosic biomass due to its rigid structure. Therefore, lignocellulosic biomass needs intensive pretreatment for producing sugars during enzymatic hydrolysis. The main aim of conducting pretreatment of biomass is to reduce the recalcitrant nature, increase the surface area, remove the lignin content hence enhance the digestibility of biomass resulting in high efficiency of biogas production. Work has been done on biogas production and to increase its production by different pretreatment methods. This study recommended that more studies are required for effective treatment of substrates to enhance biogas production. The purification of biogas is also important to supply pure biogas to the common and poor people of India.

References

Abbasi, T., Tauseef, S.M. and Abbasi, S.A. (2012). A brief history of anaerobic digestion and "biogas". *In: Biogas Energy.* Springer Briefs in Environmental Science, Vol. 2, Springer, New York, NY, pp. 11–23.

Amin, F.R., Khalid, H., Zhang, H., Rahman, S.U., Zhang, R., Liu, G. and Chen, C. (2017). Pretreatment methods of lignocellulosic biomass for anaerobic digestion. *AMB Express*, **7:** 72. doi: https://doi.org/10.1186/s13568-017-0375-4.

Bhatt, A.K., Bhatia, R.K., Thakur, S., Rana, N., Sharma, V. and Rathour, R.K. (2018). Fuel from waste: A review on scientific solution for waste management and environment conservation. *In: Prospects of Alternative Transportation Fuels*. Springer, Singapore, pp. 205–233.

Bond, T. and Templeton, M.R. (2011). History and future of domestic biogas plants in the developing world. *Energy Sustainable Dev*, **15:** 347–354.

Boyle, G. (1996). Renewable Energy: Power for a Sustainable Future. Oxford University Press.

Breitenmoser, L., Gross, T., Huesch, R., Rau, J., Dhar, H., Kumar, S., Hugi, C. and Wintgens, T. (2019). Anaerobic digestion of biowastes in India: Opportunities, challenges and research needs. *J Environ Manage*, **236:** 396–412.

Government of India. Ministry of new and renewable energy. https://mnre.gov.in/physical-progress-achievements.

Karthikeyan, O.P., Trably, E., Mehariya, S., Bernet, N., Wong, J.W. and Carrere. H. (2018). Pretreatment of food waste for methane and hydrogen recovery: A review. *Bioresource Technol*, **249:** 1025–1039. doi: https://doi.org/10.1016/j.biortech.2017.09.105.

Khalid, A., Arshad, M., Anjum, M., Mahmood, T. and Dawson, L. (2011). The anaerobic digestion of solid organic waste. *Waste Manage*. **31:** 1737–1744.

Luo, C., Lu, F., Shao, L. and He, P. (2015). Application of eco-compatible biochar in anaerobic digestion to relieve acid stress and promote the selective colonization of functional microbes. *Water Res*, **68:** 710–718.

Maurya, R., Tirkey, S.R., Rajapitamahuni, S., Ghosh, A. and Mishra, S. (2019). *Recent Advances and Future Prospective of Biogas Production. Advances in Feedstock Conversion Technologies for Alternative Fuels and Bioproducts – New Technologies, Challenges and Opportunities*. Woodhead Publishing Series in Energy, pp. 159–178. doi: https://doi.org/10.1016/B978-0-12-817937-6.00009-6.

Ministry Of New and Renewable Energy. https://mnre.gov.in/file-manager/annual-report/2016-2017/EN/pdf/1.pdf.

Pan, J., Ma, J., Qiu, L., Gou, X. and Gao, T. (2016). The performance of biochar-mediated anaerobic digestion of chicken manure. *China Environ Sci*, **36:** 2716–2721.

Parawira, W., Read, J.S., Mattiasson, B. and Bjornsson, L. (2008). Energy production from agricultural residues: High methane yields in pilot-scale two-stage anaerobic digestion. *Biomass Bioenergy*, **32:** 44–50.

Popp, J., Laknerb, Z., Harangi-Rákosa, M. and Fári, M. (2014). The effect of bioenergy expansion: Food, energy, and environment. *Renewable Sustain Energy Rev*, **32:** 559–578.

Quaschning, V.V. (1969). Renewable Energy and Climate Change. 2nd ed. John Wiley & Sons Ltd.

Roder, M. and Welfle, A. (2019). Bioenergy. *Managing Global Warming*. doi: https://doi.org/10.1016/B978-0-1.

Saladini, F., Patrizi, N., Pulselli, F.M., Marchettini, N. and Bastianoni, S. (2016). Guidelines for emergy evaluation of first, second and third generation biofuels. *Renewable Sustain Energy Rev*, **66:** 221–227.

Satyanarayan, S. and Murkute, P. (2008). Biogas production enhancement by *Brassica compestries* amendment in cattle dung digesters. *Biomass and Bioenergy*, **32:** 210–215. doi: https://doi.org/10.1016/j.biombioe.2007.09.008.

Sims, R.E. (2003). Bioenergy to mitigate for climate change and meet the needs of society, the economy and the environment. *Mitigation Adapt Strat Global Change*, **8:** 349–370.

Singh, R., Roy, S. and Singh, S. (2015). Energy recovery from waste – A review. *Impending Power Demand Innov Energy Paths*, **1:** 298–303.

Srirangan, K., Akawi, L., Young, M.M. and Chou, C.P. (2012). Towards sustainable production of clean energy carriers from biomass resources. *Appl Energy*, **100:** 172–186.

Stougie, L., Giustozzi, N., van der Kooi, H. and Stoppato, A. (2018). Environmental, economic and exergetic sustainability assessment of power generation from fossil and renewable energy sources. *Int J Energy Res*, **42:** 2916–2926.

Sunyoto, N.M.S., Zhu, M., Zhang, Z. and Zhang, D. (2016). Effect of biochar addition on hydrogen and methane production in two-phase anaerobic digestion of aqueous carbohydrates food waste. *Bioresource Technol*, **219:** 29–36.

Wagner, A.O., Lackner, N., Mutschlechner, M., Prem, E.M., Markt, R. and Illmer, P. (2018). Biological pretreatment strategies for second-generation lignocellulosic resources to enhance biogas production. *Energies*, **7:** 1797. doi: https://doi.org/10.3390/en11071797.

Zabeda, H.M., Aktera, S., Yuna, J., Zhanga, G., Awada, F.N., Qi, X. and Sahu, J.N. (2019). Recent advances in biological pretreatment of microalgae and lignocellulosic biomass for biofuel production. *Renew Sustain Energy Rev*, **105:** 105–128. doi: https://doi.org/10.1016/j.rser.2019.01.048.

Zaidi, A.A., Feng, R., Malik, A., Khan, S.Z., Shi, Y., Bhutta, Asad, J. and Shah, A.H. (2019). Combining microwave pretreatment with iron oxide nanoparticles enhanced biogas and hydrogen yield from green algae. *Processes*, **24:** 1–13. doi: https://doi.org/10.3390/pr7010024.

Zheng, Yi, Zhao, J., Xu, F. and Li, Y. (2014). Pretreatment of lignocellulosic biomass for enhanced biogas production. *Progress Energy Combust Sci*, **42:** 35–53. doi: https://doi.org/10.1016/j.pecs.2014.01.001.

Chapter 15
Study on the Effect of Load Resting Time on Recycled Brick Aggregate Cement Concrete

Pranav Sharma, Jince P. Zachariah, and Partha Pratim Sarkar

1 Introduction

The scarcity of natural stone aggregates (NSA) is becoming a common problem worldwide (Abu Noaman et al., 2018; Vancura et al., 2009). Many research studies are thus being carried out on material conservation and the use of sustainable materials in pavement construction, which is a major consumer of NSAs (Cuttell et al., n.d.; Poon and Chan, 2006). Crushed brick aggregate (CBA), reclaimed pavement waste, recycled construction waste, treated soil are some major non-conventional materials used for pavement construction purposes (Abbas et al., 2017; Arulrajah et al., 2014; Herrador et al., 2011; Mazumder et al., 2006). In Tripura, a north-eastern state of India, brick aggregate concrete is used conventionally for ordinary concrete because of the unavailability of proper resources. For the unavailability of natural aggregates, normal concrete production usually cost high in these places for the increased transportation cost of natural stone aggregates.

As discussed, brick aggregate is effectively used in concrete production in parts of the world. Generally, two different methodologies are adopted to collect the brick aggregates. In the first method, normal construction bricks are collected from the brick kilns and crushed down for the required aggregate size. Whereas, in the second, bricks were collected from construction and demolition waste and is crushed down. Waste from the demolished constructions exist in every location of the world and has been established as a growing problem. The sustainable use of this crushed brick aggregate can solve large environmental problems. These demolished construction waste can be recycled and reused as raw materials for a new application like rigid pavements. By this operation, the consumption of raw materials is reduced, and the waste is eliminated. Recycled crushed brick as fine aggregate can also be used for producing pavement concrete mix. The use of these waste materials will

P. Sharma (✉) · J. P. Zachariah · P. P. Sarkar
National Institute of Technology, Agartala, India

A. S. Kalamdhad (ed.), *Integrated Approaches Towards Solid Waste Management*, https://doi.org/10.1007/978-3-030-70463-6_15

also make the pavement more economical. The brick aggregate has proved as an efficient alternative for natural stone aggregate mixes by many researchers (Arisha et al., 2016; Arisha et al., 2018; Dey and Pal, 2013; Park, 2016). However, the utilisation of crushed brick aggregate in rigid pavement construction has not been studied deeply.

The nature of loading is having a huge role to play while analysing pavement surfaces. Fast-moving vehicles will be imparting loads quickly compared to slow-moving traffic. Hence, the loading and resting time of each load application is having a considerable impact in determining the service life of the pavement (Arisha et al., 2018; Herrador et al., 2011). This study is aimed to analyse the effect of loading and resting time in load-bearing characteristics of pavement surfaces. At shorter intervals, in every load case, a superimposed effect will be developed because of the previous load application, which affects the durability of the pavement. Loads were applied to centre, edge and corner regions of the pavement surfaces to study the effects. Different loading resting time intervals were selected for this purpose. Three-dimensional finite element analyses were carried out using ABAQUS to simulate the elliptical tire print and pavement surfaces. The details of the methodology adopted are explained in further sections.

In this study, pavement responses were studied in the centre, corner and edge regions using ABAQUS software for different loading and resting cycles. Thus, this study holds high value as it provides an in-depth study of the effect of loading and resting time in determining pavement distresses.

2 Materials and Methods

Finite element analysis using ABAQUS has been found to be very efficient in modelling rigid pavements to incorporate different realistic conditions in the analysis for obtaining stresses and deformations of pavements (Mackiewicz, 2015; Patil et al., 2013). The accuracy of finite element solutions depends on the mechanical behaviour of materials which must be selected in such a way that it better represents the characteristics of the material such as plasticity, brittleness, ductility, elasticity, etc. In this study, a 3D finite element modelling has been adopted using ABAQUS, considering the circular footprint of the loaded wheel. Table 1 shows the material properties used in the finite element analysis.

The modelling of the slab was done by using the property of material shown in Table 1. The size of the model was adopted from IRC specification

Table 1. Material characteristics

Properties	*Value*	*Unit*
Density	2.3544E-005	N/mm^3
Young's modulus	30000	N/mm^2
Poisson's ratio	0.15	–

(4.3 m × 3.5 m × 0.33 m). The subgrade always works as the medium to transfer the load from pavement to the earth. Winkler's foundation was considered for the effective representation of the surface layer- base/subgrade interaction.

Closely arranged linear springs with spring stiffness characterising the modulus of subgrade reaction were developed in ABAQUS to model the pavement system. Elliptical tyre imprints were shifted to circular areas for applying load. In the study, 0.25 m diameter circles were used for this purpose. The details of the pavement sections along with the ABAQUS models are given in Figs. 1, 2, and 3.

Different loading patterns were studied to analyse the effect of loading and resting time. Four different load patterns were selected, and the pavement responses were studied from all three critical regions of the pavement: corner, centre and edge. In all the load cycles, the loading time was kept as a constant (0.1 second), and the resting time was varied. Loading resting cycle times were selected as 0.2 seconds, 0.5 seconds, 1 second and 2 seconds. Five load applications were introduced in each case for the analysis. Peak stress and deflection at the critical regions were obtained from ABAQUS and the performance was evaluated. The details of the loading cycles are provided in Table 2. In shorter loading cycles, the effect of loading will be compounded due to the effect of the previous load application, which increases the impact of loading.

3 Results and Discussion

For efficient loading on the pavement, it is necessary that deformation and stress with respect to the time should be low, especially over the critical points of the load such as centre, edge and corner. Using ABAQUS modelling, the stress and deformation for this critical point for different load cycle and loading span were obtained. For accurate and comparative study, the stresses and deformations at the load centre, as well as the centre of the slab in each case, were recorded. A sample result output

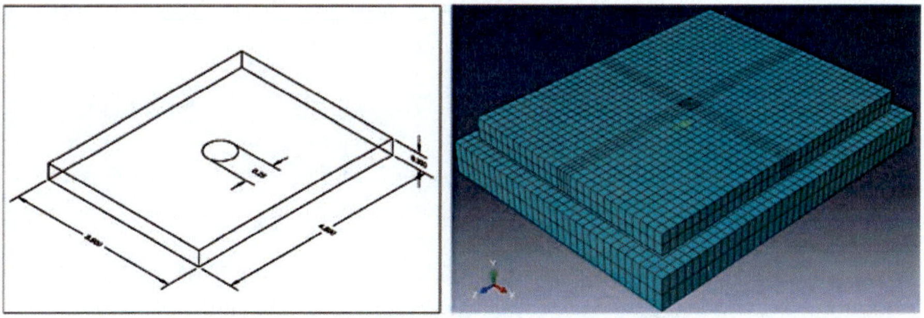

(a) (b)

Fig. 1. Central loading: (a) schematic diagram, (b) ABAQUS model.

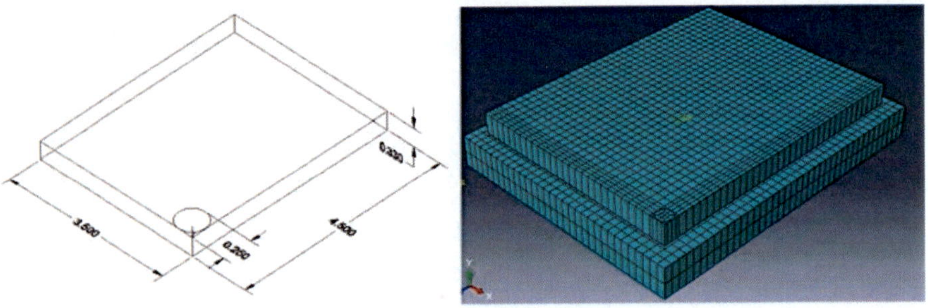

(a) **(b)**

Fig. 2. Corner loading: (a) schematic diagram, (b) ABAQUS model.

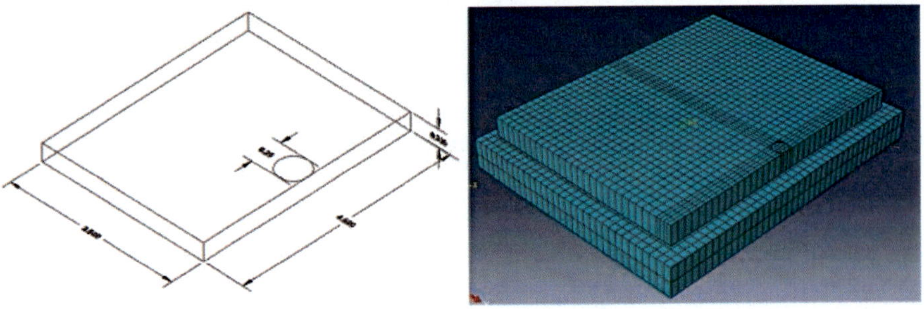

(a) **(b)**

Fig. 3. Edge loading: (a) schematic diagram, (b) ABAQUS model.

obtained from ABAQUS software is given in Fig. 4, where Fig. 4(a) shows the deformation and Fig. 4(b) shows the stresses generated.

Fig. 5 explains the behaviour of different load patterns causing deflections at the load centre. From Fig. 5, it was clearly observed that the deflection at the corner is maximum for 0.2 second loading cycle time and decreasing for later cycles of 0.5, 1.0 and 2.0 seconds. Edge loading also showed results similar to edge loading, but the deformation was observed less in the edge load case. The deformation for the centre loading has very less value as compared to the edge and corner loading since it is constrained in all four sides.

For the comparative study of deflection at one point due to all type of loading, the deflection at slab centres due to the critical point loading was plotted, as shown in Fig. 6. From Fig. 6, the plot for corner and edge loading is very less as compared to the central loading. It was further noticed that the deflection for the 0.2-second loading cycle is more than that of other loading cycles.

Another major area of focus in this study was the analysis of stresses generated in the recycled brick aggregate concrete pavement. The behaviours of stresses generated were also studied using ABAQUS models, and the results are plotted in Figs. 7 and 8. Figure 7 explains the behaviour of stresses at loading centres in the

Table 2. Loading details

S. no.	*Loading time (seconds)*	*Resting time (seconds)*	*Time for one loading cycle (seconds)*	*Time for five loading cycle (seconds)*
1	0.1	0.1	0.2	1.0
2	0.1	0.4	0.5	2.5
3	0.1	0.9	1.0	5.0
4	0.1	1.9	2.0	10.0

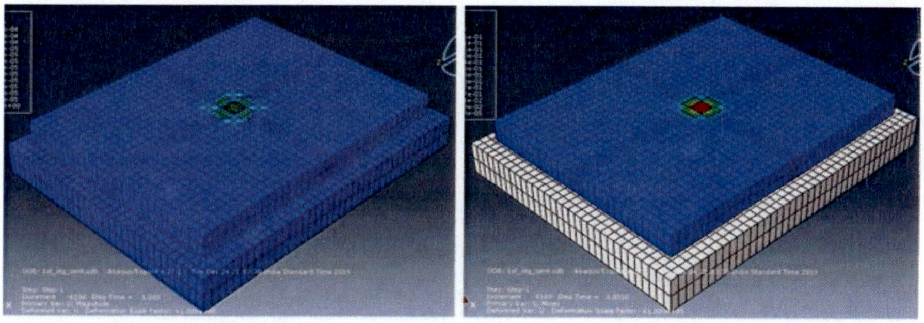

(a) Deformation **(b) Stress**

Fig. 4. Result output obtained from ABAQUS software.

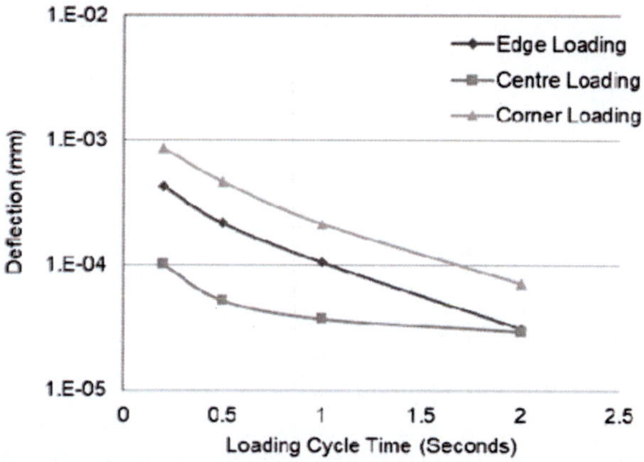

Fig. 5. Deflection at the load centres.

critical points. From the figure, it was seen that the maximum stress is generated at the corner loading, followed by edge and centre, respectively. However, only a very small change was observed in the peak stresses of corner, centre and edge loading when loading centres were analysed. It was clearly seen that the stress is maximum for 0.2-second loading cycle and later on reduced with increase in the cycle time in all the three loading cases.

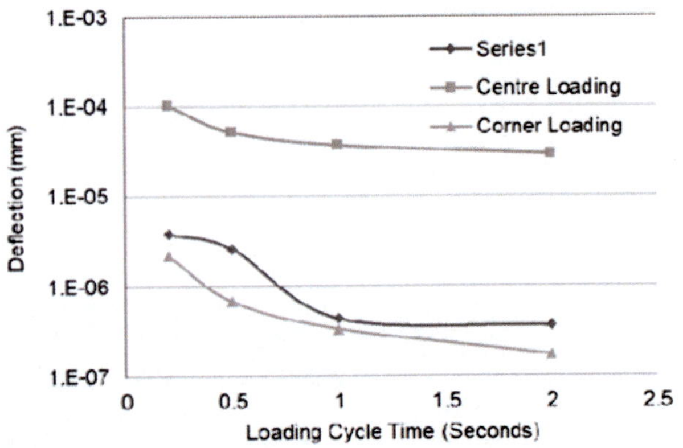

Fig. 6. Deflection at the centre of the slab.

Stresses due to the critical point loadings were plotted at the slab centres against time, as shown in Fig. 8. As expected, centre stress was found larger, followed by the stresses caused due to edge loading and corner loading, respectively. Similar to the previous cases, a decreasing trend was observed as the loading cycle time increased recording the peak values at 0.2-second load cycle time.

4 Conclusions

As per the result and graphs, it can be concluded that the load cycle time is having a considerable impact in determining the pavement performance. Peak stress and peak deformation was obtained in the corner region for all cycles. All the recorded deformations and stresses were minimal for the recycled brick aggregate pavement surfaces proving the efficiency of recycled brick aggregate as a concrete pavement aggregate. It was also observed that, as the resting time increases the magnitude of stress and deformation will be reduced as the peak stresses were recorded at 0.2-second loading cycle time. Therefore, fast-moving traffic, generally giving a smaller time gap between successive loads generate more distress to the pavement than the slow-moving traffic for the same pavement surface.

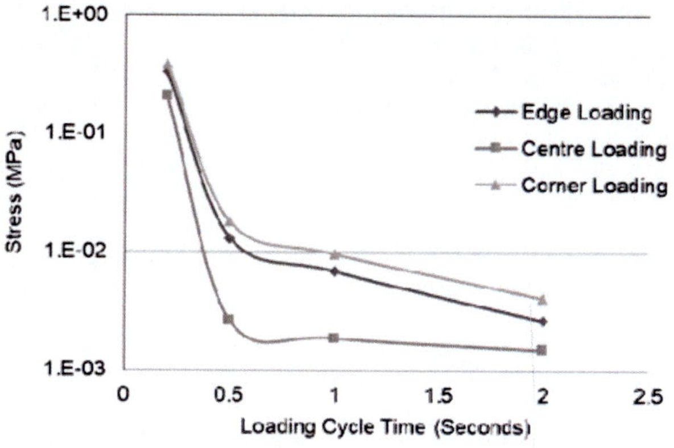

Fig. 7. Stress at the load centre.

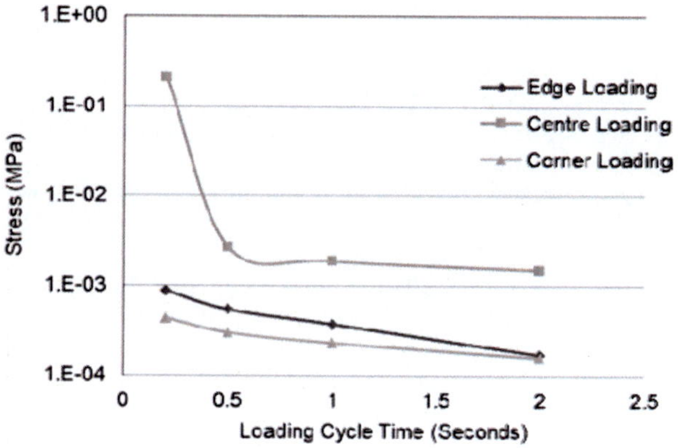

Fig. 8. Stress at the slab centre.

References

Abbas, S., Saleem, M.A., Kazmi, S.M.S. and Munir, M.J. (2017). Production of sustainable clay bricks using waste fly ash: Mechanical and durability properties. *J Build Eng*, **14:** 7–14. doi: https://doi.org/10.1016/j.jobe.2017.09.008.

Abu Noaman, M., Islam, M.N., Islam, M.R. and Karim, M.R. (2018). Mechanical properties of brick aggregate concrete containing rice husk ash as a partial replacement of cement. *J Mater Civil Eng*, **30(6):** 1–10. doi: https://doi.org/10.1061/(ASCE)MT.1943-5533.0002272.

Arisha, A., Gabr, A., El-Badawy, S. and Shwally, S. (2016). Using blends of construction & demolition waste materials and recycled clay masonry brick in pavement. *Procedia Eng*, **143(Ictg):** 1317–1324. doi: https://doi.org/10.1016/j.proeng.2016.06.148.

Arisha, A.M., Gabr, A.R., El-Badawy, S.M. and Shwally, S.A. (2018). Performance evaluation of construction and demolition waste materials for pavement construction in Egypt. *J Mater Civil Eng*, **30(2):** 1–14. doi: https://doi.org/10.1061/(ASCE)MT.1943-5533.0002127.

Arulrajah, A., Maghoolpilehrood, F., Disfani, M.M. and Horpibulsuk, S. (2014). Spent coffee grounds as a non-structural embankment fill material: Engineering and environmental considerations. *J Cleaner Prod*, **72:** 181–186. doi: https://doi.org/10.1016/j.jclepro.2014.03.010.

Cuttell, G.D., Snyder, M.B., Vandenbossche, J.M. and Wade, M.J. (n.d.). Performance of rigid pavements containing recycled concrete aggregates. *Transp Res Rec*, **971071**.

Dey, G. and Pal, J. (2013). Use of brick aggregate in standard concrete and its performance in elevated temperature. *Int J Eng Technol*, **5(4):** 523–526. doi: https://doi.org/10.7763/ijet.2013.v5.610.

Herrador, R., Pérez, P., Garach, L. and Ordóñez, J. (2011). Use of recycled construction and demolition waste aggregate for road course surfacing. *J Transp Eng*, **138(2):** 182–190. doi: https://doi.org/10.1061/(ASCE)TE.1943-5436.0000320.

Mackiewicz, P. (2015). Finite-element analysis of stress concentration around dowel bars in jointed plain concrete pavement. *J Transp Eng*, **141(6):** 1–8. doi: https://doi.org/10.1061/(ASCE)TE.1943-5436.0000768.

Mazumder, A.R., Kabir, A. and Yazdani, N. (2006). Performance of overburnt distorted bricks as aggregates in pavement works. *J Mater Civil Eng*, **18(6):** 777–785. doi: https://doi.org/10.1061/(ASCE)0899-1561(2006)18.

Park, C.E. (2016). *OOD Brickassembly: Vol. February*.

Patil, V.A., Sawant, V.A. and Deb, K. (2013). 3D finite-element dynamic analysis of rigid pavement using infinite elements. *Int J Geomech*, **13(5):** 533–544. doi: https://doi.org/10.1061/(ASCE)GM.1943-5622.0000255.

Poon, C.S. and Chan, D. (2006). Feasible use of recycled concrete aggregates and crushed clay brick as unbound road sub-base. *Constr Build Mater*, **20(8):** 578–585. doi: https://doi.org/10.1016/j.conbuildmat.2005.01.045.

Vancura, M., Khazanovich, L. and Tompkins, D. (2009). Reappraisal of recycled concrete aggregate as coarse aggregate in concretes for rigid pavements. *Transp Res Rec*, **2113:** 149–155. doi: https://doi.org/10.3141/2113-18.

Chapter 16
Energy Densification of Juice Waste Using Hydrothermal Carbonisation

Divya Gupta, Sanjay M. Mahajani, and Anurag Garg

1 Introduction

Fossil fuels have played a major role in providing energy for the industrial and economic growth ever since industrialisation began. However, continuous extraction of fossil fuels over decades has led to debates over sustainability of these fuel reserves. Most of the countries are now shifting steadily towards renewable sources of energy. Coal catered to 96% of the total global energy requirements in 1900s, which declined to less than 30% in 2000 (Ritchie and Roser, 2019). Moreover, the release of greenhouse gases (GHG) from fossil fuels can be reduced by using biogenic sources of energy since the carbon dioxide generated during combustion of biomass is equivalent to the carbon dioxide consumed during plant growth (Cao et al., 2017). According to a report, global fossil fuel reserves is equivalent to ~750 Gt of carbon out of which only ~35% can be burned to keep the increase in global temperature <2°C (Ritchie and Roser, 2019). Therefore, there is an urgent need to develop alternate and renewable energy sources to reduce dependency on fossil fuels and GHG emissions.

Municipal wet biodegradable waste is one of the potential sources for production of renewable energy since it is generated in large amounts worldwide throughout the year. India produces ~ 62 million tonnes of municipal solid waste (MSW) annually, of which >50% is composed of wet biodegradable waste (Central Public Health and Environmental Engineering Organization (CPHEEO) 2016). In the last 10 years, waste generation has increased by 2.44 times (Joshi and Ahmed, 2016) and the same

D. Gupta · A. Garg (✉)
Environmental Science and Engineering Department, Indian Institute of Technology Bombay, Mumbai, Maharashtra, India
e-mail: a.garg@iitb.ac.in

S. M. Mahajani
Department of Chemical Engineering, Indian Institute of Technology Bombay, Mumbai, Maharashtra, India

A. S. Kalamdhad (ed.), *Integrated Approaches Towards Solid Waste Management*, https://doi.org/10.1007/978-3-030-70463-6_16

trends are also expected in future. The wet waste is largely produced from hotels, restaurants, households, market and food and juice processing industries. Upon land disposal in unscientific manner, the putrescible wet waste leads to the generation of methane, leachate and odour. The present study particularly deals with the conversion of fruit pulp and peelings (generated from juice processing industries and shops) into fuel-like material.

Wet waste is usually not suitable for direct energy recovery due to its high moisture content (>70%) and low calorific value. However, hydrothermal carbonisation (HTC) process is recognised as an efficient treatment of such waste fraction for enhancing its calorific value by concentrating carbon in the more homogeneous solid phase with reduced moisture content. Due to these properties, the resulting solid material (also called as hydrochar (HC)) can be utilised as fuel. Alternatively, this material can also be used as co-fuel, adsorbent, soil amendment materials, raw material for carbon fuel cells etc. (Parshetti et al., 2014; Kasturi et al., 2016; Wang et al., 2018; Fei et al., 2019). Apart from this, a process wastewater (PW) stream is also produced, which contains high load of different organic compounds and chemical oxygen demand (COD). Traditionally, HTC is carried out under moderate temperature conditions (usually 180–260°C) and self-generated pressure in a high-pressure reactor. Prior to the pretreatment, no drying is required and water helps in the reaction as catalyst as well as solvent during the reaction (Kambo and Dutta, 2015). Nevertheless, the major reaction parameters such as temperature and duration should be optimised for different types of waste materials.

Global citrus fruit production (e.g., sweet lime, orange, malta (*Citrus sinensis*) etc.) is reported to be >88 million tonnes annually (Puccini et al., 2016). Sweet lime is the third largest fruit produced in India with a total of >3 million tonnes in the top ten states (Agricultural and Processed Food Products Export Development Authority APEDA). Food and juice processing industries produce large quantities of waste after juice or wine production including pulp, seeds and peels. Sabio et al. (2016) reported the effect of solid to moisture ratio, temperature and time on HC characteristics prepared from tomato peel waste. They found temperature to be the highest impacting factor on product characteristics. The high heating value (HHV) of HC varied in the range of 23.6–34.8 MJ/kg at different temperature and time durations of 150–230°C and 1.6–15 h. Benavente et al. (2015) studied HTC of orange waste along with olive mill and artichoke wastes at temperatures of 225°C and 250°C for 2–24 h durations. Hydrophobicity of HC was reported to enhance with increasing reaction time or temperature due to hydrolysis of water retaining cell walls and hydrogen bonds in hydroxyl moieties. Major gaseous products obtained after HTC were observed to be CO_2, CO, CH_4, C_2H_4 and H_2. Zhang et al. (2018) treated different fruit wastes such as apple pomace, rotten apple and grape pomace using HTC at temperatures of 190–260°C for 15 min. Excess water was added to maintain solids to water ratio of 1:12 along with addition of excess pressure. The HHV of HC obtained from apple pomace was highest among different wastes (~29 MJ/kg at 260°C). It was also reported that the solid yield was improved with increasing temperature due to high sugar content. The reaction temperature of 260°C showed an increase in carbon content by 20% for all the samples within 15 min of reaction.

Biomethane potential test for PW produced from HTC of orange pomace have been carried out by Erdogan et al. (2015). They also reported presence of different sugars (sucrose, glucose, fructose), organic acids, HMF and furfural. The concentration of furfural was found to be significantly higher than HMF and both of these decreased with increased temperature and reaction duration.

To summarise, all the studies have reported an increase in carbon content and HHV of raw waste after HTC. Nevertheless, there is lack of information on various types of citrus fruits which are generated in large quantities. Hence, the present study investigated the fuel characteristics of HC produced from sweet lime waste. Furthermore, the pellets from the HC were prepared and tested for strength. Moreover, chemical characteristics of PW were also determined with an intention to understand resource recovery potential. This is another major by-product from HTC pretreatment, which has not been discussed in detail in most of the previous studies.

2 Materials and Methods

2.1 Experimental Methods

Juice waste (JW) (comprised of both pulp and peel) produced during juice extraction from sweet lime was obtained from a juice shop in IIT Bombay campus. Approximately 700 g of JW (as received) was weighed and added in a 2 L high-pressure stainless-steel reactor (SS 316) supplied by Amar Equipments Pvt. Ltd., Mumbai. Due to high moisture content in JW (70–80%), no additional water was mixed during HTC reaction. The details of reactor have been reported in our previous publication (Gupta et al., 2019). To summarise, the reactor was equipped with PID (proportional integral derivative) temperature controller, cooling coil to quench the reaction immediately and pressure gauge. The reactions were carried out at a fixed temperature of 200°C for durations ranging from 1 to 5 h. The PW produced during HTC was separated from solid residue (i.e., HC) using vacuum filtration and both the streams were subjected to analysis.

2.2 Methods for Hydrochar Analysis

The substrate and HC samples were analysed for proximate analysis using USEPA Method 1684 and ASTM Method D 3174-02. Moisture was determined for waste samples by heating them at 80°C temperature for 24 h. Volatile matter was determined by heating the oven-dried samples in covered crucibles at 550°C temperature for 4 h in a muffle furnace. The mass loss was recorded as volatile matter. Different sets of oven-dried samples were kept at 950°C for 7 min holding time in open crucibles to determine ash content. The remaining mass in the crucible is reported as ash content. The fixed carbon was calculated by subtracting the sum of ash and volatile matter from total oven-dried sample mass. Elemental analysis was carried out on a Thermo Finnigan, CHNSO analyser, Italy. HHV of a HC sample was obtained using automatic bomb calorimeter (A. Kumar & Company, Mumbai).

Solid yield was calculated as percentage of the dry solids recovered compared to initial dry solids, as given in equation (1):

$$\text{Solid yield} = \frac{\text{Mass of dry hydrochar}}{\text{Mass of dry solids in JW}} \tag{1}$$

Energy related parameters such as carbon densification and energy densification were calculated using the equations (2) and (3) (Li et al., 2013):

$$\text{Carbon densification} = \frac{\%\text{C in hydrochar}}{\%\text{C in feedstock}} \tag{2}$$

$$\text{Energy densification} = \frac{\text{Measured energy of hydrochar}}{\text{Measured energy of feedstock}\left(\text{with initial moisture}\right)} \tag{3}$$

Total chlorine was determined in JW and hydrochars using a combination of bomb calorimeter method (ASTM D4208-18) and argentometric titration as described in American Public Health Association (APHA) handbook (APHA 2012).

The HC pellets were prepared in a screw type pelletiser. Drop test was done by dropping each pellet from a height of 1.5 m twice (Yilmaz et al., 2018). The pellets were dipped in water for 24 h to measure their water absorbability. The increase in mass after 24 h suggests the maximum amount of water that can be absorbed by pellets. Water resistance was measured by the number of pellets, which pass the drop test again after being submerged in water.

2.3 *Process Wastewater Analysis*

The yield of PW was calculated as a percentage of initial moisture present as given in equation (4).

$$\text{PW yield} = \frac{\text{Volume of PW obtained after filtration}}{\text{Volume of water}\left(\text{moisture}\right)\text{in JW}} \tag{4}$$

COD of PW samples was measured by closed reflux method (APHA, 2012) using a HACH COD digester (DRB200, USA). pH of the PW samples was measured with a digital pH meter (Polmon, LP-139S, India). The pH meter was calibrated using standard pH buffer solutions (pH = 4, 7 and 10) before measuring the pH of unknown samples.

5-Day biochemical oxygen demand (BOD_5) was determined according to the method specified in APHA handbook (APHA, 2012). Dissolved oxygen was measured using HACH HQ30d dissolved oxygen (DO) meter. The BOD_5 was computed using the difference between initial and final DO.

Dubois method was used to determine the total carbohydrate concentration in the liquid samples (Dubois et al., 1956) whereas ammonium nitrogen (NH_4^+-N) was measured in the wastewater using salicylate method (Willis et al., 1996). Proteins and humic-like substances (HLS) were analysed simultaneously using a response surface plot for different concentrations of both proteins and HLS as suggested by Vakondios et al. (2014). Volatile fatty acids (VFAs), i.e., acetic acid, propionic acid and butyric acid were measured in PW through extraction in methyl tertiary butyl ether (MTBE) by gas chromatography (GC) using Perkin Elmer GC column (COL ELITE FFAP 30 m × 0.25 μm × 0.25 μm) according to the method given by Banel and Zygmunt (2011). 5-Hydroxymethyl furfural (HMF) and furfural were measured by high performance liquid chromatography (HPLC) using C18 column (250 mm × 4.6 mm, 5 μm particle size) and measuring the absorbance at 285 nm (Mijares et al., 1986).

3 Results and Discussion

3.1 *Characteristics of JW*

The initial moisture present in JW was ~70% in all the samples, hence no water needs to be added for the hydrothermal reaction. The waste contained high volatile matter (~96%) and low ash content (3.8%). The carbon content in the waste was <40%. Apart from this, lower heating value (LHV) and HHV were 2.9 MJ/kg and 15.9 MJ/kg, respectively.

3.2 *Hydrothermal Carbonization Process*

After the HTC reaction, the waste could be separated into solid HC and PW. The solid and liquid fractions were separated using vacuum filtration with Grade 5 Whatman filter paper. The HC samples recovered after HTC at 200°C for different durations had a moisture content of ~60%, which were oven-dried at 80°C temperature before storing at room temperature. The solid recovery was almost similar (45–50%) at the three reaction durations as shown in Table 1. PW (filtrate) was stored at ~4°C and diluted appropriately before subjecting to detailed analyses. The PW yield was increased from 60% to 77% with increase in reaction duration (Table 2). The enhancement in PW yield with time may be attributed to the increased

Table 1 Chemical characteristics of JW and hydrochars derived from JW

		JW derived HC samples		
Parameter	*JW*	*JW_200_1*	*JW_200_3*	*JW_200_5*
Solid yield (%)	–	50	45.3	49.5
Volatile solids (%)	96.4 ± 0.9	95.8 ± 0.4	93.8 ± 0.2	93.1 ± 0.7
Fixed carbon (%)	0.4 ± 0.68	0.7 ± 0.2	1.6 ± 0.2	2.5 ± 0.7
Ash (%)	3.8 ± 0.05	3.82 ± 0.06	4.6 ± 0.09	4.4 ± 0.09
C (%)	38.2	45.8	49.8	52.7
H (%)	5.9	7.4	6.0	4.2
O (%)	50.8	40.4	37.8	37.5
N (%)	1.3	2.6	1.8	1.2
Chlorine (mg/g)	0.65	0.4	0.5	0.5
LHV (MJ/kg)	2.9	–	–	–
HHV (MJ/kg)	15.9	21.2	21.5	23.6
Carbon densification	–	1.20	1.30	1.38
Energy densification#	–	1.3	1.4	1.5
Energy densification*	–	7.3	7.4	8.1

#Calculated based on HHV of JW. * Calculated based on LHV of JW.

Table 2 Characteristics of process wastewater

	Values		
Sample	*JW_PW_200_1*	*JW_PW_200_3*	*JW_PW_200_5*
PW yield (%)	60	71.4	77
COD (g/L)	~81	~97	~92
BOD_5 (g/L)	51	56	49
Carbohydrates (g/L)	5.1	2.2	1.1
NH_4^+-N (mg/L)	204.2	181.9	338.2
Proteins (g/L)	5.8	6.5	6.7
HLS (g/L)	~31	~28	~22

dehydration reactions. The occurrence of such reactions can be supported by reduction in hydrogen and oxygen contents in HC samples at higher reaction durations.

3.3 *Characteristics of Hydrochar Derived from JW*

HC produced from JW treatment at 200°C temperature showed an increase in carbon and decrease in oxygen. The change in JW characteristics after HTC are summarised in Table 1. HTC resulted in recovery of 45–50% of the initial solids as hydrochar whereas, carbon recovery was 69%. After five hours of HTC, the carbon content in the solids increased by 14.5% while the oxygen content reduced by 13%. Hydrogen to carbon ratio was found to decrease to 0.96 from 1.9 and oxygen to carbon ratio decreased from ~1 to 0.53. The reduction in oxygen and increase in carbon content led to increase in heating value of the raw material. According to Van

Krevelen diagram (Fig. 1), the hydrochar produced after 5 h of reaction resembled to lignite and brown coal. Hence, significant improvement in fuel quality of JW was obtained after HTC for sufficient duration.

However, there was no significant change in volatile matter and ash contents. After 5 h of HTC, carbon and energy densifications were increased substantially as compared to the initial raw material. By increasing the reaction duration, HHV of JW increased from 15.9 MJ/kg to 23.6 MJ/kg, resulting in an energy densification 1.5 times or 150% compared to the HHV of JW. Meanwhile, carbon densification in HC samples was 130–150% of JW.

Chlorine content is one of the important elements since its presence in fuel may lead to generation of chlorinated organic pollutants such as dioxins and furans (Yasuhara et al., 2001). However, HTC is supposed to decrease in the chlorine content (Prawisudha et al., 2012). Although other wet wastes such as cooked food waste may have high concentrations of chlorine contributed by table salt, JW had very low chlorine in all its samples (~0.65 mg/g of dry JW). After HTC, the chlorine content was slightly reduced to 0.45–0.5 mg/g in HC samples. In addition, it has been reported in literature that inorganic chlorine can be washed from the produced HCs using distilled water, which can further reduce chlorine concentrations (Prawisudha et al., 2012).

3.4 Durability of HC Samples Derived from JW

To increase the ease of storage, handling and transportation, hydrochars were densified into pellets. The pellets were tested using the drop test as discussed earlier. The pellets were 1 cm in length with 0.5 cm diameter. Firstly, the dried HC samples were

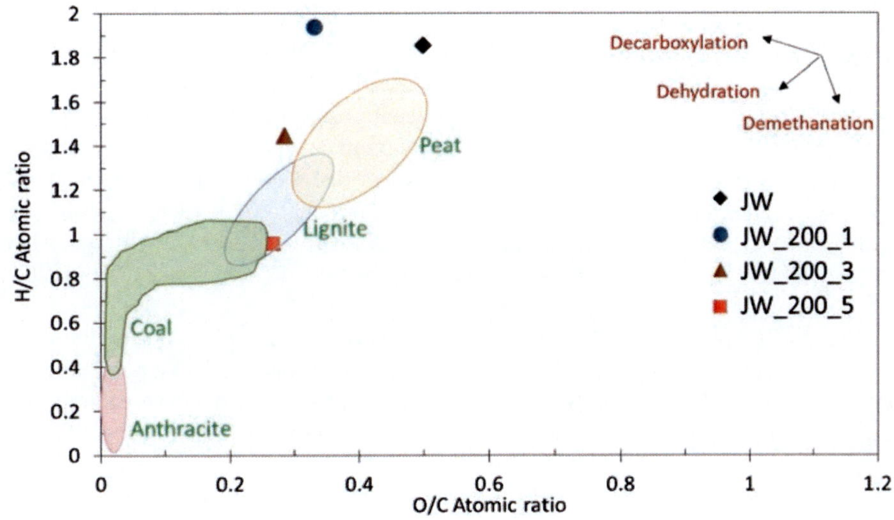

Fig. 1. Locations of JW and HCs produced after HTC process on Van Krevelen diagram

used for pellet preparation. However, these could not pass the drop tests and disintegrated on the first drop.

In the next trial, 15% moisture was added to the HC (JW_200_5) and kept overnight in covered crucibles since it is reported that ~10–15% moisture is required for proper consistency (Jiang et al., 2014). However, the pellets with 15% moisture also failed the drop test. Therefore, oil was added as a binder to the HC before pellet formation. The pellets thus prepared were able to pass the drop test.

Subsequently, the pellets were immersed in water for 24 h to observe their water absorbability and strength after being exposed to moisture. It was observed that the pellets obtained from JW_200_5 HC sample were stable and passed the drop test again. There was only slight release of pale-yellow coloured compounds from the HC pellets. Therefore, it can be inferred that structural integrity of the JW pellets formed with binder was higher.

3.5 Process Wastewater Characteristics

COD in the PW samples increased upto 3 h run time and then decreased after HTC of 5 h duration. During initial phase of HTC, solids are solubilised in the aqueous medium. Subsequently, re-deposition of the sugars on the HC takes place (Kambo et al., 2018). As a result, increase in COD is observed initially which is reduced if the reaction continues for longer duration. The change in BOD also followed similar trend. This suggests that the organic compounds released into the wastewater were biodegradable in nature. The decrease in BOD_5 may also be due to the degradation of readily biodegradable compounds such as carbohydrates, proteins and carboxylic acids.

Various PW samples were then analysed for identification and quantification of different macromolecules concentration, including proteins, HLS and carbohydrates (Table 2). HLS may be formed during HTC by Maillard's reactions occurring between proteins and carbohydrates (Jokic et al., 2001; Danso-Boateng et al., 2015). To analyse degradation of proteins, the concentration of NH_4^+-N was measured. Moreover, HMF and furfural concentrations were measured to predict the effect of HTC on carbohydrates degradation. 5-HMF is reported to be a major component of PW recovered after HTC of biomass materials (other than MSW) which is formed from cellulose hydrolysis. The compound can replace fossils to prepare furan derivatives which are listed among the top twelve value added chemicals from biomass (Kambo et al., 2018).

The concentration of carbohydrates was maximum in the PW sample obtained after 1 h HTC reaction. Extending the HTC reaction period caused degradation of HMF and reduced significantly in the PW. On the other hand, the proteins concentration continuously improved with reaction duration along with increasing NH_4^+-N concentration. This suggests that solubilisation of carbohydrates was almost complete within first hour of the reaction after which their degradation was started. However, the proteins solubilisation from JW solids was progressively increased with reaction time. HLS were detected in high concentrations of 22–30 g/L in all the PWs, hence the recovery of such compounds may be feasible.

Like carbohydrates, HMF and furfural concentrations in PWs were also maximum after one-hour HTC reaction (Fig. 2). With increase in the reaction time, these compounds reduced significantly. It shows that the solubilisation and subsequent degradation of carbohydrates and its decomposition products was faster than protein solubilisation and degradation. Moreover, the concentrations of organic acids increased with time suggesting the degradation of furfurals and carbohydrates into organic acids. The results in the present study are in accordance with those obtained during HTC of orange pomace (Erdogan et al., 2015). However, the COD levels were lower in the previous study due to addition of excess water before the reaction.

Further, VFAs formed after HTC mainly composed of acetic acid whose peak concentration (i.e., 3000 mg/L) was observed after 3 h of reaction, after which there was decrease in its concentration (Fig. 2). Similar trend was observed for propionic acid though its concentration was much lower. Butyric acid was detected only in the waste sample treated for 3 h duration HTC. The results indicate that the acids could have undergone decarboxylation reactions to form unsaturated compounds. This also corresponds with the BOD_5 values of JW samples which was also decreased in PW sample recovered after 5 h reaction.

Therefore, it can be suggested that during HTC, first solubilisation of different macromolecules takes place which is followed by hydrolysis. Subsequently, simultaneous dehydration, decarboxylation, aromatisation and polymerisation reactions occur to form variety of unsaturated compounds, phenolics and VFAs. These compounds may be polymerised to form complex structures like HLS and may redeposit into the char matrix.

To summarise, several compounds such as HLS, HMF, furfural and organic acids were detected in PW samples, which may be subjected to recovery depending on their recoverable concentrations and economic viability. Their presence in PW depends upon reaction conditions. HLS can be used as fertilisers. It has been

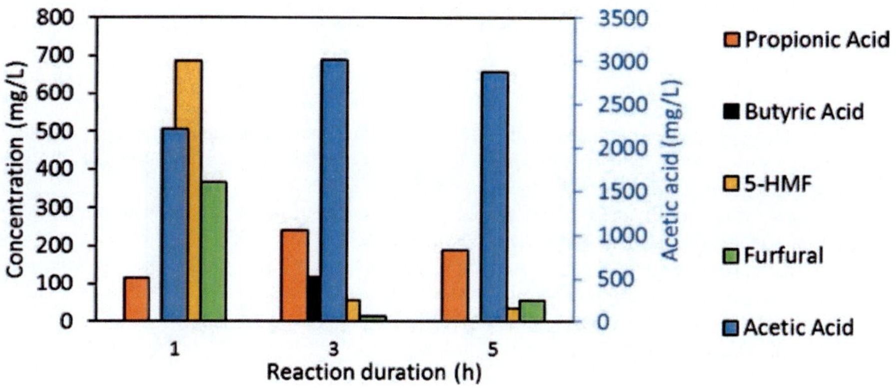

Fig. 2. Concentration of volatile fatty acids, 5-HMF and furfural in process wastewater samples (temperature = 200°C, self-pressure = 18 bar).

mentioned earlier that HMF is a promising compound for synthesis of furan derivatives which can replace fossil fuel derived compounds. The recovery of HMF has been reported earlier from model compounds (Ventura et al., 2018); however, in-depth studies on their extraction from PW generated during hydrothermal treatment of waste are lacking. Moreover, organic acids rich PW can be subjected to fermentation for producing biofuels or ethanol (Raftery and Karim, 2017).

4 Conclusions

In the present study, we investigated the effect of HTC pretreatment on JW charcateristics. HTC was recognised as an effective method to produce a good heating value fuel-like material (i.e., HC). After HTC at 200°C for 5 h duration, HC having HHV of 23.6 MJ/kg could be produced. However, the stable pellets could be formed only after adding oil as binder. Therefore, use of other binders can be explored in further studies.

In PW samples, high concentration of HLS (upto 30 g/L) was obtained, which may be recovered and used for soil amendment. Among other organic compounds detected in aqueous phase, 5-HMF and furfural were found in high concentrations after HTC treatment of 1 h. These compounds are toxic for anaerobic microbes. During extended reaction (run time = 5 h), these were almost completely degraded. In future study, efforts should be made to recover various useful organic compounds from PW after which the residual organic acids rich liquid stream can be subjected to anaerobic treatment or fermentation to produce biogas or ethanol.

References

APEDA. Indian Production of sweet orange (malta, mosambi). Taken from: http://agriexchange.apeda.gov.in/India%20Production/India_Productions.aspx?cat=fruit&hscode=1065. Accessed on 2 January 2020.

APHA (2012). Standard methods for the examination of water and wastewater. Washington DC, USA.

Banel, A. and Zygmunt, B. (2011). Application of gas chromatography-mass spectrometry preceded by solvent extraction to determine volatile fatty acids in wastewater of municipal, animal farm and landfill origin. *Water Sci Technol*, **63:** 590–597.

Benavente, V., Calabuig, E. and Fullana, A. (2015). Upgrading of moist agro-industrial wastes by hydrothermal carbonization. *J Anal Appl Pyrol,* **113:** 89–98.

Cao, L., Zhang, C., Chen, H., Tsang, D.C., Luo, G., Zhang, S. and Chen, J. (2017). Hydrothermal liquefaction of agricultural and forestry wastes: State-of-the-art review and future prospects. *Bioresource Technol*, **245:** 1184–1193.

Central Public Health and Environmental Engineering Organization, Part II: The Manual, Municipal Solid Waste Management Manual (CPHEEO) (2016). Ministry of Urban Development, New Delhi.

Danso-Boateng, E., Shama, G., Wheatley, A.D., Martin, S.J. and Holdich, R.G. (2015). Hydrothermal carbonisation of sewage sludge: Effect of process conditions on product characteristics and methane production. *Bioresource Technol*, **177:** 318–327.

Dubois, M., Gilles, K.A., Hamilton, J.K., Rebers, P.T. and Smith, F. (1956). Colorimetric method for determination of sugars and related substances. *Anal Chem*, **28:** 350–356.

Erdogan, E., Atila, B., Mumme, J., Reza, M.T., Toptas, A., Elibol, M. and Yanik, J. (2015). Characterization of products from hydrothermal carbonization of orange pomace including anaerobic digestibility of process liquor. *Bioresource Technol*, **196:** 35–42.

Fei, Y.H., Zhao, D., Cao, Y., Huot, H., Tang, Y.T., Zhang, H. and Xiao, T. (2019). Phosphorous retention and release by sludge-derived hydrochar for potential use as a soil amendment. *J Environ Qual*, **48:** 502–509.

Gupta, D., Mahajani, S.M. and Garg, A. (2019). Effect of hydrothermal carbonization as pretreatment on energy recovery from food and paper wastes. *Bioresource Technol*, **285** (available online).

Jiang, L., Liang, J., Yuan, X., Li, H., Li, C., Xiao, Z., Huang, H., Wang, H. and Zeng, G. (2014). Co-pelletization of sewage sludge and biomass: The density and hardness of pellet. *Bioresource Technol*, **166:** 435–443.

Jokic, A., Frenkel, A.I., Vairavamurthy, M.A. and Huang, P.M. (2001). Birnessite catalysis of the Maillard reaction: Its significance in natural humification. *Geophys Res Lett*, **28:** 3899–3902.

Joshi, R. and Ahmed, S. (2016). Status and challenges of municipal solid waste management in India: A review. *Cogent Environ Sci*, **2:** 1139434.

Kambo, H.S. and Dutta, A. (2015). A comparative review of biochar and hydrochar in terms of production, physico-chemical properties and applications. *Renew Sust Energ Rev*, **45:** 359–378.

Kambo, H.S., Minaret, J. and Dutta, A. (2018). Process water from the hydrothermal carbonization of biomass: A waste or a valuable product? *Waste Biomass Valori* **9:** 1181–1189.

Kasturi, P.R., Selvan, R.K. and Lee, Y.S. (2016). Pt decorated *Artocarpus heterophyllus* seed derived carbon as an anode catalyst for DMFC application. *RSC Adv*, **6:** 62680–62694.

Li, L., Diederick, R., Flora, J.R. and Berge, N.D. (2013). Hydrothermal carbonization of food waste and associated packaging materials for energy source generation. *Waste Manage*, **33:** 2478–2492.

Mijares, R.M., Park, G.L., Nelson, D.B. and McIVER, R.C. (1986). HPLC analysis of HMF in orange juice. *J Food Sci*, **51:** 843–844.

Parshetti, G.K., Chowdhury, S. and Balasubramanian, R. (2014). Hydrothermal conversion of urban food waste to chars for removal of textile dyes from contaminated waters. *Bioresource Technol*, **161:** 310–319.

Prawisudha, P., Namioka, T. and Yoshikawa, K. (2012). Coal alternative fuel production from municipal solid wastes employing hydrothermal treatment. *Appl Energy*, **90:** 298–304.

Puccini, M., Licursi, D., Stefanelli, E., Vitolo, S., Galletti, A.M.R. and Heeres, H.J. (2016). Levulinic acid from orange peel waste by hydrothermal carbonization (HTC). *Chem Eng Trans*, **50:** 223–228.

Raftery, J.P. and Karim, M.N. (2017). Economic viability of consolidated bioprocessing utilizing multiple biomass substrates for commercial-scale cellulosic bioethanol production. *Biomass Bioenergy*, **103:** 35–46.

Ritchie, H. and Roser, M. (2019). Fossil Fuels. Taken from: https://ourworldindata.org/fossil-fuels. Accessed on 2 January 2020.

Sabio, E., Álvarez-Murillo, A., Román, S. and Ledesma, B. (2016). Conversion of tomato-peel waste into solid fuel by hydrothermal carbonization: Influence of the processing variables. *Waste Manage*, **47:** 122–132.

Vakondios, N., Koukouraki, E.E. and Diamadopoulos, E. (2014). Effluent organic matter (EfOM) characterization by simultaneous measurement of proteins and humic matter. *Water Res*, **63:** 62–70.

Ventura, M., Dibenedetto, A. and Aresta, M. (2018). Heterogeneous catalysts for the selective aerobic oxidation of 5-hydroxymethylfurfural to added value added products in water. *Inorg Chim Acta*, **470:** 11–21.

Wang, T., Zhai, Y., Zhu, Y., Gan, X., Zheng, L., Peng, C., Wang, B., Li, C. and Zeng, G. (2018). Evaluation of the clean characteristics and combustion behaviour of hydrochar derived from food waste towards solid biofuel production. *Bioresource Technol*, **266:** 275–283.

Willis, R.B., Montgomery, M.E. and Allen, P.R. (1996). Improved method for manual, colorimetric determination of total Kjeldahl nitrogen using salicylate. *J Agric Food Chem*, **44:** 1804–1807.

Yasuhara, A., Katami, T., Okuda, T., Ohno, N. and Shibamoto, T. (2001). Formation of dioxins during the combustion of newspapers in the presence of sodium chloride and poly (vinyl chloride). *Environ Sci Technol*, **35:** 1373–1378.

Yilmaz, E., Wzorek, M. and Akcay, S. (2018). Co-pelletization of sewage sludge and agricultural wastes. *J Environ Manage*, **216:** 169–175.

Zhang, B., Heidari, M., Regmi, B., Salaudeen, S., Arku, P., Thimmannagari, M. and Dutta, A. (2018). Hydrothermal carbonization of fruit wastes: A promising technique for generating hydrochar. *Energies*, **11:** 2022.

Chapter 17
Terracotta Membrane-based Microbial Fuel Cell with Algal Biocathode: A Low-Cost Alternative to Dairy Wastewater Treatment Coupled Electricity and Biomass Production

Smriti Mehrotra, V. Kiran Kumar, K. Man Mohan, S. Gajalakshmi, and Bhawana Pathak

1 Introduction

The 21st century is marked by the massive global shortages of water and energy throughout the world. Consumption of food and dairy products at a higher rate is considered a major contributor to it. Among them, the dairy industry is distinctive to this count due to the release of a large quantity of wastewater yielded through its various milk processing entities, cleaning and sanitation units (Kumar et al., 2016). Dairy industry on average discharges 2.5 L of wastewater while processing 1 L of milk (Ramasamy et al., 2004; Elakkiya and Matheswaran, 2013). Its wastewater is marked by high chemical oxygen demand (COD), nitrate and phosphate content due to the fats, high lactose content, various dissolved and suspended solids and the use of nitric acid, caustic soda used in its cleansing operations (Al-Shammari et al., 2015). Various traditional physicochemical wastewater treatment techniques are practiced in industries to meet environmental discharge standards. While these techniques are less effective and more costly owing to the energy-intensive processes and the use of chemical coagulants (Sharma, 2014). The efficient management of wastewater resources and their use for energy generation can prove to be a judicious solution to it (Venkata Mohan et al., 2010). Microbial fuel cells (MFCs) are such bioelectrochemical systems, contributing to this attempt in a cost-effective and sustainable manner. This prototype serves the dual purpose of energy generation and

S. Mehrotra · B. Pathak (✉)
School of Environment and Sustainable Development, Central University of Gujarat, Gandhinagar, Gujarat, India
e-mail: bhawana.pathak@cug.ac.in

V. Kiran Kumar · K. Man Mohan · S. Gajalakshmi
Bioelectrochemical Systems Lab, Centre for Pollution Control and Environmental Engineering, Pondicherry University, Pondicherry, India

A. S. Kalamdhad (ed.), *Integrated Approaches Towards Solid Waste Management*, https://doi.org/10.1007/978-3-030-70463-6_17

wastewater treatment with insignificant negative environmental impacts (Mathuriya et al., 2018). Typical MFCs are composed of an anode and cathode compartment containing electrodes alienated by a cation exchange membrane (CEM). The anodic compartment is fed by the wastewater and sludge, where organic matter oxidises into carbon dioxide in anaerobic conditions through the metabolism of exoelectrogenic bacteria releasing electrons and protons. These electrons pass through the external circuit while the protons cross the CEM to enter the cathodic region, where they combine with an electron acceptor such as oxygen to produce water (González Del Campo et al., 2013). Although these traditional MFCs have proved to be promising in wastewater treatment techniques and energy generation their economic and technical feasibility especially the cost of separators, the kinetics of reduction reaction and cost of aeration in the cathodic chamber have proved to be an obstacle in its practical applications. Therefore, the scientific community is now inclining its efforts towards the use of biocathode and cheaper ion-exchange membranes. Microalgae and terracotta membranes are considered to be the purpose of solvers to these efforts (Salar-García et al., 2016).

Incorporating microalgae in the cathode of MFCs serves the dual purpose of carbon sequestration and electricity generation due to its efficient photosynthetic mechanism (Chandra et al., 2019). Utilising light and the carbon dioxide, released by respiration of anodic bacteria, these algae release oxygen, which serves as the terminal electron acceptor in the cathodic compartment thus promoting the oxidation–reduction reaction for the generation of electricity (Lin et al., 2013; Nayak and Ghosh, 2019). In comparison to traditional MFCs, which underperformed due to the penetration of wastewater organics through proton exchange membrane, these organisms proliferate and utilise its micronutrients as a medium of their growth (Tice and Kim, 2014). Furthermore, the scalability of A-MFCs can be enhanced by the use of cheap and efficient separators, which is one of the major challenges in commercialising it. Use of Nafion membranes, have though increased the efficiency but its stability and cost is the major issue in its practical application (Winfield et al., 2016). The sulfonic acid group present in these membranes reacts with ammonia present in wastewater making it highly unstable (Rabaey and Verstraete, 2005). While other separators have critical issues of high cost, low electron recovery and lower mechanical strength (Li et al., 2010). Behera et al. (2010a) identified the unique ability of the terracotta membrane as a separator in MFCs due to its porous wall structure, which promotes ionic mobility from anodic to the cathodic chamber. While Ghadge and Ghangrekar (2015) studied the properties of Montmorillonite and Kaolinite clay as efficient cationic exchangers. These act as efficient separators due to their resistivity in extreme pressure and sustenance in extremely acidic or alkaline environments (Gajda et al., 2015).

The present study is an attempt to construct biocathode assisted dual chambered A-MFC setup using terracotta pot and recycled transparent plastic pickle bottles for dairy wastewater treatment, green energy generation and algal production.

2 Materials and Methods

2.1 Microalgal Strain

The microalgal freshwater strain of *Chlorella sorokiniana* is purchased from the culture collection laboratory of National Collection of Industrial Microorganisms (NCIM), Pune, India. These cultures were maintained in controlled conditions at 25 ± 2°C temperature, under continuous illumination of 2500–3000 lux with light photoperiod of 18 h light/6 h darkness in BG 11 medium maintained at pH 7. Cultures were manually shaken daily to maintain homogeneity. The dense culture of 10th day was weighed and 0.5 g/L was used as biocatalyst in the cathode fed with BG 11 medium ($NaNO_3$, $MgSO_4{\cdot}7H_2O$, K_2HPO_4, $CaCl_2{\cdot}2H_2O$, $ZnSO_4{\cdot}7H_2O$, $CuSO_4{\cdot}5H_2O$, citric acid, EDTA, Na_2CO_3, H_3BO_3, $NaMoO_4{\cdot}2H_2O$, $MnCl_2{\cdot}4H_2O$, $Co(NO_3)_2{\cdot}6H_2O$) (Pandit et al., 2017).

2.2 MFC Configuration

Three identical dual chambered MFCs (Fig. 1) were constructed using terracotta pots being placed inside large recycled plastic pickle bottles. Components of each experimental setup is given in Table 1. Each MFC consists of a terracotta pot with a working volume of 800 ml which acts as an anodic chamber cum separator/ion exchange membrane, and a transparent recycled plastic bottle with a working volume of 1500 ml composes the cathodic chamber of the reactor. Stainless steel (SS) mesh of each 300 cm^2 projected area is wrapped in the inner and outer

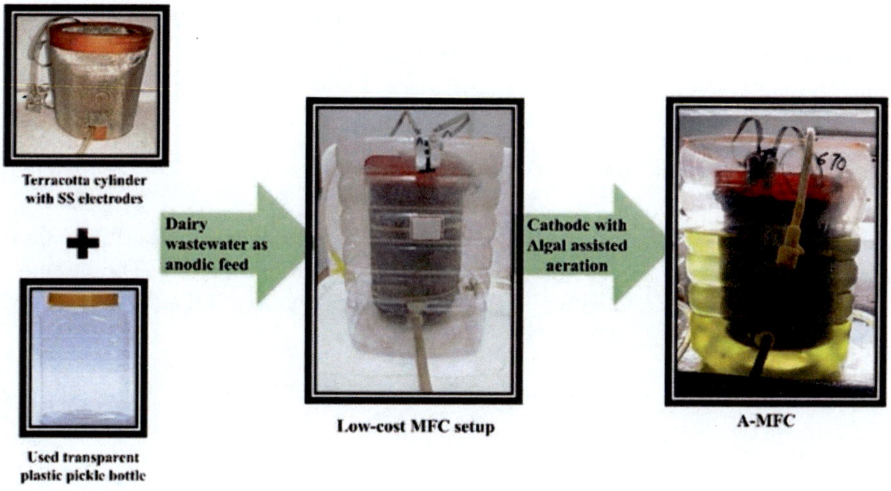

Fig. 1. Low-cost microbial fuel cell construction.

Table 1. Microbial fuel cell configuration

Reactor	*Anolyte*	*Catholyte*	*Anode*	*Cathode*
Reactor 1	Dairy wastewater + pretreated anaerobic sludge	BG 11 medium with only aeration	SS mesh	SS mesh
Reactor 2 (Fig. 1)	Dairy wastewater + pretreated anaerobic sludge	BG 11 medium inoculated with algae	SS mesh	SS mesh
Reactor 3	Dairy wastewater + pretreated anaerobic sludge	BG 11 medium without algae and aeration	SS mesh	SS mesh

circumference of the terracotta cylinder forms the electrodes of the MFC. The intact SS wire of the SS mesh electrodes is used as the current collector.

The real-time dairy wastewater from the Pondicherry Dairy plant is collected and diluted to the desired COD (1000 mg L^{-1}) and is used as anolyte. Anaerobic sludge from Pondicherry municipal wastewater treatment plant is used as the inoculum in the anode. Before inoculation, the sludge was pretreated at 110°C for 15 min as explained by Jadhav and Ghangrekar (2009) to suppress the methanogen activity. BG 11 Medium with and without 0.5 g L^{-1} of *Chlorella sorokiniana* was used as catholyte.

2.3 Data Analysis of MFCs

Hourly working voltages were recorded manually through a Digital Multimeter. The polarisation study was done by varying resistances from 10 kΩ to 10 Ω. Current and Power were measured by Ohm's law calculations (Logan, 2007). Power and current density were calculated by using the projected cathodic surface area. Chemical oxygen demand (COD) was obtained by following the titrimetric open reflux method. The pH and electrical conductivity were measured by digital pH and EC meter. Dissolved oxygen was measured by a digital DO meter.

2.4 Data Analysis of Microalgal Biomass

Microalgal growth in the cathode chamber was monitored by measuring optical density on alternate days at 700 nm in UV/Vis Spectrophotometer. The biomass concentration was calculated by using the calibration curve of OD_{700} by following equation (Yadav et al., 2019):

$$\text{Biomass concentration} = 0.44 \times OD_{700}$$

3 Results and Discussion

3.1 Voltage Generation and Fuel Cell Behaviour

For acclimatisation, the MFCs were run at the open circuit for the initial one week after which all the MFCs were connected to 1000 Ω external resistance and the hourly working voltage was used for the analysis. The maximum open-circuit voltage (OCV) of 568, 443 and 368 mV (data not shown) were observed in the MFCs with algal assisted aeration (MFC-2), mechanical aeration (MFC-1) and MFC without aeration (MFC-3) respectively. For the seven days of operation, the average working voltage of 236.3 mV, 191.3 mV and 130.3 mV was achieved by MFC-2, MFC-1 and MFC-3. Figure 2 depicts the voltage performances of the three MFCs, where MFC-1 with mechanical aeration has produced a stable voltage output of 170 ± 20 mV from the initial to the final day of operation. The voltage output of MFC-2 was gradually increased from day 1 to day 3 after which stable output was achieved; this was coinciding with the dissolved oxygen levels in the cathode (Fig. 5). DO concentration in the cathode is directly related to the kinetics of reduction reaction and power production in MFCs (Xiao et al., 2012). Similar trends were observed by Shukla and Kumar (2018) and Commault et al. (2017).

A polarisation study was conducted on the 7th day to determine the fuel cell behaviour of MFCs. Figure 3 shows the polarisation and power density curves of the MFCs. The peak power density of 28.8 mW m^{-2} was observed in the reactor with algal assisted aeration followed by reactors with aeration (20.4 mW m^{-2}) and without aeration (10.6 mW m^{-2}). Table 2 compares the power density of the previously reported A-MFCs. Power density exhibited by A-MFC and aerated MFC in this

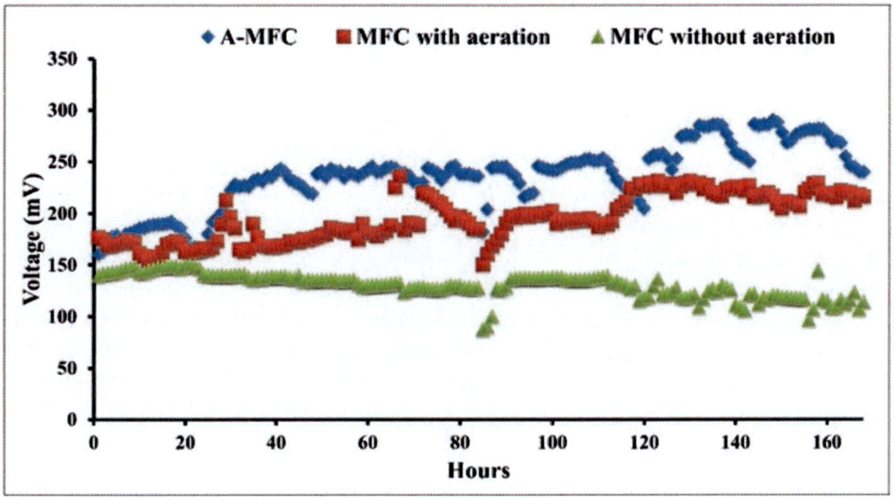

Fig. 2. Voltage generation in the different MFCs.

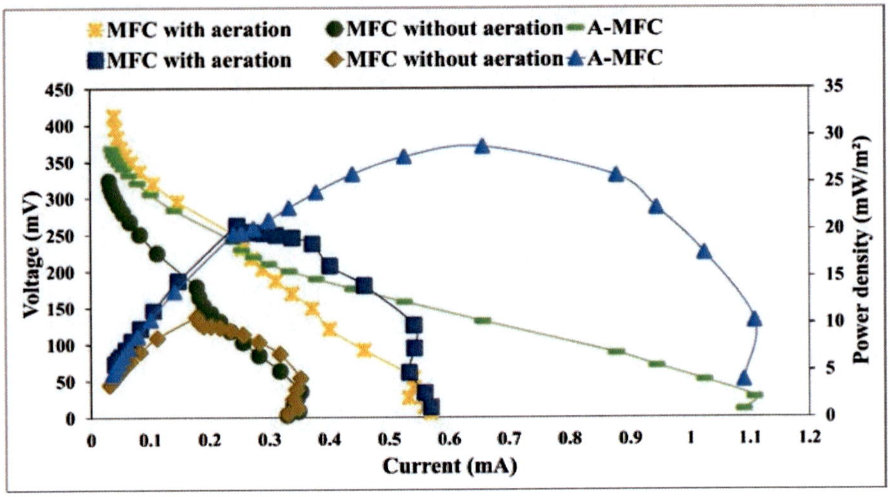

Fig. 3. Polarisation and power density curves of MFCs.

research is higher than the previous study of Behera et al. (2010b) and Neethu and Ghangrekar (2017), where they used earthen pot MFC with abiotic and biotic cathode to treat rice mill wastewater, and photo sediment MFC (sMFC) with algal biocathode. Yadav et al. (2019) reported 54.48 mW m^{-2} power density with the algal biocathode equipped earthen pot MFC, where they used carbon felt cathode with a surface area of 2152.18 cm^2 and wastewater mixed algal biomass as anolyte. More power density in their research can be attributed to the base material and surface area of the cathode.

3.2 *Wastewater Treatment*

Initial and final COD concentrations in the three MFCs were given in Fig. 4. The Microbial consortium present in the anode and its affinity towards the substrate will determine the COD reduction rate and in turn determine the Coulombic efficiency of the system (Khandelwal et al., 2018). The inoculum used in this study is efficient and the maximum COD removal efficiency of 73.08%, 77.40% and 63.30% was achieved by the MFC-1, MFC-2 and MFC-3. COD removal efficiency is following the previously reported studies of Naina Mohamed et al. (2020) and Yadav et al. (2019). Naina Mohamed et al. (2020) reported the removal efficiency of 68.5% and 73.5% with the initial COD concentration of 5000 and 2500 mg L^{-1} of kitchen wastewater. Yadav et al. (2019) reported average removal efficiencies between 70% and 75% with artificial wastewater and algae as anolyte with the initial concentration of 1100 mg L^{-1}.

Table 2. Comparison of algal MFC performance vis-a-vis our study

Microalgal species	*MFC configuration*	*Substrate*	*Anode*	*Cathode*	*Maximum power density*	*Maximum biomass concentration ($g L^{-1}$)*	*Dissolved oxygen concentration ($mg L^{-1}$)*	*Reference*
Mixed microalgae	Sediment MFC	Sediment and effluent of aquaculture the pond	Graphite plates	Non-catalysed carbon felt	22.19 mW/m^2	0.8	14.2	(Neethu and Ghangrekar, 2017)
Pseudokirchneriella subcapitata	Integrated tubular MFC installed in a glass beaker	Synthetic wastewater	Carbon brush	Carbon cloth with Pt/C catalysts	2.2 W/m^3	0.128	20	(Xiao et al. 2012)
Desmodesmus* sp. *A8	Two chambered Plexiglass MFC	Artificial wastewater	Graphite felt	Graphite felt	103.88 mW/m^2	–	13.2	(Wu et al., 2014)
Scenedesmus obliquus	H-shaped DMFC	Wastewater	Plain carbon paper	Platinum-coated carbon paper	153 mW/m^2	–	15.7	(Kakarla and Min, 2014)
Golenkinia* sp. *SDEC-16)	Tubular dual chambered microbial fuel cell (DMFC)	Anaerobically digested effluent from kitchen waste	Carbon cloth	Carbon cloth	2.8 W/m^3	0.94	–	(Pei et al., 2018)
***Chlorella* sp.**	DMFC (earthen pot placed in glass reactor)	Artificial wastewater	SS mesh	Carbon felt	54.48 mW/m^2	1.34	11.2	(Yadav et al., 2019)
Scenedesmus quadricauda SDEC-8	ABMFC (algae biofilm microbial fuel cell)	Domestic wastewater	Carbon cloth	Carbon cloth	62.93 mW m^{-2})	–	–	(Yang et al. 2019)

(continued)

Table 2. (continued)

Microalgal species	*MFC configuration*	*Substrate*	*Anode*	*Cathode*	*Maximum power density*	*Maximum biomass concentration ($g\ L^{-1}$)*	*Dissolved oxygen concentration ($mg\ L^{-1}$)*	*Reference*
Chlorella sorokiniana	Dual-chambered flat plate microbial carbon capture cell	Synthetic wastewater	Carbon felt	Carbon felt	2.32 W/m^3	0.812	-	(Varanasi et al., 2020)
Golenkinia* sp. *SDEC-16	Tubular PMFC	Anaerobically digested effluent from food waste (ADE-FW)	Carbon cloth	Carbon cloth	3.5 W/m^3	0.7	20 mg/L	(Yang et al., 2018)
Chlorella vulgaris (FACHB-26)	H-type MFC	Synthetic wastewater	Carbon brushes	Carbon brushes	466.9 mW/m^3	0.376	17 mg/L	(Wang et al., 2019)
Chlorella sorokiniana	Dual chambered microbial fuel cell (DMFC)	Synthetic wastewater	Carbon felt	Carbon felt	3.2 W/m^3	2.45	7 mg/L	(Neethu et al., 2018)
Scenedesmus abundans	H-type Dual chambered photosynthetic microbial fuel cell (PMFC)	Pharmaceutical wastewater	Graphite rods	Graphite rods	838.68 mW/m^2	0.97	–	(Nayak and Ghosh 2019)
Chlorella sorokiniana	DMFC (earthen pot placed in used plastic bottles)	Dairy wastewater	SS mesh	SS mesh	28.8 mW/m^2	1.8	9.2	This study

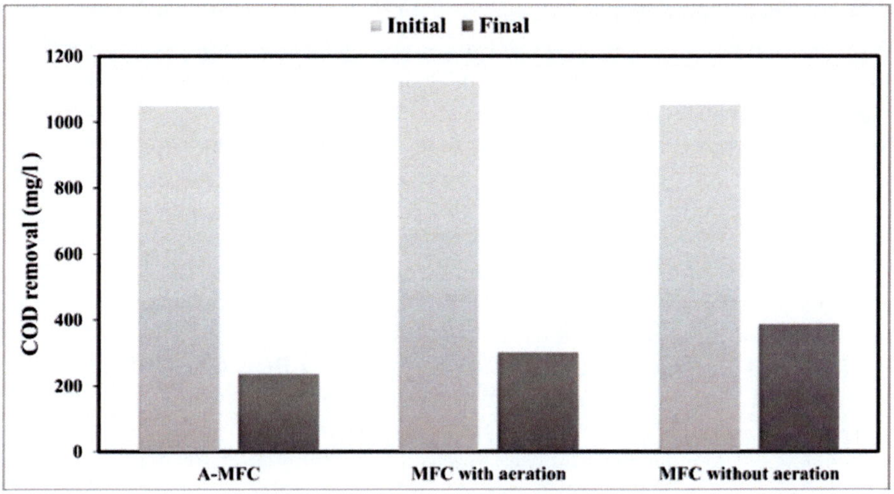

Fig. 4. COD removal in A-MFC, MFC with aeration, and MFC without aeration.

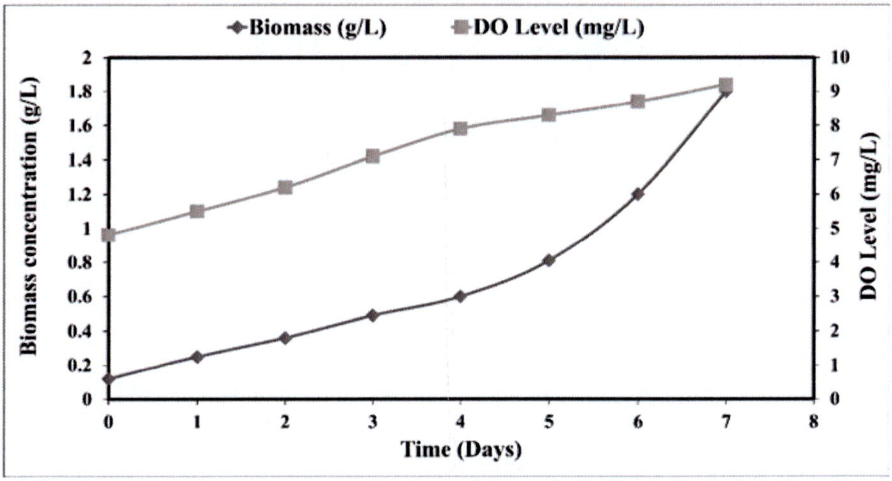

Fig. 5. Profile of algal biomass concentration and DO level in the cathodic chamber of A-MFC.

3.3 *Algal Biomass and DO Levels in the Cathode Chamber*

The time profile of algal biomass concentration and DO level in the cathode chamber of A-MFC is given in Fig. 5. Throughout the study, the DO concentration was between 4.8 mg L^{-1} to 9.2 mg L^{-1} and the levels were increased with elevated biomass concentration (0.12–1.8 g L^{-1}). A positive correlation between power density and dissolved oxygen in biocathode MFC was observed by Yadav et al. (2019)

achieving a maximum DO concentration of 11.2 mg L^{-1} using *Chlorella vulgaris*. This concomitant increase in DO levels may be attributed to the substantial oxygen production through algal photosynthetic activity under illuminating conditions which in turn enhanced reduction at cathode resulting in higher electrogenesis. Higher electrogenicity is contributed due to the oxidation–reduction reaction (ORR) occurring at the cathode. Oxygen being a terminal electron acceptor in the cathodic chamber increases the conduction of electrons that in turn increases the voltage and power density of the system, which is a critical indicator of its performance (Hou et al., 2020). In comparison to mechanical aeration, a 32% higher DO value of 19.57 mg L^{-1} and maximum power density of 154 mW m^{-2} was obtained by Kakarla and Min (2014). This additional advantage is due to its ability to utilise the diffused substrate as a nutrient of growth which normally is a limitation in other MFCs (Neethu et al., 2018). Yang et al. (2018) found a distinctive relation between the algal growth and DO concentration. During the exponential growth stage of algae, an increase in DO concentration was observed from 8 mg L^{-1} to 18 mg L^{-1} while a decrease was observed between 9 and 10 mg L^{-1} during the stationary growth rate of algae. Hence, it gives clear evidence of the direct relationship of the effect of algal growth concentration on DO and power density values.

4 Conclusion

In this research, the holistic approach of waste recycling was used to construct and operate cost-effective MFCs for industrial wastewater treatment, green energy generation and algal production. Chassis made of recycled plastic bottles and terracotta pot along with SS mesh electrodes has cut down the cost of MFC construction whereas the use of algal biocathode has reduced the cost of aeration with the additional benefit of biomass production. In A-MFC, dairy industry wastewater was treated successfully with 77.4% COD removal efficiency and the algal biocathode resulted in 18.6% and 44.8% more voltage output than the MFCs with aerated and non-aerated controls. In the entire operation, the power output of A-MFC (average power density of 18.6 mW m^{-2}) outperformed the two controlled MFCs (MFC-1 – 12.7 mW m^{-2}, and MFC-3 – 5.6 mW m^{-2}). In this research, the photosynthetic efficiency of *Chlorella sorokiniana* was successfully utilised to run the cathodic reduction reaction, and the microalgae were able to generate the DO levels of 9.2 mg L^{-1}. The results obtained will further encourage the researchers to develop low-cost energy generating systems which are important for sustainable development.

Acknowledgements Smriti Mehrotra is thankful to the Central University of Gujarat for the support and financial assistance through the UGC-non-NET fellowship. Kiran Kumar V. and Man Mohan K. are thankful to Pondicherry University for their support and financial assistance through the UGC-non NET fellowship.

References

Al-Shammari, S.B., Bou-Hammad, S., Al-Saffar, A. et al. (2015). Treatment of dairy processing wastewater using integrated submerged membrane microfiltration system. *Journal of Environmental & Analytical Toxicology*, **05:** 1-6.

Behera, M., Jana, P.S. and Ghangrekar, M.M. (2010a). Performance evaluation of low cost microbial fuel cell fabricated using earthen pot with biotic and abiotic cathode. *Bioresource Technol*, **101:** 1183–1189.

Behera, M., Jana, P.S., More, T.T. and Ghangrekar, M.M. (2010b). Rice mill wastewater treatment in microbial fuel cells fabricated using proton exchange membrane and earthen pot at different pH. *Bioelectrochemistry*, **79:** 228–233.

Chandra, R. and Ghosh, Amit, U.K. (2019). Effects of various abiotic factors on biomass growth and lipid yield of *Chlorella minutissima* for sustainable biodiesel production. *Environmental Science and Pollution Research*, **26:** 3848–3861.

Commault, A.S., Laczka, O., Siboni, N., et al. (2017). Electricity and biomass production in a bacteria-Chlorella based microbial fuel cell treating wastewater. *Journal of Power Sources*, **356:** 299–309.

Elakkiya, E. and Matheswaran, M. (2013). Comparison of anodic metabolisms in bioelectricity production during treatment of dairy wastewater in microbial fuel cell. *Bioresource Technology*, **136:** 407–412.

Gajda, I., Stinchcombe, A., Greenman, J., et al. (2015). Ceramic MFCs with internal cathode producing sufficient power for practical applications. *International Journal of Hydrogen Energy*, **40:** 14627–14631.

Ghadge, A.N. and Ghangrekar, M.M. (2015). Development of low cost ceramic separator using mineral cation exchanger to enhance performance of microbial fuel cells. *Electrochimica Acta*, **166:** 320–328.

González Del Campo, A., Cañizares, P., Rodrigo, M.A., et al. (2013). Microbial fuel cell with an algae-assisted cathode: A preliminary assessment. *Journal of Power Sources*, **242:** 638–645.

Hou, Q., Yang, Z., Chen, S. and Pei, H. (2020). Using an anaerobic digestion tank as the anodic chamber of an algae-assisted microbial fuel cell to improve energy production from food waste. *Water Research*, **170:** 115305.

Jadhav, G.S. and Ghangrekar, M.M. (2009). Performance of microbial fuel cell subjected to variation in pH, temperature, external load and substrate concentration. *Bioresource Technology*, **100:** 717–723.

Kakarla, R. and Min, B. (2014). Photoautotrophic microalgae *Scenedesmus obliquus* attached on a cathode as oxygen producers for microbial fuel cell (MFC) operation. *International Journal of Hydrogen Energy*, **39:** 10275–10283.

Khandelwal, A., Vijay, A., Dixit, A. and Chhabra, M. (2018). Microbial fuel cell powered by lipid extracted algae: A promising system for algal lipids and power generation. *Bioresource Technology*, **247:** 520–527.

Kumar, R.V., Goswami, L., Pakshirajan, K. and Pugazhenthi, G. (2016). Dairy wastewater treatment using a novel low cost tubular ceramic membrane and membrane fouling mechanism using pore blocking models. *Journal of Water Process Engineering*, **13:** 168–175.

Li, F., Sharma, Y., Lei, Y. and Li, B. (2010). Microbial fuel cells: The effects of configurations, electrolyte solutions, and electrode materials on power generation. *Appl Biochem Biotechnol*, **160:** 168–181.

Lin, C.C., Wei, C.H., Chen, C.I., et al (2013). Characteristics of the photosynthesis microbial fuel cell with a Spirulina platensis biofilm. *Bioresource Technology*, **135:** 640–643.

Logan, B.E. (2007). Microbial Fuel Cells. https://doi.org/10.1002/9780470258590.

Mathuriya, A.S., Jadhav, D.A. and Ghangrekar, M.M. (2018). Architectural adaptations of microbial fuel cells. *Applied Microbiol Biotechnol*, **102:** 9419–9432.

Naina Mohamed, S., Ajit Hiraman, P., Muthukumar, K. and Jayabalan, T. (2020). Bioelectricity production from kitchen wastewater using microbial fuel cell with photosynthetic algal cathode. *Bioresource Technology*, **295:** 122226.

Nayak, J.K. and Ghosh, U.K. (2019). Post treatment of microalgae treated pharmaceutical wastewater in photosynthetic microbial fuel cell (PMFC) and biodiesel production. *Biomass and Bioenergy*, **131:** 105415.

Neethu, B., Bhowmick, G.D. and Ghangrekar, M.M. (2018). Enhancement of bioelectricity generation and algal productivity in microbial carbon-capture cell using low cost coconut shell as membrane separator. *Biochemical Engineering Journal*, **133:** 205–213.

Neethu, B. and Ghangrekar, M.M. (2017). Electricity generation through a photo sediment microbial fuel cell using algae at the cathode. *Water Science and Technology*, **76:** 3269–3277.

Pandit, P.R., Fulekar, M.H. and Karuna M.S.L. (2017). Effect of salinity stress on growth, lipid productivity, fatty acid composition, and biodiesel properties in *Acutodesmus obliquus* and *Chlorella vulgaris*. *Environmental Science and Pollution Research*, **24:** 13437–13451.

Pei, H., Yang, Z., Nie, C., et al. (2018). Using a tubular photosynthetic microbial fuel cell to treat anaerobically digested effluent from kitchen waste: Mechanisms of organics and ammonium removal. *Bioresource Technology*, **256:** 11–16.

Rabaey, K. and Verstraete, W. (2005). Microbial fuel cells: Novel biotechnology for energy generation. *Trends in Biotechnology*, **23:** 291–298.

Ramasamy, E.V., Gajalakshmi, S., Sanjeevi, R., et al. (2004). Feasibility studies on the treatment of dairy wastewaters with upflow anaerobic sludge blanket reactors. *Bioresource Technology*, **93:** 209–212.

Salar-García, M.J., Gajda, I., Ortiz-Martínez, V.M., et al. (2016). Microalgae as substrate in low cost terracotta-based microbial fuel cells: Novel application of the catholyte produced. *Bioresource Technology*, **209:** 380–385.

Sharma, D. (2014). Treatment of dairy waste water by electro coagulation using aluminum electrodes and settling, filtration studies. *International Journal of ChemTech Research*, **6:** 591–599.

Shukla, M. and Kumar, S. (2018). Algal growth in photosynthetic algal microbial fuel cell and its subsequent utilization for biofuels. *Renewable and Sustainable Energy Reviews*, **82:** 402–414.

Tice, R.C. and Kim, Y. (2014). Influence of substrate concentration and feed frequency on ammonia inhibition in microbial fuel cells. *Journal of Power Sources*, **271:** 360–365.

Varanasi, J.L., Prasad, S., Singh, H. and Das, D. (2020). Improvement of bioelectricity generation and microalgal productivity with concomitant wastewater treatment in flat-plate microbial carbon capture cell. *Fuel*, **263:** 116696.

Venkata Mohan, S., Mohanakrishna, G., Chiranjeevi, P., et al. (2010). Ecologically engineered system (EES) designed to integrate floating, emergent and submerged macrophytes for the treatment of domestic sewage and acid rich fermented-distillery wastewater: Evaluation of long term performance. *Bioresource Technology*, **101:** 3363–3370.

Wang, Y., Lin, Z., Su, X., et al. (2019). Cost-effective domestic wastewater treatment and bioenergy recovery in an immobilized microalgal-based photoautotrophic microbial fuel cell (PMFC). *Chemical Engineering Journal*, **372:** 956–965.

Winfield, J., Gajda, I., Greenman, J. and Ieropoulos, I. (2016). A review into the use of ceramics in microbial fuel cells. *Bioresource Technology*, **215:** 296–303.

Wu, Y. Cheng, Wang, Z.J., Zheng, Y., et al. (2014). Light intensity affects the performance of photo microbial fuel cells with *Desmodesmus* sp. A8 as cathodic microorganism. *Applied Energy*, **116:** 86–90.

Xiao, L., Young, E.B., Berges, J.A. and He, Z. (2012). Integrated photo-bioelectrochemical system for contaminants removal and bioenergy production. *Environmental Science and Technology*, **46:** 11459–11466.

Yadav, G., Sharma, I., Ghangrekar, M. and Sen, R. (2019). A live bio-cathode to enhance power output steered by bacteria-microalgae synergistic metabolism in microbial fuel cell. *Journal of Power Sources*, **449:** 227560.

Yang, Z., Pei, H., Hou, Q., et al. (2018). Algal biofilm-assisted microbial fuel cell to enhance domestic wastewater treatment: Nutrient, organics removal and bioenergy production. *Chemical Engineering Journal*, **332:** 277–285.

Yang, Z., Nie, C., Hou, Q. et al. (2019). Coupling a photosynthetic microbial fuel cell (PMFC) with photobioreactors (PBRs) for pollutant removal and bioenergy recovery from anaerobically digested effluent. *Chemical Engineering Journal*, **359:** 402–408.

Chapter 18
Upcycling Textile Waste Towards Green Nanocomposites

Zunjarrao Kamble and Bijoya Kumar Behera

1 Introduction

The twenty-first century will be named in its history as a highly destructive period of the planet if we fail to take proper action and to aware our people about efficient energy and water utilisation, pollution control and waste management. According to the World Bank report (Kaza et al., 2018), the projected global municipal solid waste generation in 2030 and 2050 is 2.59 and 3.40 billion tonnes respectively. However, the waste collection rates for high income, upper-middle-income, lower-middle-income and low-income countries are 96, 82, 51 and 39% respectively. The surprising fact is almost 40% of global waste is disposed of in landfills and about 19% of waste is recovered through recycling and composting. Textile waste is one of the important categories of waste and needs special attention. Textiles are the second most important thing for a human being to live and be protected from different weather conditions. However, with increasing population, the need for textiles is also increased. The developments in regenerated and synthetic fibres, new functional finishes and finishing techniques, range of new shades, stitching styles and fashion demands, developments and automation of textile manufacturing machines for higher production contributed in increased consumption of textiles (Wang, 2006). The average consumption of textiles per person has been increased from 7 kg in 1992 to 13 kg in 2013. As consumption of textiles increased, the corresponding textile waste also increased. It has been anticipated that total fashion waste in 2030 will be 148 million tonnes, which would be equivalent to an annual waste of 17.5 kg per capita across the planet. Also, more than 150 million tonnes of clothing would be landfilled or burned in 2050 (Ellen Macarthur Foundation, 2014). Textile wastes are broadly classified as pre-consumer waste and post-consumer waste. The

Z. Kamble (✉) · B. K. Behera
Department of Textile and Fibre Engineering, Indian Institute of Technology Delhi, New Delhi, Delhi, India
e-mail: ttz178482@textile.iitd.ac.in

A. S. Kalamdhad (ed.), *Integrated Approaches Towards Solid Waste Management*, https://doi.org/10.1007/978-3-030-70463-6_18

pre-consumer wastes are generated during textile manufacturing and which includes fibre waste during spinning, fabric wastes generated at different stages of manufacturing etc. The post-consumer wastes are those who have served their useful life (Wang, 2006).

Variety of products such as sound insulation material (Seddeq et al., 2013), thermal insulation material (Hassanin et al., 2018; Wazna et al., 2019), nonwoven fabrics as agricultural mulching (Abidi et al., 2019), yarns, fabrics to develop school blazer fabrics, drapes, doormats, prayer rugs, blankets, bedsheets etc. (Handique, 2010), can be developed by using textile wastes. Textile waste reinforced composite is one of the unique products to efficiently realise the true potential of waste textiles (Baillie et al., 2011; Meng et al., 2020). Further, the mechanical properties of the epoxy composites get improved upon the addition of the graphite oxide as a filler (Liu et al., 2012; Tang et al., 2013; Wan et al., 2014; Wei et al., 2015; Tshai et al., 2016). Rafiee et al. (2009) report a 40% increase in tensile strength of epoxy nanocomposites at 0.1 wt% of graphene platelets loading. However, the textile waste reinforced thermoset composites filled with graphite oxide filler are not reported much. Furthermore, textile waste reinforced composites can be used for applications such as items of furniture materials, automotive components etc. Thus, the cotton textile waste reinforced composites help to reduce the tree cuttings and virgin materials consumption. Therefore, this study is focused on the development of composite materials reinforced with cotton textile wastes and filled with rGO nanoparticles. The present research reports extraction of cotton shoddy from waste textiles, production of carded web and then developing thermoset composites using the carded web as reinforcement material. No other study than authors previous study (Kamble and Behera, 2020) reports a similar type of composites development process as per the best of author's knowledge. Further, this study reported the addition of rGO content on the weight of fibres in the composites. Whereas many other studies reported the addition of rGO content on the total weight or total volume of the composite. Further, these composites were characterised by mechanical properties such as tensile strength and modulus, flexural strength and modulus, izod impact strength and thermogravimetric analysis.

2 Materials and Methods

2.1 Materials

The cotton shoddy was used as reinforcement material for composite making. The cotton shoddy is a fibrous form obtained by tearing away cotton fabric waste by using the rag-tearing machine. The rag-tearing machine consists of rollers clothed with fine pins or wires which cuts and open the fabric pieces to fibre form. The cotton waste textiles are firstly cut into small pieces and then fed to the rag-tearing machine. The cotton shoddy obtained after the rag-tearing process and its microscopic view are shown in Fig. 1. A thermoset matrix material epoxy resin

(CHS-EPOXY 520) and hardener (Telalit 0492) having a mixing ratio of 100:32 by weight as specified by the supplier, was purchased from SPOLCHEMIE, Czech Republic. The reduced graphite oxide (rGO) was purchased from TECHINSTRO, India and it was used as supplied. The particle size of rGO particles sonicated in hardener was measured by using particles size analyser. It has been observed that 90% of the rGO particles have an average size of 445 nm. The thickness of rGO particles is 8-10 nm.

2.2 *Development of Thermoset Nanocomposites*

2.2.1 Preparation of Carded Web

The shoddy was processed on roller carding machine at 200 rpm of the cylinder to produce fibre web. The carding speed was optimised by the authors in their earlier study to get maximum tensile strength (Kamble and Behera, 2020). The single-layer of fibre web, produced by using the carding machine as shown in Fig. 2(a), has an average areal density of 27 grams per square meter. The web areal density is measured by using ASTM D3776-96 standards. The fibre web delivered by a carding

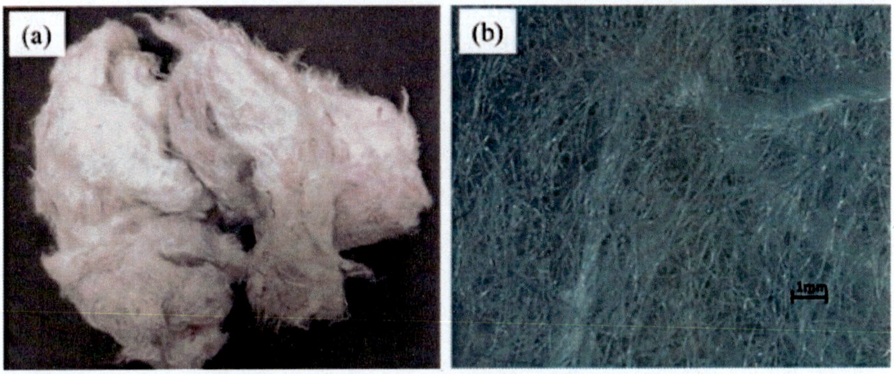

Fig. 1. Cotton shoddy obtained after rag tearing (a) and its microscopic view (b).

Fig. 2. A piece of a single layer of the fibre web (a), multi-layered carded web (b) and its cross-sectional view (c).

machine was subsequently wound onto a wooden roller to get a multilayer web. The top and the cross-sectional view of the multi-layered carded web is shown in Fig. 2(b) and (c) respectively. The multilayer web was used as a preform for thermoset composite specimen production.

2.2.2 Development of Thermoset Nanocomposite Specimens

The composite development process as shown in Fig. 3 is divided into three stages namely carded web preparation, rGO particles exfoliation into the epoxy resin and nanocomposite development. The stainless steel mould, having dimension 25 cm × 20 cm × 0.3 cm was fabricated. According to the mould's dimension, the multi-layer fibre web was first cut. The web of desired weight as per fibre volume fraction desired in the composite was taken. Here, it must be noted that the carded web areal density is independent of composite fibre volume fraction. The weighted multi-layer web was then divided into six parts having an equal number of layers. To incorporate the rGO nanoparticles into the composite, a weighed quantity of rGO nanoparticles was first sonicated in hardener for one hour. The hardener was then mixed with epoxy resin as shown in Fig. 4. The resin and hardener were further mixed by hand stirring and then degassed to alleviate any air bubbles before applying it on to the fibre web. The composites were produced by using compression moulding technique and the composites making process was same as explained above. The thermoset nanocomposites with four different rGO contents namely 0.1,

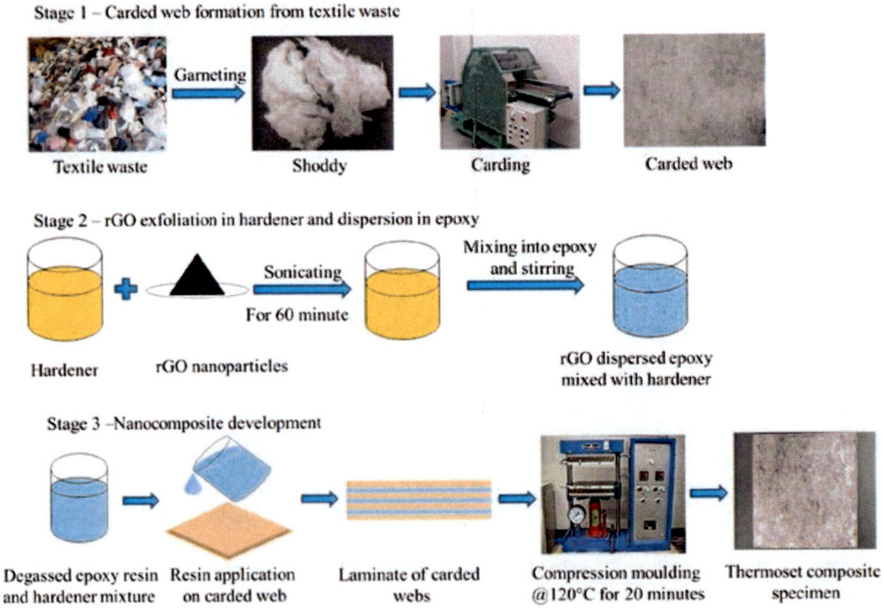

Fig. 3. Cotton shoddy reinforced epoxy nanocomposite development process.

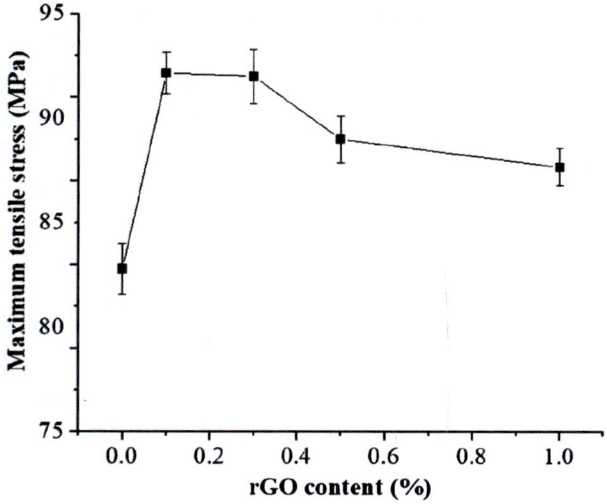

Fig. 4. Maximum tensile stress of composites with different rGO contents.

0.3, 0.5 and 1.0% on the weight of fibres in the composite were developed. The fibre volume fraction and thickness of all nanocomposites were 0.3 and 3 mm respectively.

2.3 Characterisation of Composite Specimens

2.3.1 Characterisation of Mechanical Properties

The composite specimens were characterised by tensile, three-point bending and notched izod impact strength properties. These tests were performed to assess the suitability of these composite materials for applications such as items of furniture materials and automotive components etc. The tensile properties of composite specimens were characterised according to ASTM D3039/ASTM D3039M standards (ASTM International 2017). The composite specimen for a tensile strength test has dimension 250 mm × 25 mm × 3 mm, which was tested on Autograph universal strength tester. The maximum tensile stress and Young's modulus were determined. The composite specimens were characterised by flexural strength according to ASTM D7264/ASTM D7264M-15 (ASTM International 2007) standards in a three-point bending mode on Autograph universal strength tester. The standard span-to-thickness ratio of 32:1 was chosen. Therefore, the span length was 96 mm and the dimensions of the flexural test specimen were 115 mm × 13 mm × 3 mm. The test speed was maintained at 2 mm/min. The maximum flexural stress was determined. Further, the composite specimens were characterised by notched izod impact strength test according to ISO 180:2000 standards (Indian standards 2018). The tests were performed on Toyoseiki digital impact tester with specimen dimensions 80 mm × 10 mm × 3 mm. The five specimens each for tensile, flexural strength and notched izod impact strength are tested to determine the average value.

2.3.2 Thermogravimetric Analysis (TGA)

The thermogravimetric analysis of composite specimens was performed on the thermogravimetric analyser to understand the thermal response of composites. The composite specimens of weight 6–9 mg were used for the test. The tests were performed for the temperature range of room temperature to 600°C at a constant heating rate of 20°C per minute in a nitrogen atmosphere.

2.3.3 Scanning Electron Microscopy (SEM)

The tensile and izod impact fractured surfaces of the composite specimens were analysed for fibre and matrix interface and the failure modes of the composites by using a Zeiss EVO18 scanning electron microscope at 20 kV acceleration voltage.

3 Results and Discussion

Effect of rGO content on mechanical properties of the nanocomposites

3.1 Tensile Test

It has been observed that the addition of rGO nanoparticles improves the maximum tensile stress of the composites. Rafiee et al. (Rafiee et al., 2009) reported a crack deflection mechanism of tensile strength improvement in epoxy composites reinforced with graphene platelets. The crack deflection is the process in which the initial crack gets tilted or twisted due to the presence of nanoparticles during tensile loading. However, the crack deflection mechanism alone is not justifiable when there is an interphase zone, fibres and chemical bonding between filler and matrix are also present in the composite. Monteserín et al. (2017) reported that the functional groups present on the rGO surface enhance its uniform dispersion within the matrix. The chemical interaction of these functional groups with epoxy resin as shown in Fig. 5, increases the interphase strength. Therefore, the rGO nanoparticles along with the cotton fibres bear the load transferred by a matrix which in turn enhances the mechanical properties of the nanocomposite. As shown in Fig. 4, the maximum tensile stress for composite without rGO and with rGO content of 0.1, 0.3, 0.5 and 1% was 82.79, 92.14, 92, 89 and 87.67 MPa respectively. With an increase in rGO content from 0 to 0.1, 0.1 to 0.3, 0.3 to 0.5 and 0.5 to 1%, the maximum tensile stress on the composite increases by 11.3, 11.1, 7.5 and 5.8% respectively. The decrease in the improvement of maximum tensile stress with an increase in rGO content is due to agglomeration rGO particles. The exfoliated and agglomerated rGO sheets observed in TEM image are shown in Fig. 6. This agglomeration of

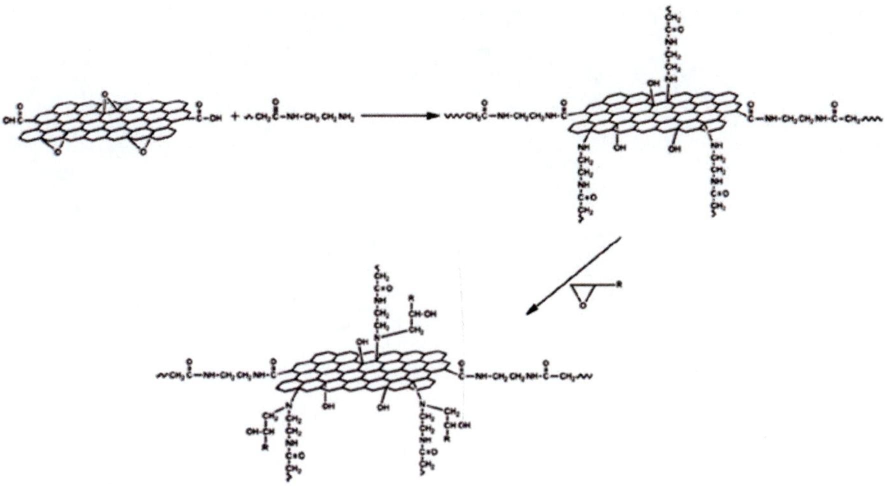

Fig. 5. Proposed curing mechanism of GO- and rGO-modified epoxy systems (Monteserín et al., 2017).

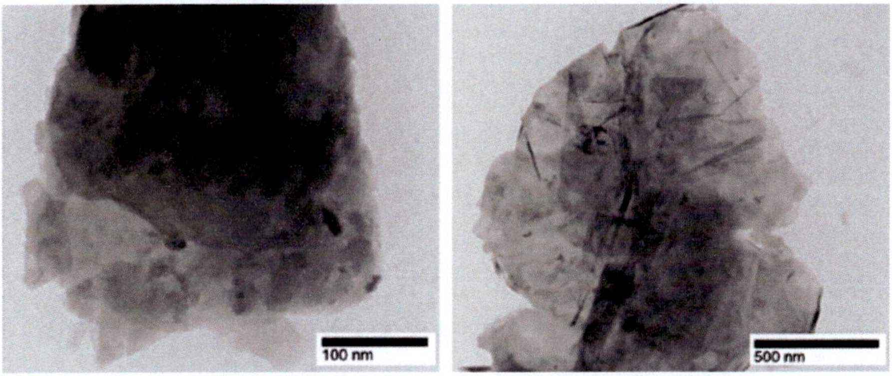

Fig. 6. TEM image of agglomerated rGO sheets (a) and exfoliated rGO sheets (b).

rGO nanoparticles results into decrease in interface area and it also acts as stress concentration point (Tshai et al., 2016). The tensile modulus of the composites without filler and filled with 0.1, 0.3, 0.5 and 1% rGO were 3.39, 3.87, 3.86, 3.30 and 3.1 GPa respectively. Further, the scanning electron microscope images of tensile fractured specimens are shown in Fig. 7 and which indicates that the tensile failures of the nanocomposite are mainly due to fibre matrix de-bonding, fibre pull-out and fibre breakage.

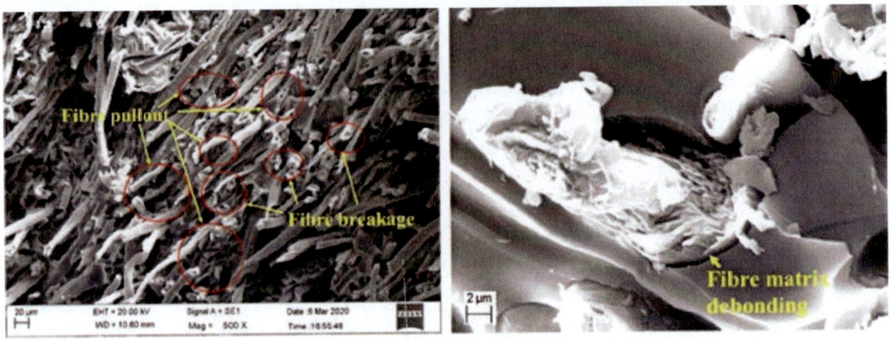

Fig. 7. Scanning electron microscope image of tensile fractured specimen.

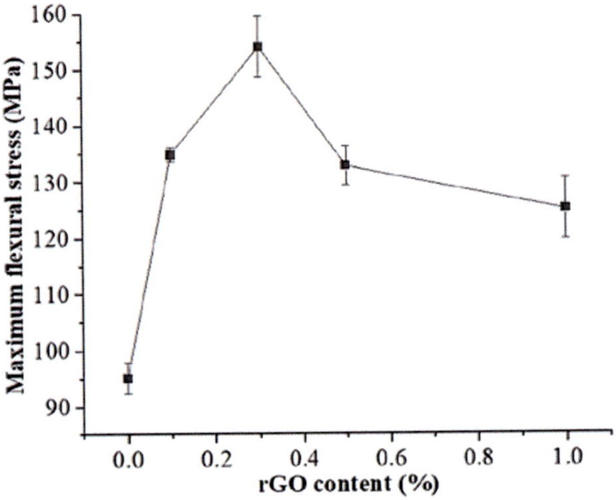

Fig. 8. Maximum flexural stress of composites with different rGO contents.

3.2 *Flexural Strength*

As shown in Fig. 8, with an increase in rGO content, increases from 0 to 0.3% the maximum flexural stress increases from 95.2 MPa to 154.02 MPa respectively and then decreases for 0.5 and 1% rGO content. The maximum flexural stress for composite with 0.1, 0.5 and 1% rGO content were 134.79, 132.75 and 125.06 MPa respectively. With an increase in rGO content from 0 to 0.1, 0.1 to 0.3, 0.3 to 0.5 and 0.5 to 1%, the maximum flexural strength of the composite increases by 41.6, 61.8, 39.3 and 31.3% respectively. The addition of rGO nanoparticles leads to an increased matrix interface, which results in higher load transfer from the matrix to the reinforcement. However, due to an increase in the agglomeration of rGO nanoparticles for 0.5 and 1% content, the interface area of rGO with the epoxy matrix decreases and thus the flexural stress decreases (Tshai et al., 2016).

3.3 *Izod Impact Strength*

As shown in Fig. 9, as rGO content increases from 0 to 0.3% the izod impact strength increases from 10.13 to 13.88 kJ/m^2 respectively. The izod impact strength was decreased for rGO content of 0.5% and 1%. This may be attributed to the agglomeration of rGO nanoparticles within the matrix. The izod impact strength for composite with 0.1, 0.5 and 1% rGO content were 11.16, 11.45 and 10.83 kJ/m^2 respectively. The improvement in izod impact strength with increase in rGO content from 0 to 0.1, 0.1 to 0.3, 0.3 to 0.5 and 0.5 to 1% is 10.2, 36.8, 13 and 6.9% respectively. The failure of the composites under impact load was majorly due to fibre matrix de-bonding, fibre pull-out and fibre breakage as shown in Fig. 10. The

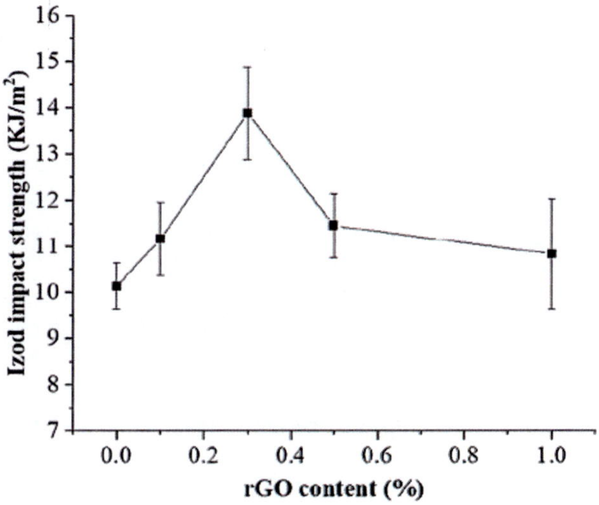

Fig. 9. Izod impact strength of composites with different rGO.

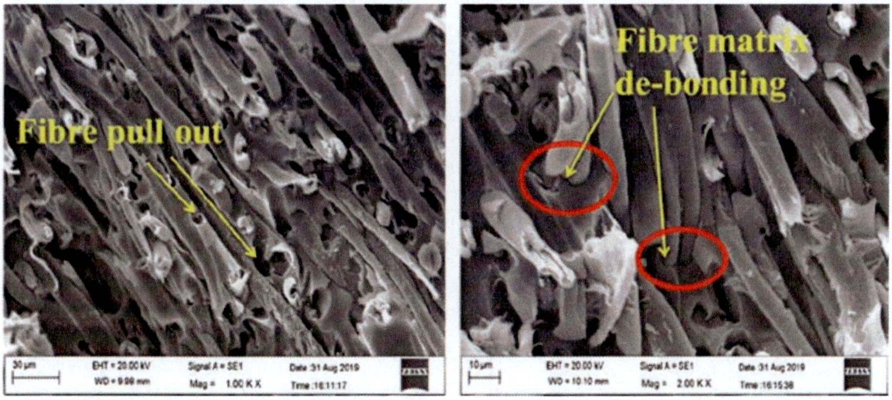

Fig. 10. Scanning electron microscope image of izod impact fractured specimen.

mechanical properties of the composites without filler and filled with rGO nanoparticles were found comparable to that of industrial particle board and medium-density fibreboard products (Ross and USDA Forest Service, 2010) and rubberwood epoxy composites (Prakash et al., 2019). A composite to be used as automotive dashboard panel should have density, Young's modulus and a tensile strength greater than 1.18 gm/cm^3, 2.3 GPa and 25 MPa respectively (Sapuan et al., 2011). The composites developed in the present research shows mechanical properties higher than required for an automotive dashboard panel and therefore these composites are potential materials for automotive dashboard panel manufacturing. Furthermore, typical values of the mechanical properties for automotive applications (Müssig, 2008) and values obtained from the developed composites are shown in Table 1.

3.4 Thermogravimetric Analysis (TGA) of Composites Filled with Different Graphite Oxide Content

As shown in Fig. 11, the composites filled with rGO and without rGO shows a nearly similar trend of weight loss. The onset temperatures for composite without rGO and filled with 0.1, 0.3, 0.5 and 1% rGO were 319, 319, 316, 321 and 281°C respectively. The half weight loss temperature of composites without rGO and filled with 0.1, 0.3, 0.5 and 1.0% rGO were 379, 375, 376, 383 and 372°C. The half weight loss temperature decreases with an increase in rGO content. This is attributed to the fact that the rGO nanoparticles have low specific heat and high thermal conductivity as compared to cotton fibres and epoxy matrix (Wan and Zarins, 2003; Bastiurea et al., 2015), as a result, the heat absorbed by rGO nanoparticles causes

Table 1. Comparison of mechanical properties of composites for automotive applications

	Unit	*Typical values of natural fibre composites used in the automotive industry*	*Cotton shoddy reinforced epoxy composites*	*Cotton shoddy reinforced epoxy composites filled with 0.3 wt% of rGO*
Processing technique		Hot pressing	Hot pressing	Hot pressing
Natural fibre content	Mass%	65 (Bast fibre)	38 (Extracted waste cotton fibres)	38 (Extracted waste cotton fibres)
Polymer	Mass%	35 (epoxy)	62 (epoxy)	62 (epoxy)
Density	g/cm^3	0.8–0.85	1.23	1.23
Tensile strength	MPa	40–50	82.79	92
Flexural strength	MPa	50–70	95.2	154
Impact strength	mJ/mm^2	14–20	10.13	13.88

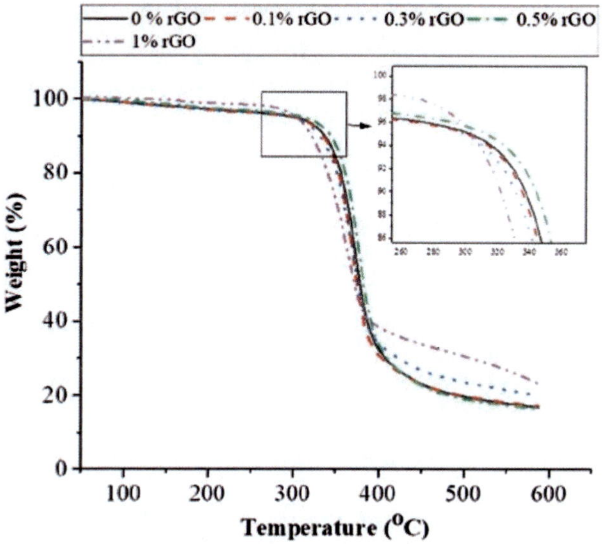

Fig. 11. Thermogravimetric analysis of nanocomposites reinforced with different rGO content.

degradation of fibres and epoxy matrix surrounded by it. The maximum weight loss rate for all composites was observed between temperatures 250°C to 450°C. This is attributed to thermal decomposition of cellulose in the fibres which occurs between temperatures 255°C and 340°C (Silva et al., 2012). The half weight loss temperature of the composite with 0.5% rGO is higher than others, which may be due to uneven distribution rGO within the composite. The thermogravimetric analysis of composite specimens confirms that the composites developed are enough thermally stable.

4 Conclusions

The cotton fibres extracted from textile waste can be successfully employed to reinforce the composite materials. The addition of rGO nanoparticles improves the mechanical properties of the composites. The composites developed in the present research shows required mechanical properties to be used in the items of furniture material and to develop some visible and non-visible automotive components. The composites developed with and without rGO nanoparticles are thermally enough stable. The improvement in even distribution of rGO nanoparticles within the composite specimen will further improve the mechanical properties of the composites. Furthermore, the techno-economic study of the developed composites for proposed applications is required to promote these composites in the market.

Declaration of Conflicting Interests The author(s) declared no potential conflicts of interest with respect to the research, authorship, and/or publication of this article.

Funding This research did not receive any specific grant from funding agencies in the public, commercial, or not-for-profit sectors.

References

Abidi, H., Rana, S., Chaouch, W., Azouz, B., Aissa, I.B., Hassen, M.B. and Fangueiro, R. (2019). Accelerated weathering of textile waste nonwovens used as sustainable agricultural mulching. *J Ind Text*, doi: https://doi.org/10.1177/1528083719855326.

ASTM International (2017). ASTM-D3039/D3039M-17 standard test method for tensile properties of polymer matrix composite materials.

ASTM International (2007). ASTM-D7264/D7264M-07 standard test method for flexural properties of polymer matrix composite materials. *Annu B ASTM Stand*, **1–11**. doi: https://doi.org/10.1520/D7264.

Baillie, C., Matovic, D., Thamae, T. and Vaja, S. (2011). Waste-based composites - Poverty reducing solutions to environmental problems. *Resour Conserv Recycl*, doi: https://doi.org/10.1016/j.resconrec.2011.05.006.

Bastiurea, M., Rodeanu, M.S., Andrei, G., Dima, D. and Cantaragiu, A. (2015). Correlation between graphene oxide/graphite content and thermal properties of polyester composites. *Dig J Nanomater Biostructures*, **110:** 1569–1579.

Ellen Macarthur Foundation (2014). Towards the circular economy : Accelerating the scale-up across global supply chains.

Handique, M. (2010). https://www.livemint.com/Companies/x9tnCSRrdT3RSn3B XfRtON/Old-clothes-spin-a-new-yarn-in-India.html.

Hassanin, A.H., Candan, Z., Demirkir, C. and Hamouda, T. (2018). Thermal insulation properties of hybrid textile reinforced biocomposites from food packaging waste. *J Ind Text*, doi: https://doi.org/10.1177/1528083716657820.

Indian standards (2018). Plastics – Method of testing; Part 5: Mechanical properties; Section 4: Determination of izod impact strength.

Kamble, Z. and Behera, B.K. (2020). Mechanical properties and water absorption characteristics of composites reinforced with cotton fibres recovered from textile waste. *J Eng Fiber Fabr*, **15:** 1–8. doi: https://doi.org/10.1177/1558925020901530.

Kaza, S., Yao, L., Bhada-Tata, P. and Van, W.F. (2018). What a waste 2.0: A global review of solid waste management to 2050.

Liu, Q., Zhou, X., Fan, X., Zhu, C., Yao, X., Liu, Q., Zhou, X., Fan, X., Zhu, C. and Yao, X. (2012). Mechanical and thermal properties of epoxy resin nanocomposites reinforced with graphene oxide. *Polym Plast Technol Eng*, **51:** 251–256. doi: https://doi.org/10.1080/03602559.2011.625381.

Meng, X., Fan, W., Ma, Y., Wei, T., Dou, H., Yang, X., Tian, H., Yu, Y., Zhang, T. and Gao, L. (2020). Recycling of denim fabric wastes into high-performance composites using the needle-punching nonwoven fabrication route. *Text Res J*, **90:** 695–709. doi: https://doi.org/10.1177/0040517519870317.

Monteserín, C., Blanco, M., Aranzabe, E., Aranzabe, A., Laza, J.M., Larrañaga-Varga, A. and Vilas, J.L. (2017). Effects of graphene oxide and chemically-reduced graphene oxide on the dynamic mechanical properties of epoxy amine composites. *Polymers (Basel)*, **9:** 1–16. doi: https://doi.org/10.3390/polym9090449.

Müssig, J. (2008). Cotton fibre-reinforced thermosets versus ramie composites: A comparative study using petrochemical- and agro-based resins. *J Polym Environ*, **16:** 94–102. doi: https://doi.org/10.1007/s10924-008-0089-4.

Prakash, V., Bera, T. and Acharya, S.K. (2019). Mechanical and erosive wear behavior of rubber wood particulate reinforced epoxy composite. *Mater Today Proc*, doi: https://doi.org/10.1016/j.matpr.2019.06.708.

Rafiee, M.A., Rafiee, J., Wang, Z., Song, H., Yu, Z. and Koratkar, N. (2009). Enhanced mechanical properties of nanocomposites at low graphene content. *Am Chem Soc*, **3:** 3884–3890.

Ross, R.J., USDA Forest Service. FPL (2010). *Wood Handbook : Wood As An Engineering Material.*

Sapuan, S.M., Kho, J.Y., Zainudin, E.S., Leman, Z., Ahmed Ali, B.A. and Hambali, A. (2011). Materials selection for natural fiber reinforced polymer composites using analytical hierarchy process. *Indian J Eng Mater Sci*, **18:** 255–267.

Seddeq, H.S., Aly, N.M., Marwa, A.A. and Elshakankery, M.H. (2013). Investigation on sound absorption properties for recycled fibrous materials. *J Ind Text.* doi: https://doi.org/10.1177/1528083712446956.

Silva, C.G., Benaducci, D. and Frollini, E. (2012). Lyocell and cotton fibers as reinforcements for a thermoset polymer. *BioResources*, **7:** 78–98.

Tang, L.C., Wan, Y.J., Yan, D., Pei, Y.B., Zhao, L., Li, Y.B., Wu, L.B., Jiang, J.X. and Lai, G.Q. (2013). The effect of graphene dispersion on the mechanical properties of graphene/epoxy composites. *Carbon N Y.* doi: https://doi.org/10.1016/j.carbon.2013.03.050.

Tshai, K.Y., Khalili, P., Kong, I., Yeoh, C.H. and Tshai, K.H. (2016). Synthesization of graphene and its incorporation into natural fiber reinforced thermosetting nanocomposite. *ARPN J Eng Appl Sci*, **11:** 121–127.

Wan, P.J. and Zarins, Z.M. (2003). Specific heats of cottonseed and its co-products. *J Am Oil Chem Soc*, **80:** 123–126. doi: https://doi.org/10.1007/s11746-003-0663-7.

Wan, Y.J., Tang, L.C., Gong, L.X., Yan, D., Li, Y.B., Wu, L.B., Jiang, J.X. and Lai, G.Q. (2014). Grafting of epoxy chains onto graphene oxide for epoxy composites with improved mechanical and thermal properties. *Carbon N Y*, **69:** 467–480. doi: https://doi.org/10.1016/j.carbon.2013.12.050.

Wang, Y. (2006). *Recycling in Textiles*. Woodhead Publ. Limited, England, pp. 1–230.

Wazna, M.E.L., Gounni, A., Bouari, A.E.L., Alami, M.E.L. and Cherkaoui, O. (2019). Development, characterization and thermal performance of insulating nonwoven fabrics made from textile waste. *J Ind Text*. doi: https://doi.org/10.1177/1528083718757526.

Wei, J., Atif, R., Vo, T. and Inam, F. (2015). Graphene nanoplatelets in epoxy system: Dispersion, reaggregation, and mechanical properties of nanocomposites. *J Nanomater*, **1–12**. doi: https://doi.org/10.1155/2015/561742.

Chapter 19
Reduction of Significant Aspects and Enhancement of Non-Significant Aspects for Hazardous Wastes in a Medical Electronics Manufacturing Firm

M. Aishwarya, Rajesh Gopinath, L. R. Phanindra, K. Clarina, Rashmi R. Kagawad, and S. G. Ananya

1 Introduction

Urbanisation in its unplanned inorganic rampant form has led to more perils than boon for the natural environment and civilisation at large (Gopinath et al., 2017; Vrinda and Gopinath, 2018). In this context, one among the most serious global consequences faced is the repercussions of untreated discharge of hazardous materials onto the natural environment (Nandini et al., 2012). Any substance which possesses either of the characteristics such as flammable, explosive, reactive, corrosive, toxic, radio-active and ignitable is termed as hazardous (Ramya et al., 2014). Certain urban day-today examples for these include heavy metal laden paints, pesticides, insecticides, weedicides, motor oils, batteries, cleaning and polishing solvents etc. (Venugopal and Gopinath, 2012).

Improper management of these at residential and industrial sectors is hence a matter of major concern (Balasubramanya and Gopinath, 2018). Off late, there has been an increase in awareness amidst manufacturing firms towards reducing hazardous wastes produced within its premises as an outcome of its operations (Gopinath et al., 2014).

On a corporate level, this effort is incepted by the implementation of an 'EMS', abbreviated as Environmental Management System (Mohamed et al., 2015). It may be elucidated as a set of procedures, practices, techniques and methods that allows an establishment to minimise its environmental impacts and magnify its efficiency of operation. ISO 14001 specifies the standard for the establishment and maintenance of an EMS. The basic principles of ISO 14001 are based on the well-known

M. Aishwarya · R. Gopinath (✉) · K. Clarina · R. R. Kagawad · S. G. Ananya
Department of Civil Engineering, BMS Institute of Technology and Management, Bengaluru, Karnataka, India

L. R. Phanindra
Wipro GE Healthcare Pvt. Ltd., Bengaluru, Karnataka, India

A. S. Kalamdhad (ed.), *Integrated Approaches Towards Solid Waste Management*, https://doi.org/10.1007/978-3-030-70463-6_19

Plan-Do-Check-Act (PDCA) cycle, which had been developed by Edwards Deming and is depicted in Fig. 1 (Deming, 1986).

As expressed in the official documentation for Environmental Management Systems of ISO 14001:2015; Identification and Evaluation of Environmental Aspects and their Environmental Impacts form the most significant component. An Environmental Aspect is defined as any constituent element of an establishment's procedures, products, protocols or services that can interact with its surrounding environment.

An Environmental Impact is defined as an alteration to the environment, which may be direct or indirect, affecting the environment either wholly or partially, whether adverse or beneficial, as a resultant of an establishment's products, protocols, processes, activities or services (ISO 14001, 2015). While few industries have envisaged EMS as decision-making tool to ascertain its viability as profit or loss component (Kevin et al., 2014); most others have employed it to reduce the environmental costs (Goh et al., 2006) and environmental damages (Snezana et al., 2013). Further, researches have also pointed that although implementation of EMS strategies may incur initial additional expenditures; it however over a period enhances monetary benefits and competitive edge (Kevin et al., 2014).

Mostly the researches encompassed have usually restricted their discussions to determination of environmental aspects and their subsequent impact on the environment (Davor and Goran, 2011); and have not delved upon solutions or recommendations (Sc Alan et al., 2013). Also, no study across globe to-date has been carried out to address the implementation of EMS or its viability for a medical-electronics firm, which especially deals extensively with hazardous, radioactive and e-wastes; each of which has negative bearing upon a growing city like Bengaluru, which caters to both manufacturers, consumers and recyclers (Gopinath and Vinod, 2020).

The present study hence attempts to plug these gaps in literature, and also provide holistic sustainable corrective and preventive remedial measures. Of late, there

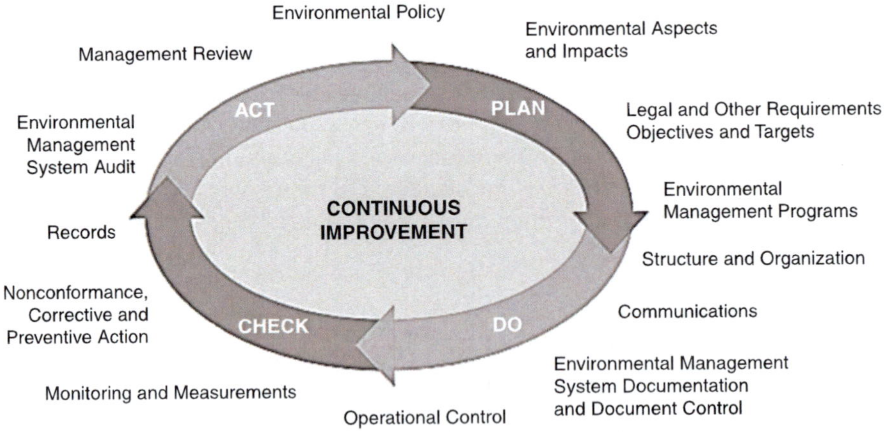

Fig. 1. Plan-Do-Check-Act cycle (Deming, 1986).

has been an increase in awareness amidst manufacturing firms towards reducing hazardous wastes produced within its premises as an outcome of its operations (Gopinath et al., 2014). On a corporate level, this effort is incepted by the implementation of an 'EMS', abbreviated as Environmental Management System (Mohamed et al., 2015). While few industries have envisaged EMS as decision-making tool to ascertain its viability as profit or loss component (Kevin et al., 2014); most others have employed it to reduce the environmental costs (Goh et al., 2006) and environmental damages (Snezana et al., 2013).

Further, researches have also pointed that although implementation of EMS strategies may incur initial additional expenditures; it however over a period enhances monetary benefits and competitive edge (Kevin et al., 2014). Mostly the researches encompassed have usually restricted their discussions to determination of environmental aspects and their subsequent impact on the environment (Davor and Goran, 2011); and have not delved upon solutions or recommendations (Sc Alan et al., 2013).

Also, no study across globe to-date has been carried out to address the implementation of EMS or its viability for a medical-electronics firm, which especially deals extensively with hazardous, radioactive and e-wastes; each of which has negative bearing upon a growing city like Bengaluru which caters to both manufacturers, consumers and recyclers (Gopinath and Vinod, 2020). The present study hence attempts to plug these gaps in literature, and also provides holistic sustainable corrective and preventive remedial measures.

2 Materials and Methods

The present study has been carried out during 2018 in the Bengaluru based 'X-ray division of Wipro GE Healthcare Pvt. Ltd.'. The industry essentially manufactures X-ray machines, CT machines and MR systems for imaging modalities that are used in medical imaging procedures; as detailed and showcased in the company website at https://www.gehealthcare.in. The scope of the research is to mainly enhance the effectiveness of the existing EMS viz. identification of the non-conformities and unidentified aspects that are a resultant of the existing processes and also due to procedures adopted based on outcome of previous environmental audits. Further, the study strengthens the prevalent EMS by delineating the Significant Aspects. Finally, it delves upon recommendation of corrective/preventive measures and their viability.

The methodology adopted for the conduction of this study is based on the Plan-Do-Check-Act (PDCA) framework (Deming, 1986). In this study, the methods of materials identification, compliance with legal requirements and process flowchart method along with the Life Cycle Assessment have been used to identify Environmental Aspects and their Impacts. The methodology adopted comprises of the sequential steps of Extensive Literature Survey, Reconnaissance Survey, Intensive Data Collection, Data Analysis and Interpretation, and Recommendation of Corrective and Preventive measures to reduce Significant Aspects to Non-Significant ones. Keeping this in view, the PDCA framework has accordingly been modified to suit the study. The same has been represented in Chart I and elaborated further on. To attain the objectives of study, an extensive literature survey was carried out which involved the review of relevant literatures and research articles pertaining to the project carried out at international, national and regional levels in order to understand the process and gaps. A Reconnaissance Survey was carried out to understand the procedures, protocols and safety aspects across the company. Further company's environmental policy, objectives, targets, previous EMS documents were reviewed thoroughly.

Processes across each area were studied and flowcharts developed to identify all the various possible aspects, namely emissions to air; release to water and land; use of natural resources and energy and release of energy; waste and by-products (Laarifi et al., 2017).

Also, a structured survey was carried out with professionals at the management level, supervisors, engineers, operators and workers on-site, to collect direct and indirect pertinent data. Information from the surveys and process flowcharts were then subjected to SWOT Analysis, for determination of all possible aspects.

For each product, service or activity, all actual, potential, positive and negative impacts were consequently identified (Davor and Goran, 2011). These impacts were then assigned scores based on (i) Severity of Impact, (ii) Scale of Impact, (iii) Frequency or Likelihood of the associated Environmental Impacts and (iv) Probability of Occurrence of the activity. These scores were then multiplied for each indicator across the relevant life cycle stages.

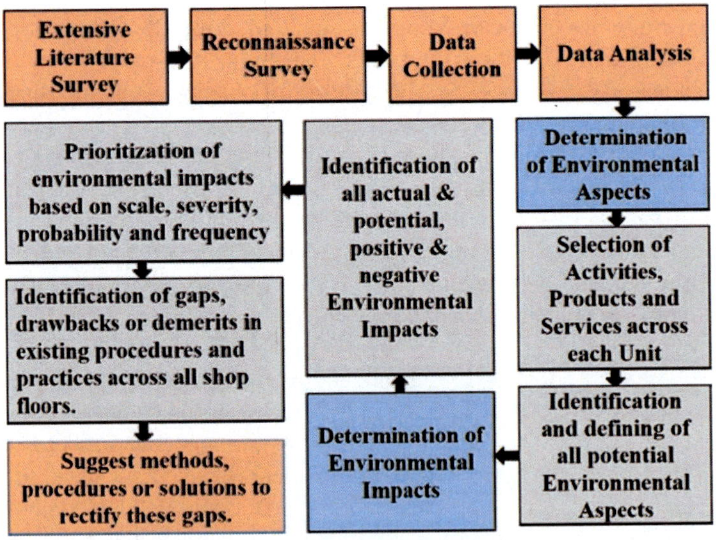

Chart I. Research methodology.

The total score of an indicator was finally categorised as (i) positive (or low) impact on the environment (for a score of 8 to 16); (ii) non-significant (or moderate) impact on the environment (for a score of 18 to 46) and (iii) significant (or severe) impact on the environment (for a score greater than 48). Further environmental aspects and their respective impacts were hence identified and rated across all units. Ratings of the pre-identified aspects were also re-reviewed and revised. Some aspects that were irrelevant were eliminated (Dagmara and Thomas, 2009). These revisions were then made to the existing Aspect Impact registers of the respective unit. Also, along with this, the existing processes and procedures were closely studied in order to identify the gaps, demerits or drawbacks persistent in them across all shop floors and units. Accordingly, recommendations were made, which were inclusive of ecofriendly practices, non-conventional practices, innovative concepts and modern technologies that could be adopted and practiced feasibly and economically by the industry in order to strengthen its environmental performance and thereby improve their Environmental Management System in the due course. The recommendations proposed were with respect to ecology and environment, waste management, energy, water conservation, wastewater reuse etc. The further portions of paper detail the outcome of the Aspect Impact Studies and Recommendations.

3 Results and Discussion

For the study, the industry was divided into three areas, namely Administration, Facility and Assembly-Testing areas. It was observed that the firm was practicing EMS with ISO 14001-2015, with regular monitoring by the Environmental Health and Safety (EHS) department. From the Structured Survey and SWOT Analysis, 107 'new' aspects were identified. For example, in the facility area it was identified that there was an excess loading on the STP due to the back washing process of the filter press adding onto quantity of the wastewater. This was a new aspect in the Aspect Impact Register as the filter press was introduced after the previous risk assessment was carried out.

Also, 28 of the existing aspects that were not relevant were eliminated. For example, in the assembly area, emission of lead fumes during soldering was an aspect that was deleted as the lead-based solder usage was shifted to a lead-free solder. Also, the ratings of 53 existing aspects were revised. For example, earlier the consumption of toner in the admin area was rated as 16 with scores of 2 for scale, Severity, Probability of Occurrence and Duration.

However, with the incorporation of digital documentation in the admin area, the rate of printing has considerably increased in comparison to the previous year. Thereby, the rating for occurrence of consumption of toner when increased with score of 4 for probability of occurrence, its rating was revised to 32. Upon reviewing a total of 385 aspects were updated in the Aspect Impact Register from the study, which is significant considering the fact that the total count before conduction of the study was 306. Chart II unveils the comparison in the total count of aspects 'before'

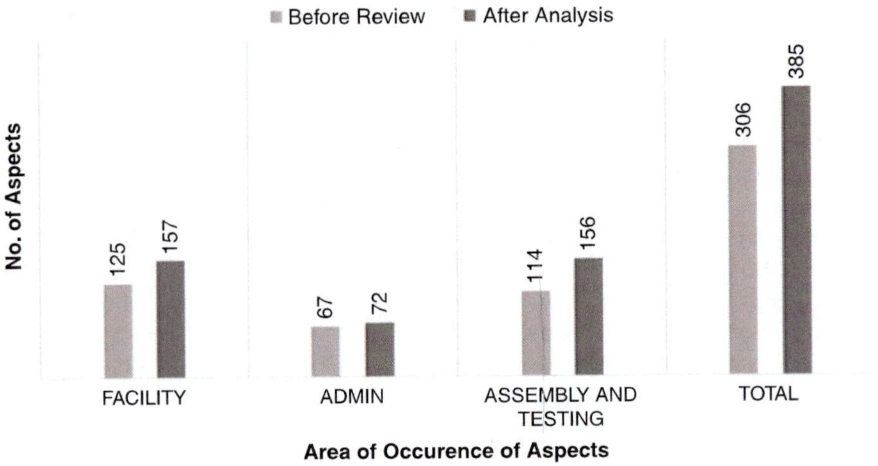

Chart II. Comparison of aspect count 'before' and 'after' study.

(provided from documentation) and 'after' (outcome of present research) across the three areas in the entire firm. Distinguishable detection of newer aspects was made in Assembly-Testing and also the Facility areas. However, there was not much difference observed for the Admin area.

Out of the 385 aspects, 121 (i.e. 31.43%) were found to be Significant, i.e., aspects which are of immediate concern. The same have been further addressed in this research with recommendations to manage them. Fourteen of the total aspects (i.e. 3.63%) were with Positive Impacts and contributed well to the conservation of environment. For example, reuse of packaging material of individual parts such as foam boards and carton boxes in the packing of the assembled equipments leads to positive impact as it saves resources. The rest of the aspects with count of 250 (i.e. 64.93%) were deemed Non-Significant Aspects, which although may not be of immediate concern; however, may turn out to become major concern in the longer run as the stringency of environmental targets to be achieved increases.

For example, consumption of paper in the company was rated 16 with Severity, Scale of Impact and Probability of Occurrence, duration being rated 2. This was the lowest rated aspect, and is considered non-significant as it can cause due-damage to the environment if the rate and quantity of use of paper increases in the industry.

Chart III depicts the distribution of these aspects identified across each area. The count of non-significant aspects was highest across all the three areas with the numbers of 37, 109 and 104 for Administration, Facility and Assembly-Testing areas respectively. In comparison, the count for positive aspects was lowest with 8, 4 and 2 in Administration, Facility and Assembly-Testing areas respectively. Also, the number of significant aspects, (in total) were found to be as high as 50 in the Assembly-Testing area while they added up to 44 and 27 in the Facility and Administration areas respectively. The aspect which garnered the highest rating was Consumption of Power in the Facility area for the running of DG Sets, STP and

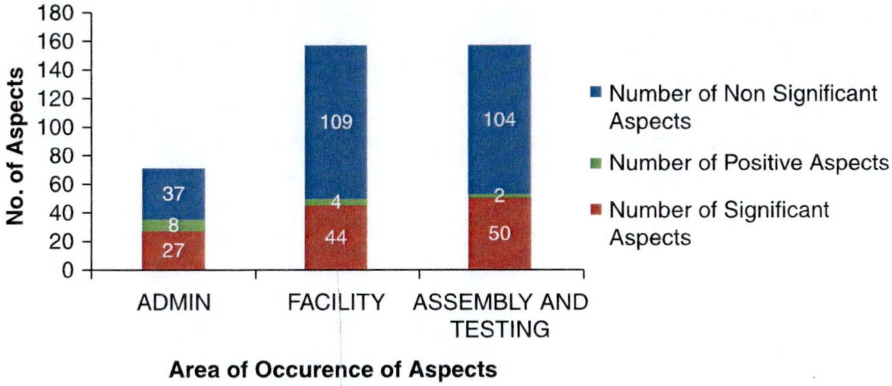

Chart III. Delineation of different types of aspects across all the three areas.

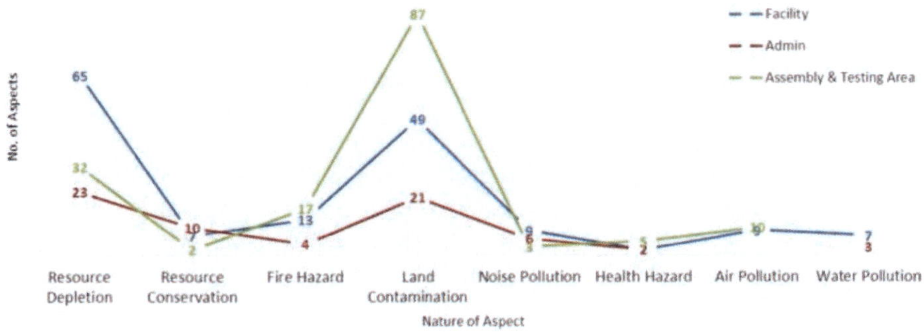

Chart IV. Delineation of aspects in the Administration, Facility and Assembly-Testing areas.

other utilities; with a rating of 144 (owing to scale and severity of impact as 3 and probability of occurrence and duration as 4). This aspect secured the highest rating as huge amount of power is consumed daily across the facility area for the smooth functioning of the entire industry, thereby making it a significant aspect leading to resource depletion. All these aspects were further deduced for their impact on environmental attributes such as natural resource depletion and conservation; pollution of land, air and water; hazards related to health and fire etc.

Chart IV divulges the aforementioned findings about all the three areas. It can be observed that among all areas, the maximum aspects leading to Land Contamination, was with contributions of 87, 49 and 21 from Assembly and Testing, Facility and Administration areas respectively. The prospect of land contamination is high due to disposal of oil-soaked cotton wastes, oil-filters, oil spillage while transferring it for storage and while lubrication from the DG set area.

In a similar manner, maximum resource depletion was found at Facility (65), followed by Assembly and Testing areas (32) and Administration (23). For instance, in the Administration area, consumption of new paper for printing, documentation and labelling operations contributed to vast natural resource depletion; in Assembly and Testing area it was due to wide usage of power, paper and chemicals like isopropyl alcohol (IPA) etc. Water pollution and air pollution was found not to exist in Assembly and Testing area, and Administration area respectively. The threats of fire hazard existed more in facility and especially Assembly and Testing area wherein there is a potential risk of fire during the testing process due to the possibility of static discharge.

Resource conservation was found to be least in Assembly and Testing area and quite commendable in Administration area. In the facility area where for instance the treated wastewater from STP was being reused for gardening purposes, led to enhanced conservation measures of raw water.

On the overall, it may be observed that while 41% of aspects resulted in land contamination, 31% resulted in resource depletion. On the lop-side, only about ½% of the aspects actually contributed to resource conservation, due to advent of application of motion sensing light and water sensors in the washrooms.

On the basis of these aforementioned results, varying strategic recommendations were made for the most Significant Aspects. However, within the scope of this paper, only that w.r.t. Hazardous Waste Management is discussed in this paper. From the reconnaissance survey, verification of records, visual inspections and survey; it was observed that the various types of hazardous wastes generated across the Administration, Facility and Assembly-testing areas were e-wastes, oil filters, oil-soaked cotton wastes, discarded used containers, adhesives and residues and lead-tin compounds. Among these, the most significant aspects with the recommendations have been discussed further, alongside their resultant outcome that had improved the environmental performance of the company.

Oil-soaked cotton waste generated in the lubrication of equipments in the Facility and Assembly and Testing areas, was found to cause major land contamination. The same was found to be disposed to a waste collection vendor approved by Karnataka State Pollution Control Board. The remedial measure provided for this was to adopt pelletisation of oil-soaked cotton waste, which involves converting them into fuel pellets. Upon implementation, it has helped in reducing the waste generation by 34.2% and the cost of disposal by 56%. During the manufacturing, oil is a major resource that is utilised for lubrication, and hence the production of used-oil is inevitable. The potential effect of leakage or improper storage of it on land thereby becomes a Significant Aspect. The quantity of used-oil had shown an increasing trend from 140 litres in 2016 to 550 litres in 2018 on account of filtering the transformer oil used in the testing area. The remedial measure suggested to combat this was to eliminate the filtration process in manufacturing and a changing over to a new 'Diesel Generator (DG) set', in utilities of higher efficiency to result in low used-oil generation.

Oil filters are also potentially hazardous to the environment if not disposed properly. The oil filter numbers had increased from 22 in 2017 to 25 in 2018 due to the frequent maintenance of the ageing diesel generator. Upon recommendation, a new energy efficient DG set was commissioned, which coincidentally had also resulted in reduced production of used-oil. Discarded containers of toxic cleaning agents such as IPA in the hazardous waste storage area were also a major concern. This was especially due to the fact that previously the cleaning of the entire assembled unit was being accomplished with such toxic chemicals. This was addressed by bringing about a change in the Life Cycle Process of the Gantry Covers that were assembled at site. Herein, adoption of touch-up for only scratches on the assembled units with chemical instead of cleaning the entire unit drastically reduced the consumption of chemicals that thereby reduced the quantity of chemical containers as well.

This process modification has led to a cost savings of 34% as the chemical acquisition also reduced along with the reduction in the cost of disposal of chemical containers. Also, the implementation upon recommendation to adopt reservoir printers in place of LaserJet or inkjet printers was successful in terms of saving paper, and especially ink, since these cartridges need not be disposed and merely required replacement of ink tanks. These ink tanks could print up to 7000–10,000 black and white pages and over 10,000 pages in colour in one refilling. Conventional cartridges could churn only upto 1000 pages. The company had 12 printers and the cartridges were being replaced once in 4 months. Therefore, with the adoption of reservoir printers, the waste generated was merely 33% of the original. Usage and disposal of printer ink, toner and cartridges is another Significant Aspect. Also, with the recommended adoption of password activated printing process, the company had saved 15% of paper waste from being generated, alongside conservation of the printing ink.

Upon implementation of aforementioned recommendations, significant reduction was observed in the generation of the hazardous wastes as presented in Table 1. The % reduction achieved for used-oil, oil-soaked cotton waste, oil filters and empty chemical containers were computed to be 59.1%, 34.2%, 60.0% and 41.1% respectively. Also, it was observed that the remedial measures had reduced the cost of handling the used-oil by about 36%.

4 Conclusions

The major outcome of this study has resulted in identification of 28 pre-existing aspects as irrelevant, and further the ratings of 53 pre-existing aspects being revised. Out of the total aspects, 121 (31.43%) were distinguished as significant, and 14 were proved to have a positive impact on the environment. The study recommends

Table 1. Variation of hazardous waste generation over the years

	Rate of waste generation		*Before the study*			*After the study*
Sl. no.	*Type of waste (unit)*	*Quantity*	*2016*	*2017*	*2018*	*2019*
1	Used-oil	litres	140.00	200.00	550.00	255.00
2	Oil-soaked cotton waste	kg	1079.50	650.40	662.40	436.19
3	Oil filters	numbers	26.00	22.00	25.00	10.00
4	Empty chemical containers	kg	399.70	369.80	379.00	223.30

both corrective and preventive measures to combat the issue of hazardous wastes and high environmental cost.

The recommendation of pelletisation of oil-soaked cotton waste had reduced the waste generation by 34.2% and cost reduction by 56% in terms of disposal. To reduce the generation of used-oil and the oil filters, it was recommended to adopt an energy efficient DG Set. This had reduced the used-oil generation from 550 to 255 litres, and the number of oil filters used from 25 to 10. These measures had reduced the overall cost of handling the used-oil by about 36%. Reservoir printers were recommended as a replacement to laser and inkjet printers. In these printers, cartridges need not be disposed, as the ink tanks could be refilled. Hence, waste generated was reduced to merely 33% to that in comparison to that of conventional inkjet printers.

Also, with the recommendation and adoption of password activated printing process, the company was able to save 15% of paper waste from being generated, as well conserve the printing ink used for the same. The successful implementation of the recommendations had not only led to reduction in the significant aspects to 21.4%, but had also shown a rise in the positive aspects to 9.1%. The results from analysis and evaluation of environmental performance and compliance, have improved the readiness of the industry for certification process towards risk assessment via internal audits and management reviews. It has aided in improvement of environmental attributes such as air, water, land, natural resources with the implementation of given recommendations. The present study hence served as a guideline to strengthen the prevalent EMS for other similar industrial users, as there was no other pre-existing research, for this type of industry across the nation and globe. The study could further be extended to identify aspects according to the latest ISO 45001 document, which considered occupational health and safety along with quality and environment.

Acknowledgements We would like to thank the management and staff at GE Healthcare, Bangalore for their cooperation and support, during the course of this study.

References

Balasubramanya, N. and Gopinath, R. (2018). Environmental Science & Engineering. 1st edn. Cengage Learning India Private Limited. ISBN-13:978-93-87511-56-9.

Dagmara, N. and Thomas, P. (2009). Finding the connection: Environmental management systems and environmental performance. *J Cleaner Prod*, **17:** 601–607.

Davor, L. and Goran, S. (2011). Possibility of Environmental Aspects and Impacts Evaluations as per ISO 14001 standard on the example of an Academic Institution. *Holistic Approach Environ*, **1:** 75–84.

Deming, W.E. (1986). Out of the Crisis. Massachusetts Institute of Technology, Center for Advanced Engineering Study, Cambridge, MA, p. 88. ISBN 978-0911379013. OCLC 13126265.

Goh, E.A. et al. (2006). A study on the impact of environmental management system (EMS) certification towards firms' performance in Malaysia. *Manage Environ Qual*, **17:** 73–93.

Gopinath, R., Lokesh, Y. et al. (2014). A study on the effective management of E-waste ceramic by its utilization as a replacement to aggregates in SDBC grade-2 mix. *Int J Innov Appl Stud*, **7:** 981–984.

Gopinath, R. et al. (2017). Real-time thermal mapping for heat & cool archipelagos of Bengaluru, India. *Civil Environ Eng*, **13(2):** 106–111. doi: https://doi.org/10.1515/CEE-2017-0014.

Gopinath, R., Vinod, B.R. et al. (2020). Analysis of E-waste ceramics as a fine aggregate for rigid pavements for replacement of M-sand and river sand. Recent developments in waste management. *Lect Notes Civil Eng*, **57:** 385–396. Springer, Singapore. doi: https://doi.org/10.1007/978-981-15-0990-2_31.

https://www.gehealthcare.in/.

ISO 14001 (2015). Environmental Management Systems – Requirements with Guidance for Use. 3rd edn. **1:** 1–32. ISO, Geneva. ISBN 978-92-67-10648-9.

Kevin, W. et al. (2014). Impact of environmental management system implementation on financial performance. *Manage Environ Qual*, **15:** 622–628.

Laarifi, A. et al. (2017). Impact of the ISO 14001 certification on the environmental performance: Case study of two Moroccan companies. *Int J Agric Environ Sci*, **2:** 43–48.

Mohamed, M.A. et al. (2015). Environmental management system: Environmental aspects & productivity. *Int J Business Manage*, **10:** 107–121.

Nandini, V., Varun, M., Gopinath, R. and Yatish, Y. (2012). Performance evaluation of Cr (VI) removal by using activated carbon and water hyacinth. *Int J Chem Sci*, **10(2):** 1389–1396.

Ramya, R., Gopinath, R. and Satish, K. (2014). Water pollution arising from laboratory facilities of educational institutions indulging in hazardous chemicals. *Int J Scientific Eng Res*, **5(11):** 980–983.

Snezana, Z. et al. (2013). The improvement of environmental performances by applying ISO 14001 standard – A case study. *Chem Ind Chem Eng Q*, **19:** 541–552.

Sc Alan, L. et al. (2013). Determining environmental aspects in company economic D.O.O. Vitez, by using a flowchart diagram method. *J Eng Sci*, **1:** 28–38.

Venugopal, N. and Gopinath, R. (2012). Isotherm studies for adsorption of Cr (VI) using activated carbon. *Ultra-Chem*, **8(3):** 333–340.

Vrinda, T. and Gopinath, R. (2018). Artificial neural network trained 'simultaneous extent analysis' as a logical tool in computation of urban heat island intensity. *Sci Technol Asia*, **23:** 18–22. doi: https://doi.org/10.14456/scitechasia.2018.28.

Chapter 20
The Potential of Biogas Production from *Water Hyacinth* by Using Floating Drum Biogas Reactor

Suryateja Pottipati, K. D. Yadav, and A. S. Kalamdhad

1 Introduction

Among the listed aquatic weeds *Water hyacinth* (*Eichhornia Crassipes*) can double its growth in 6 to 28 days, by forming a mat-like structure on the water obstructing the sunlight this plant affects the aquatic ecosystem (Malik, 2007). Any freshwater body has a potential threat from this type of aquatic weed. Several control measures were added to the scientific journals among them one of the methods is using this weed as a feedstock for the anaerobic digestion to produce biogas (Moorhead and Nordstedt 1993; Singhal et al., 1997; Nguyen et al., 2011; Ivan et al., 2012). *Water hyacinth* is suitable for the anaerobic digestion process (Jagadish et al., 2014; Shahabaldin et al., 2015; Ershad et al., 2016). Using weeds as a feedstock and converting them to useful bio-fuel may reduce the load on the conventional power generation processes from the air, water, and solar energy sources (Rao et al., 2010). The biogas producing capacity of the biodegradable wastes like weeds and agricultural wastages estimated to be 40,734 m^3 year in India (Rao et al., 2010). By using the mixture of cattle dung and *Water hyacinth* will lead to better biogas production (Kumar, 2005). The present study summons the usage of *Water hyacinth* in the anaerobic digestion process. For a better yield of biogas, cow dung was used in combination with *Water hyacinth* (Onwuliri et al., 2013; Yuhelsa et al., 2014).

S. Pottipati (✉) · A. S. Kalamdhad
Indian Institute of Technology Guwahati, Guwahati, Assam, India

K. D. Yadav
Sardar Vallabhbhai National Institute of Technology Surat, Surat, Gujarat, India

A. S. Kalamdhad (ed.), *Integrated Approaches Towards Solid Waste Management*, https://doi.org/10.1007/978-3-030-70463-6_20

2 Materials and Methods

2.1 *Biogas Setup*

A continuous type of anaerobic biogas reactor unit was set up on the terrace of the CED (civil engineering department) building at SVNIT, Surat. Plastic is modified by connecting the pipes for feeding and sampling port and outlet for the biogas reactor, gas outlet for floating drum. The materials like PVC pipes and couplers are purchased from a local shop in a nearby shopping centre. A floating drum design was chosen for the biogas reactor unit. A 50 L plastic drum as a biogas reactor and a 30 L plastic drum as a floating drum was used. In a biogas reactor drum, provisions for feeding the stock and to remove the slurry were made. A nozzle was provided on the floating drum for taking the gas out of the drum (Fig. 1).

2.2 *Sample Collection*

For this study, *Water hyacinth* is collected from a pond near Vesu, Surat. Fresh cow dung was collected from the nearby vicinity of the SVNIT campus. The details of the proximate analysis are given in Table 1.

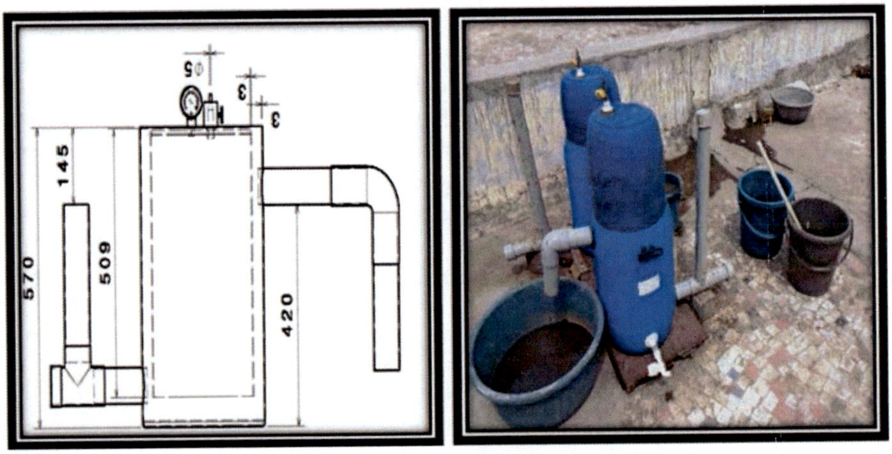

Fig. 1. Biogas setup dimensional view and biogas reactor setup.

Table 1. Proximate analysis data of *Water hyacinth* and cow dung

Parameter	*Unit*	*Water hyacinth*	*Cow dung*
Moisture content	%	90.15	70
Volatile matter	%	88.59	70
Ash (residual after burning)	%	7.43	5.39
Fixed carbon	%	0.01	0.97

2.3 *Slurry Preparation*

From raw *Water hyacinth*, roots were separated from its plant. A garden cutter is used for making the plant into small pieces and a hand-operated juice maker was used to prepare the slurry by crushing the stems and leaves of *Water hyacinth*. The feedstock was prepared by mixing the crushed *Water hyacinth* and cow dung. A constant amount of water is used to make the slurry in all the reactors and for continuous study, the amount was the same as of the batch process.

2.4 *Anaerobic Digestion*

The quantity of material is maintained on the dry weight basis and accordingly, the quantity of fresh material was taken to prepare the different combinations. In all the reactors, 1 kg of solids were mixed with 20 L of water. Quantity of dry weight of the material (*Water hyacinth* and cow dung) were calculated by removing the water content. A total of 11 number of combinations are prepared to know the best combination of *Water hyacinth* and cow dung to produce more amount of biogas. The typical combination of *Water hyacinth* and cow dung is given in Table 2 (Fig. 2).

3 Results and Discussion

All 11 reactors are established at the civil engineering department of the SVNIT campus. In this study, the reactors that got maximum production of biogas results are given. The production of biogas started from day 2 and attains a peak daily production on the 4–6th day for all the reactors.

Table 2. Typical combination of *Water hyacinth* and cow dung

Combination	*The ratio of Water hyacinth and cow dung in % (in terms of dry solids)*
1	0:100
2	10:90
3	20:80
4	30:70
5	40:60
6	50:50
7	60:40
8	70:30
9	80:20
10	90:10
11	100:0

Fig. 2. Biogas reactor setup.

Results of daily and cumulative biogas produced along with the combination of materials are explained below for both 90% *Water hyacinth* and 10% cow dung and 100% *Water hyacinth* and 0% cow dung.

3.1 *Combination 10–90%* Water hyacinth*: 10% Cow Dung*

In this reactor, the combination was of 900 g of *Water hyacinth* solids and 100 g of cow dung solids. To prepare this combination 9000 g of wet *Water hyacinth* and 330 g fresh cow dung were taken and mixed with 20 L of water and fed to the reactor (Fig. 3).

3.2 *Combination 11–100%* Water hyacinth*: 0% Cow Dung*

In this reactor, the combination was of 1000 g of *Water hyacinth* solids and 0 g of cow dung solids. To prepare this combination 10, 000 g of wet *Water hyacinth* and 100 g of fresh cow dung as inoculum was taken and mixed with 20 L of water and fed to the reactor (Fig. 4).

From Table 3, there is a reduction in the value of initial and final pH. The pH of the biogas reactor will decrease initially when organic material is first loaded into the biogas reactor and volatile acids are produced. However, as the methane-producing bacteria consume the acids, alkalinity is produced. In a properly operating anaerobic biogas reactor, a pH of 6.8 to 7.2 occurs as volatile acids are converted to methane and carbon dioxide.

The initial and final temperature values in Table 4 are dependent on the ambient atmospheric temperatures because the biogas reactors are kept open to atmospheric

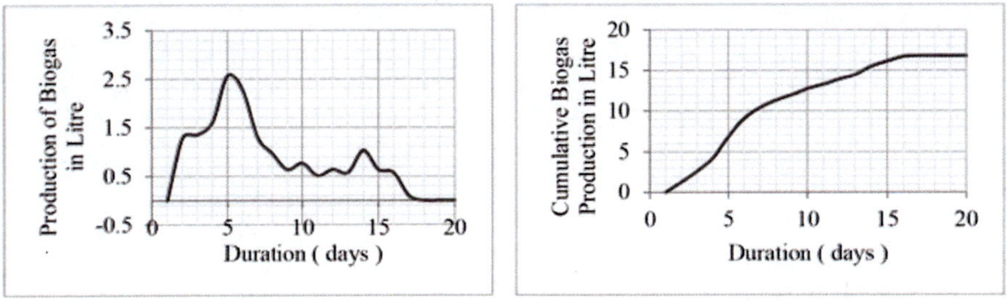

Fig. 3. Daily biogas production and cumulative production of biogas of 90% *Water hyacinth* and 10% cow dung biogas reactor.

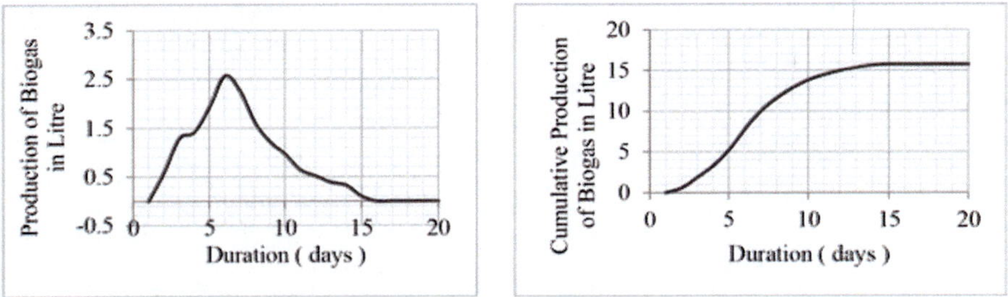

Fig. 4. Daily biogas production and cumulative production of biogas of 100% *Water hyacinth* and 0% cow dung biogas reactor.

Table 3. Best combinations

Combination	*The ratio of water hyacinth and cow dung in %*	*Quantity of feed material on dry (in terms of total solids) and wet (in terms of raw weight) basis*
Combination 10	90:10	900g *Water hyacinth* + 100 g cow dung (dry weight) 9000 g *Water hyacinth* + 332 g cow dung (wet weight)
Combination 11	100:0	1000 g *Water hyacinth* + 0 g cow dung + inoculum (dry weight) 10,000 g *Water hyacinth* + 0 g cow dung (wet weight)

conditions. From the values of alkalinity, we can infer that there is a production of alkalinity due to bacteria consuming the acids produced in the reactor.

Total solids in all the reactors are decreased by 60–65%. This is because of the digestion of volatile solids into gases by the anaerobic bacteria (Fig. 5).

Volatile fatty acids are decreased by 70% on average in all the reactors. The dominating volatile fatty acid in anaerobic digestion is acetic acid which intern converts into biogas. From Fig 6, it can be seen that in the combinations 10 and 11 there is a reduction by 85–87% of VFA.

Table 4. Summary of typical parameters and biogas production

Combination		*pH*	*Temperature*	*Alkalinity*	*Total solids*	*Volatile fatty acids*	*Gas produced*
		–	°C	*mg/L as* $CaCO_3$	*mg/L*	*mg/L*	*L*
1	Initial	7.23	27.6	5000	50250	130	8.480
	Final	7.10	32.5	8000	20,154	56.95	
2	Initial	6.95	27.5	5000	50,250	331.38	11.499
	Final	6.75	32.4	6000	21,524	79.53	
3	Initial	7.2	32.3	4500	50,250	450.98	10.664
	Final	6.94	32.5	4000	16,698	121.76	
4	Initial	7.14	27.4	5000	50,250	512.04	13.362
	Final	6.70	32.4	4000	22,361	153.04	
5	Initial	6.45	32.4	4500	50,250	582.31	11.371
	Final	7.23	32.5	4000	26,526	133.93	
6	Initial	7.00	32.2	5000	50,250	650.32	14.005
	Final	6.12	32.5	6000	21,856	227.61	
7	Initial	6.32	32.4	5000	50,250	690.35	8.608
	Final	7.10	32.5	6000	22,272	276.14	
8	Initial	7.21	32.4	5000	50,250	715.13	10.022
	Final	6.32	32.5	6000	23,894	250.29	
9	Initial	7.29	32.3	4500	50,250	865.09	9.380
	Final	6.66	32.5	5000	20,328	302.78	
10	Initial	7.03	32.4	4500	50,250	904.76	16.703
	Final	6.70	32.5	5000	18,762	110.36	
11	Initial	7.7	27.4	4500	50,250	950.88	15.675
	Final	6.43	32.5	4000	20,326	120.62	

From Fig. 6, it can be seen that the maximum production of biogas is from the combination 10 and 11, which are 90% *Water hyacinth* with 10% cow dung and 100% *Water hyacinth*.

3.3 Flam Test Results

The rubber bladder is used to collect gas from the reactor and tested by igniting a matchstick at its mouth and gas was released slowly and then the ignition of gas can be seen in Fig. 7.

The flame test resulted in a blue flame which indicates a minimum 50–70% of methane gas and also an indication of the complete burning of gas. Biogas is no more flammable if the methane content is less than 40% in the biogas composition.

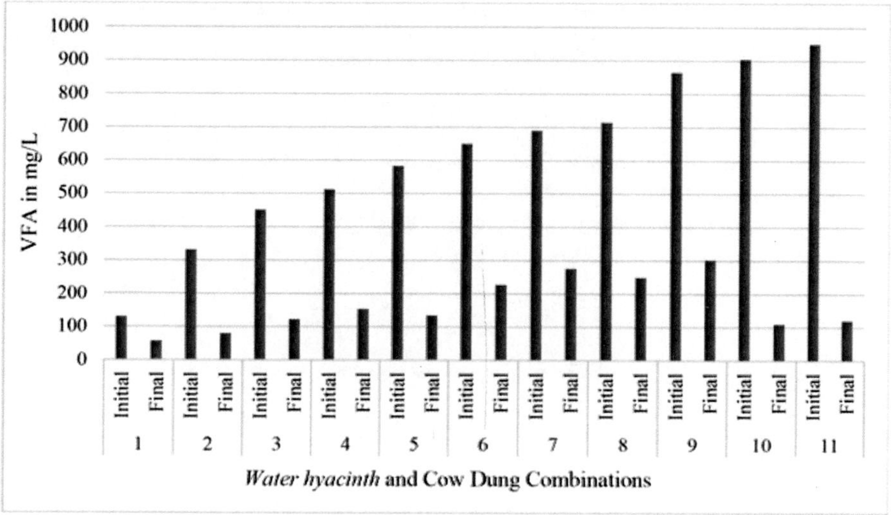

Fig. 5. Reduction of VFA in the biogas reactors of batch process.

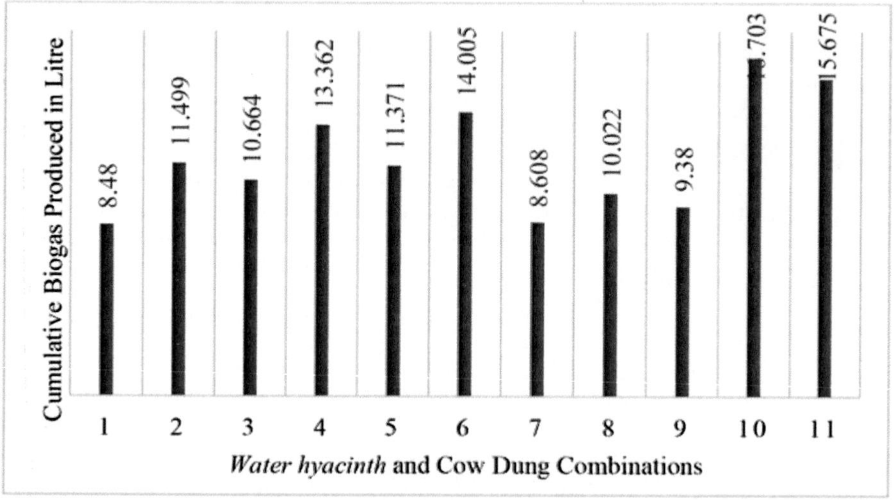

Fig. 6. Cumulative biogas produced in all combinations.

4 Conclusions and Suggestions

- In the batch process, it was seen that the maximum production of biogas is in combination 10, which is 90% *Water hyacinth* and 10% cow dung.
- The complete digestion of the batch process is in the mesophilic range (20–45°C).
- The complete digestion process is purely dependent on ambient atmospheric temperature.

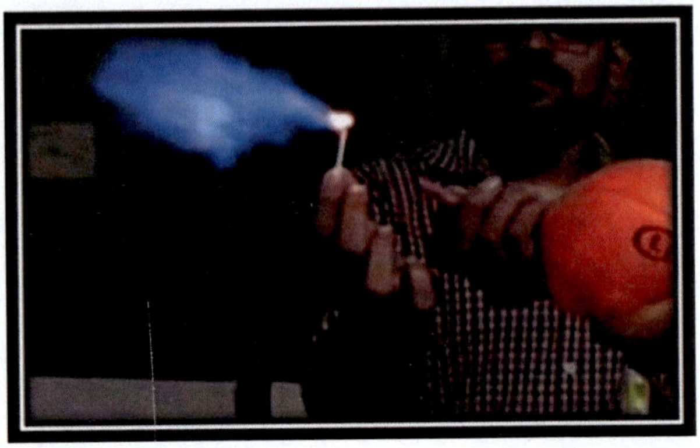

Fig. 7. Flame test of produced biogas.

- Fire test results produced a blue flame which indicates the complete burning of biogas.
- No significant difference was obtained in the continuous feeding process for 90% and 100% *Water hyacinth* biogas reactor, both reactors in the continuous feeding process produced methane which is more than 40–45%.
- The daily biogas production attains a constant value from the 11th day for continuous process.
- 10% of cow dung quantity along with *Water hyacinth* is helping to increase the quantity of biogas in the batch process.
- However, in the case of the continuous process *Water hyacinth* alone was capable of producing biogas.
- During the study, it was observed that a higher quantity of *Water hyacinth* and a lower quantity of cow dung are giving better results.
- The production of biogas from 90% and 100% biogas reactors is in the range of 20–25 L per 1 kg of *Water hyacinth* solids.

References

Bergier, I., Salis, S.M., Miranda, C.H.B., Ortega, E. and Luengo, C.A. (2012). Biofuel production from *Water hyacinth* in the Pantanal wetland. *Ecohydrol Hydrobiol*, 77–84.

Buchauer, K. (1997). A comparison of two simple titration procedures to determine volatile fatty acids in influents to waste-water and sludge treatment processes. *Water SA*, 24.

Kumar, S. (2005). Studies on efficiencies of bio-gas production in anaerobic digesters using water hyacinth and night-soil alone as well as in combination. *Asian J Chem*, **17:** 934–938.

Malik, A. (2007). Environmental challenge vis a vis opportunity: The case of water hyacinth. *Environ Int*, **33:** 122–138.

Moorhead, K.K., Nordstedt and R.A. (1993). Batch anaerobic digestion of *Water hyacinth*: Effects of particle size, plant nitrogen content, and inoculum volume. *Bioresource Technol*, 71–76.

Nguyen, V.C.N., Le Hoang, V., Nguyen, D.C. and Nguyen, H.P. (2011). Biogas Production of Pig Manure with *Water hyacinth* Juice from Batch Anaerobic Digestion. Springer Science, chapter 20.

Onwuliri, F.C., Onyimba, I.A. and Nwaukwu, I.A. (2013). Generation of biogas from cow dung. *Bioremed Biodegrad*, S18:002.

Patila, J.H., Raja, A., Shankar, B.B., Shettyc, M.K. and Pradeep Kumar, P.B. (2014). Anaerobic co-digestion of *Water hyacinth* and sheep waste. *Energy Procedia*, 572–578.

Putra, Y., Murni, D. and Kurniawan, T. (2014). Study of biogas production rate from *Water hyacinth* by hydrothermal pretreatment with buffalo dung as a starter. *Waste Technol*, **2(2):** 26–30.

Rao, P.V., Baral, S.S., Dey, R. and Mutnuri, S. (2010). Biogas generation potential by anaerobic digestion for sustainable energy development in India. *Renewable Sustain Energy Rev*, **14(7):** 2086–2094.

Rezania, S., Ponraj, M., Md Din, M.F., Songip, A.H., Md Sairan, F. and Chelliapan, S. (2015). The diverse application of *Water hyacinth* with main focus on sustainable energy and production of new era: An overview. *Renewable Sustain Energy Rev*, **41(C):** 943–954.

Singhal, V. and Rai, J.P.N. (2003). Biogas production from *Water hyacinth* and channel grass used for phytoremediation of industrial effluents. *Bioresource Technol*, **86**.

Ullah Khan, E. and Martin, A.R. (2016). Review of biogas reactor technology in rural Bangladesh. *Renewable Sustain Energy Rev*, **62:** 247–259.

Chapter 21
Influence of Initial pH on Bioleaching of Selected Metals from e-Waste Using *Aspergillus niger*

Amber Trivedi and Subrata Hait

1 Introduction

The information revolution has shown augmented growth in the production of information technology (IT) equipment like personal computers and mobile phones in addition to other electrical and electronic equipment (EEE) from the last two decades. Due to the higher obsolescence rate of EEE, waste EEE (WEEE) or electronic waste (e-waste) is presently growing at a faster rate than the municipal waste streams (Grossman, 2006). Baldé et al. (2017) estimated the global WEEE generation of about 44.7 million tonnes (Mt) in 2016 and expected to grow to 52.2 Mt by 2021. According to the recent joint study of the ASSOCHAM-NEC (2018), India generates a total of 2.0 Mt of e-waste. Each EEE contains a printed circuit board (PCB) as the main working component with heterogeneous elemental content, which includes base, toxic and precious metals. Studies have reported that obsolete computer PCB contains metal in the range of 20–24% Cu, 1–3% Ni, 0.6–6.3% of Pb, which are comparatively higher than the mineral ores. Because of their high content of base and heavy metals besides some amount of precious metals, make it valuable metal resource (Bandyopadhyay, 2008; Priya and Hait, 2018a). Recycling of metals from e-waste could save up to 90% of natural resources (Zhou et al., 2009). However, unscientifically PCB disposal in the environment creates environmental problems due to the presence of toxic metals (Sayilgan et al., 2009). Hence, metal recycling from waste PCB is the main obligation for economic development as well as for environmental protection (Deng et al., 2007). Metal recycling techniques like pyrolysis and hydrolysis are expensive, energy-intensive and generate secondary pollution (Ilyas et al., 2010; Pant et al., 2012; Priya and Hait, 2017). In the case of pyrometallurgy, heat treatment is given to wastes for metal recovery and

A. Trivedi · S. Hait (✉)
Department of Civil and Environmental Engineering, Indian Institute of Technology, Patna, Bihar, India
e-mail: shait@iitp.ac.in

A. S. Kalamdhad (ed.), *Integrated Approaches Towards Solid Waste Management*, https://doi.org/10.1007/978-3-030-70463-6_21

the process is fast and efficient but emits toxic gases and requires high energy for operation. In hydrometallurgy, chemicals like mineral acids, ligand are used to recover metals efficiently from e-waste, however toxic nature of these chemicals is a major concern for their applicability (Priya and Hait, 2017). Effluent generated from these conventional metallurgical processes, which further needs treatment to dispose of into the environment safely (Priya and Hait, 2017). To overcome these aforementioned drawbacks, nowadays researchers are focusing on the metal recycling process, which is efficient as well as environmentally friendly.

Biohydrometallurgy or bioleaching by employing microorganisms is an environmentally friendly and inexpensive alternative to conventional recycling techniques. Bioleaching is a process in which metals present in the PCB matrix are solubilised employing microbes and the metabolites excreted by them. Microorganisms involved in the bioleaching process are diverse in the range including acidophilic bacteria and fungi in which *Acidithiobacillus ferrooxidans* and *Aspergillus niger* are one of the most widely used species from each kingdom, respectively (Brandl et al., 2001; Priya and Hait, 2018b). Metal bioleaching from e-waste and other wastes by employing autotrophic bacteria are well explored. Autotrophic bacteria efficiently work in acidic medium (Priya and Hait, 2017). In past, bioextraction or bioleaching of base metals like Cu, Ni, Zn, Al and Pb from WPCBs have been investigated by employing bacterial species of *Acidithiobacillus ferrooxidans*, *Acidithiobacillus thiooxidans* and fungal species of *Aspergillus niger* and *Penicillium simplicissimum* by various researchers (Arshadi and Mousavi, 2014; Brandl et al. 2001; Faraji et al. 2018).

In recent works, La recovery by employing *Aspergillus niger* at three different pulp densities of 1%, 3% and 5% of spent fluid catalytic cracking catalyst has shown 63%, 52% and 33% La recovery, respectively (Mouna and Baral, 2019). Various factors have been optimised, which affects bioleaching of metals, i.e., Zn, Ni and Cu from e-waste including gone and multiple step process and various pulp densities to improve the bioleaching efficiency (Faraji et al., 2018). Another study from Muddanna and Baral (2019) have found that the predominantly produced biogenic acids by *Aspergillus niger* are attributed to metal leaching from spent fluid catalytic cracking catalyst. Further, previous researches have suggested that the fungal species can grow from a broad range of pH from acidic to alkaline medium (Xu et al., 2014). Xia et al. (2018) have found around 73% of Cu extraction from waste PCBs at different pulp densities using a fungal strain *Aspergillus niger* in a shake flasks mode. However, heterotrophic fungi are required to explore in different pH medium for bioleaching of metals from e-waste. In addition, metal extraction employing bioleaching are mainly dependent on composition, pH and nature of bioleaching medium. In this context, the influence of initial pH on the selected metal extraction from WPCB of an obsolete computer employing a pure culture of *Aspergillus niger* in fungal bioleaching is assessed in the present study. *Aspergillus niger* has been chosen due to its ability to produce organic acids, which presumably help in metal extraction from the solid matrix.

2 Materials and Methods

2.1 e-Waste Collection and Sample Preparation for Bioleaching Study

WPCB of obsolete computers was brought from regional electronics service centres of Bihta, Patna in the raw form (Fig. 1). Mounted parts on PCB were removed by tools like pliers and hammers followed by size cutting using a milling machine (SM200, Retsch GmbH, Germany). Comminuted PCB sample was then sieved and fragmented in the size ranging from 38 to 1000 μm by using a sieve shaker. Selected metal quantification and bioleaching experiment were performed on the prepared sample.

2.2 Metal Quantification of Waste PCB

Waste PCB sample was analysed for selected metal viz., Cu and Zn in order to compute the bioleaching efficiency. For metal characterisation, 200 mg of pulverized PCB sample along with 9 mL of nitric acid and 3 mL of hydrofluoric acid was placed in a Teflon vessel for digestion in a microwave digester (Ethos Easy, Milestone, Italy) in accordance to the USEPA 3052 acid digestion method (1995). Upon digestion, the sample volume was then made up to 50 mL and subsequently filtered by 0.22 μm filter paper. Suitable dilution was made to analyse the samples for metal content in an inductively coupled plasma mass spectrometer (ICP-MS) (7800, Agilent, USA). Metal quantification was done in the multiplication of three and average values reported.

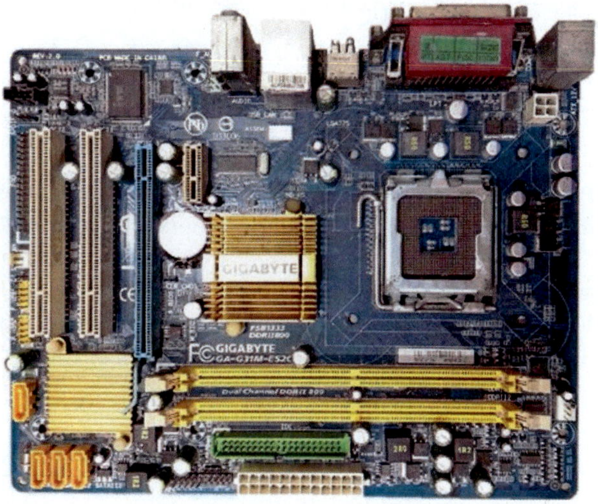

Fig. 1. Typical obsolete computer PCB used in the study.

2.3 *Bioleaching Experiment Using* Aspergillus niger

Fungal species *Aspergillus niger* was used in the present study. To assess the influence of medium initial pH on bioleaching of selected metals from e-waste using *Aspergillus niger*, an experiment was performed at three different initial pH of 5, 7 and 9. Potato dextrose broth (PDB) medium was used for fungal growth and bioleaching experiments. Fungal culture was grown in a conical flask for 7 days and 1 mL spore suspension added to the bioleaching medium as inoculum. Apart from the bioleaching medium, the other experimental parameters were kept like the volume of the reactor 100 mL, pulp density 2.5 g/L, rpm 170 and temperature. Control reactors were also run with the same bioleaching experimental conditions except addition of fungal inoculum. Samples were withdrawn at every 3 days interval for 33 days experimental period. Samples were filtered using a 0.22 μm porous filter. followed by the dilution and analysis using inductively coupled plasma mass spectrometer (ICP-MS) (7800, Agilent, USA) to analyse the metal content in the leachate. All samples were analysed in the multiplication of three and average values reported.

3 Results and Discussion

3.1 *Metallic Content of Comminuted WPCB*

The metal content of the comminuted PCB particle was quantified to calculate bioleaching efficiency. Results revealed that the Cu was present in abundance with a content of 20% and Zn with 0.10% in the waste PCB samples. With the rich metal content of selected metals like Cu and Zn, the obsolete computer waste PCB can be considered as a secondary metal resource.

3.2 *Variations in pH During Fungal Bioleaching*

Bioreactor pH values have shown variations in 33 days of the experiment, decreasing in pH can be attributed to the production of organic acids and raising of pH value to alkaline pH of waste PCB. Values shown in figures are the resultant of both organic acid and alkaline nature of the waste PCB. As shown in Fig. 2, pH increased initially followed by the sharp declination to 2.42, 2.34 and 5.61 in 33 days of experiment duration from an initial pH of 5, 7 and 9, respectively.

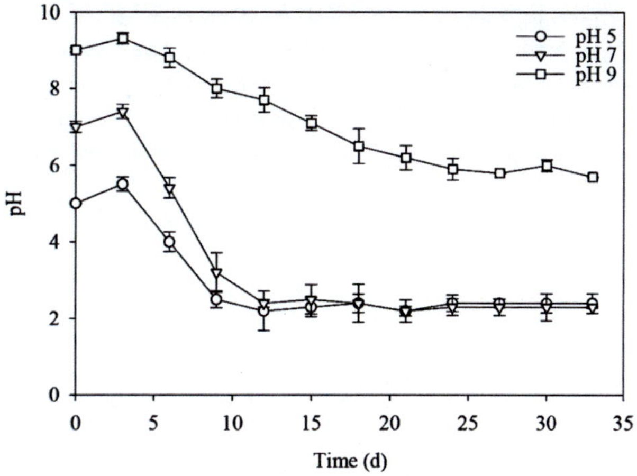

Fig. 2. Variations in pH during bioleaching study.

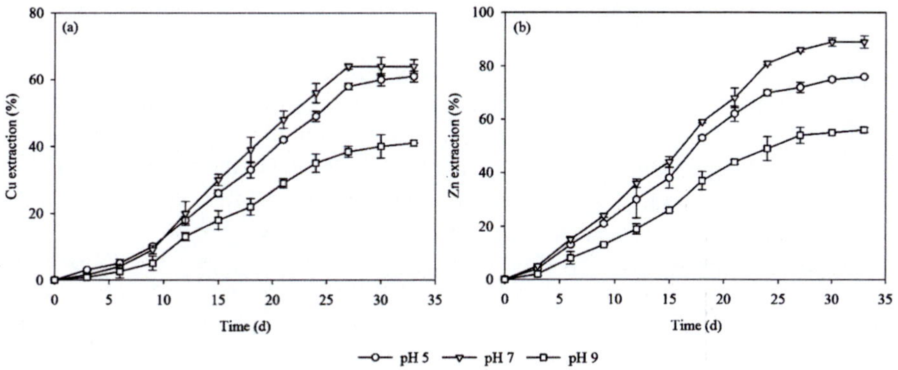

Fig. 3. At different initial pH, metal extraction efficiency of (a) Cu and (b) Zn employing *Aspergillus niger*.

3.3 *Bioleaching of Selected Metals Using Aspergillus niger*

Various factors like pH, temperature, the medium composition can affect bioleaching efficiency (Gadd, 2004). In this experimental study, the effect of initial pH viz. 5, 7 and 9 on metal bioleaching from waste PCB of an obsolete computer has been assessed. Mechanisms like acidolysis and complexolysis are mainly involved in the fungal bioleaching process which depended on the pH of the bioleaching medium (Le et al., 2006). Acidolysis and complexolysis are contributed at low and near to neutral pH, respectively. The H^+ ions concentration drops in the basic condition where complexolysis causes the metal dissolution process (Le et al., 2006).

Results from the study conducted on various initial pH have revealed that metals Cu and Zn were extracted maximum at neutral pH followed by pH 5 and 9. It can be observed from Fig. 3, maximum metal leaching were at an initial pH of 7. Fungal bioleaching led to the maximum extraction of 64% Cu and 89% Zn at pH 7 in

33 days duration. At pH 5, a slight decrement in metal extraction was observed from waste PCB with 61% Cu and 76% Zn followed by at pH 9 with 41% and 59% of Cu and Zn, respectively. Both the selected metals Cu and Zn have shown similar extraction trends initially with a slow extraction rate until 12 days followed by rapid extraction until 27 days. The increased fungal bioleaching of the metals from the comminuted particles of waste PCB under the initial neutral pH condition can be attributed to the combined actions by complexolysis and acidolysis upon the fungal growth followed by the secretion of metabolites like organic acids and amino acids (Burgstaller and Schinner, 1993). Both types of acids provide proton and complexation ability to solubilise metals (Asghari et al., 2013).

4 Conclusions

In this study, the effect of initial bioleaching medium pH has been assessed during fungal bioleaching of selected metals Cu and Zn from obsolete computer waste PCB by employing the pure culture of *Aspergillus niger*. Maximum bioleaching has occurred at pH 7 followed by pH 5 and 9, respectively. With maximum extraction efficiency, 64% Cu and 89% Zn were extracted at an initial pH of 7. The increased metal bioleaching at neutral pH condition was perhaps due to the combined actions of complexolysis and acidolysis upon the fungal growth followed by the secretion of metabolites like organic acids and amino acids. Further, it is noteworthy that fungal species *Aspergillus niger* used in this study, not only can grow but is also able to leach metals at alkaline pH of 9. As e-waste is alkaline in nature, it is an important observation from the study that metals can be extracted from alkaline waste by fungal bioleaching using *Aspergillus niger* at neutral as well as alkaline pH. However, the maximisation of metal extraction efficiency can be done by optimising different operational parameters like medium composition, inoculum density with initial neutral and alkaline pH employing *Aspergillus niger* along with few other fungal species. Molecular mechanisms involved in fungal bioleaching at different medium pH will be also investigated intensively. Further, other metal-chelating agents working at neutral and alkaline pH can also be employed in combination with bioleaching to assess the hybrid bioleaching condition.

References

Arshadi, M. and Mousavi, S.M. (2014). Simultaneous recovery of Ni and Cu from computer-printed circuit boards using bioleaching: Statistical evaluation and optimization. *Bioresource Technol*, **174:** 233–242.

Asghari, I., Mousavi, S.M., Amiri, F. and Tavassoli, S. (2013). Bioleaching of spent refinery catalysts: A review. *J Ind Eng Chem*, **19(4):** 1069–1081.

Baldé, C.P., Forti, V., Gray, V., Kuehr, R. and Stegmann, P. (2017). The global e-waste monitor 2017: Quantities, flows and resources. United Nations University, International Telecommunication Union, and International Solid Waste Association.

Brandl, H., Bosshard, R. and Wegmann, M. (2001). Computer-munching microbes: Metal leaching from electronic scrap by bacteria and fungi. *Hydrometallurgy*, **59(2–3):** 319–326.

Bandyopadhyay, A. (2008). A regulatory approach for e-waste management: A cross-national review of current practice and policy with an assessment and policy recommendation for the Indian perspective. *Int J Environ Waste Manage*, **2(1–2):** 139–186.

Burgstaller, W. and Schinner, F. (1993). Leaching of metals with fungi. *J Biotechnol*, **27(2):** 91–116.

Deng, W.J., Zheng, J.S., Bi, X.H., Fu, J.M. and Wong, M.H. (2007). Distribution of PBDEs in air particles from an electronic waste recycling site compared with Guangzhou and Hong Kong, South China. *Environ Int*, **33(8):** 1063–1069.

Faraji, F., Golmohammadzadeh, R., Rashchi, F. and Alimardani, N. (2018). Fungal bioleaching of WPCBs using Aspergillus niger: Observation, optimization and kinetics. *J Environ Manage*, **217:** 775–787.

Gadd, G.M. (2004). Microbial influence on metal mobility and application for bioremediation. *Geoderma*, **122(2–4):** 109–119.

Grossman, E. (2006). High Tech Trash: Digital Devices, Hidden Toxics, and Human Health. Island Press.

Ilyas, S., Ruan, C., Bhatti, H.N., Ghauri, M.A. and Anwar, M.A. (2010). Column bioleaching of metals from electronic scrap. *Hydrometallurgy*, **101(3–4):** 135–140.

Le, L., Tang, J., Ryan, D. and Valix, M. (2006). Bioleaching nickel laterite ores using multi-metal tolerant Aspergillus foetidus organism. *Miner Eng*, **19(12):** 1259–1265.

Mouna, H.M. and Baral, S.S. (2019). A bio-hydrometallurgical approach towards leaching of lanthanum from the spent fluid catalytic cracking catalyst using *Aspergillus niger*. *Hydrometallurgy*, **184:** 175–182.

Muddanna, M.H. and Baral, S.S. (2019). Leaching of nickel and vanadium from the spent fluid catalytic cracking catalyst by reconnoitering the potential of *Aspergillus niger* associating with chemical leaching. *J Environ Chem Eng*, **7(2):** 103025.

Pant, D., Joshi, D., Upreti, M.K. and Kotnala, R.K. (2012). Chemical and biological extraction of metals present in E waste: A hybrid technology. *Waste Manage*, **32(5):** 979–990.

Priya, A. and Hait, S. (2017). Comparative assessment of metallurgical recovery of metals from electronic waste with special emphasis on bioleaching. *Environ Sci Pollut Res*, **24(8):** 6989–7008.

Priya, A. and Hait, S. (2018a). Comprehensive characterization of printed circuit boards of various end-of-life electrical and electronic equipment for beneficiation investigation. *Waste Manage*, **75:** 103–123.

Priya, A. and Hait, S. (2018b). Extraction of metals from high grade waste printed circuit board by conventional and hybrid bioleaching using *Acidithiobacillus ferrooxidans*. *Hydrometallurgy*, **177:** 132–139.

Xia, M., Bao, P., Liu, A., Wang, M., Shen, L., Yu, R., Liu, Y., Chen, M., Li, J., Wu, X., Qiu, G. and Zeng, W. (2018). Bioleaching of low-grade waste printed circuit boards by mixed fungal culture and its community structure analysis. *Resources Conserv Recycling*, **136:** 267–275.

Xu, T.J., Ramanathan, T. and Ting, Y.P. (2014). Bioleaching of incineration fly ash by *Aspergillus niger*–precipitation of metallic salt crystals and morphological alteration of the fungus. *Biotechnol Rep*, **3:** 8–14.

Sayilgan, E., Kukrer, T., Civelekoglu, G., Ferella, F., Akcil, A., Veglio, F. and Kitis, M. (2009). A review of technologies for the recovery of metals from spent alkaline and zinc–carbon batteries. *Hydrometallurgy*, **97(3–4):** 158–166.

USEPA (1995). Microwave assisted acid digestion of siliceous and organically based matrices USEPA method 3052, 3rd edn. United States Environmental Protection Agency, Washington, DC

Zhou, H.B., Zeng, W.M., Yang, Z.F., Xie, Y.J. and Qiu, G.Z. (2009). Bioleaching of chalcopyrite concentrate by a moderately thermophilic culture in a stirred tank reactor. *Bioresource Technol*, **100(2):** 515–520.

Chapter 22
Column Leaching of Metals from PCB of End-of-Life Mobile Phone Using DTPA Under Oxidising Condition

Nipoon Gupta, Amber Trivedi, and Subrata Hait

1 Introduction

Global digitisation and availability of affordable and advanced electrical and electronic equipment (EEE) have aggravated the management problem of waste electrical and electronic equipment (WEEE) or electronic waste (e-waste) around the world. According to a report published by Baldé et al. (2017), global electronic waste (e-waste) generation in the year 2016 is estimated to be 44.7 Mt and projected to grow to 52.2 Mt by 2021. EEE has been categorised into two categories: (a) Information technology and telecommunication (IT and telecom) equipment and (b) consumer electrical and electronics (e-waste management rules, 2016). End-of-life mobile phones (Eol-MPs) are contributing a major portion in e-waste content because of its high obsolescence rate and ever-increasing consumer base and year-wise sales as shown in Fig. 1 (Beigl et al., 2012; Kasper et al., 2011; Monteiro et al., 2007; Osibanjo and Nnorom, 2008; Robinson, 2009). Printed circuit board (PCB) is the core component of any EEE so as of end-of-life mobile phone (EoL-MP). Metal recycling from the PCBs of EoL-MP as a secondary resource has become a necessity both from an economical and environmental point of view considering the abundant presence of base, precious and toxic metals in them.

Conventional recycling techniques applied for the metal recycling from e-waste viz. pyro-/hydro- and bio-metallurgical process are effective in metal recycling but has some limitations. Pyrometallurgy process required high energy for incineration, melting of waste to recover metals. In addition, process also suffered with high toxins releases in the surroundings (Lee et al., 2007; Priya and Hait, 2017). Biometallurgy or bioleaching is a process in which metal dissolution is done by employing microorganisms. Despite high metal extraction efficacy and ecofriendly

N. Gupta · A. Trivedi · S. Hait (✉)
Department of Civil and Environmental Engineering, Indian Institute of Technology, Patna, Bihar, India
e-mail: shait@iitp.ac.in

A. S. Kalamdhad (ed.), *Integrated Approaches Towards Solid Waste Management*, https://doi.org/10.1007/978-3-030-70463-6_22

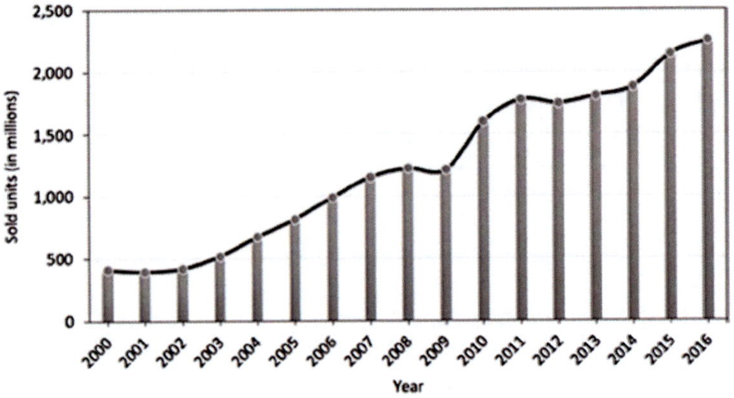

Fig. 1. Worldwide mobile terminal sales for year 2000–2016.

technique, bioleaching is a slow process (Mishra et al., 2005). In hydrometallurgy, inorganic acids like HNO_3, H_2SO_4, HCl are used to solubilise metals from PCB (Huang et al., 2009; Li et al., 2007; Priya and Hait, 2017). Further to increase the leaching efficacy with decrease in leaching time, previous studies have also showed the application of strong oxidisers like hydrogen peroxide with mineral acids for hydrometallurgical leaching of metals from e-waste (Deveci et al., 2010; Yang et al., 2011). The impediment of generation of toxic effluent in hydrometallurgical technique for metal extraction from e-waste can be curbed by using safer chemicals like organic acids or chelators having metal leachability properties. Chelators is term used for chemicals, which have the ability to form metal-ligand soluble complexes (Chauhan et al., 2015). These metal–ligand complexes do not participate in any chemical reaction like uncomplexed ions.

In previous studies, organic ligands like ethylene diamine tetra-acetic acid (EDTA) and diethylene triamine pentaacetic acid (DTPA), have shown strong metal chilating ability (Means et al., 1980; Pastor et al., 2007). Very few studies have been shown metal extraction from waste PCBs employing EDTA and DTPA as a chelator (Jadhao et al., 2016; Verma and Hait, 2019). Verma and Hait (2019) have employed safer lixiviant DTPA under oxidising condition to extract Cu and Zn from waste PCB of EoL computer. However, the previously conducted experiments are limited to shake flask mode. Therefore, there is a need to explore the different mode of experiments to scale up the process. Following the optimised conditions for metal extraction obtained from shake flask leaching studies using response surface methodology (RSM), column leaching of metals from EoL-MP PCB was performed under oxidising condition in the present study using DTPA as a safer lixiviant.

2 Materials and Methods

2.1 *Collection of End-of-Life Mobile Phone PCBs and Sample Preparation*

EoL-MPs were collected from electronics repairing shops in and around Bihta, Patna, Bihar, India. EoL-MP was dismantled and its PCB was taken out. Mounted parts were removed using plier and mechanical tools from waste PCB followed by the size cutting using a milling machine (SM200, Retsch, Germany). Comminuted PCB sample was then sieved and fragmented in the size ranging from 0.038 to 1 mm by using a sieve shaker (AS200, Retsch, Germany). Comminuted PCB samples were utilised for metal quantification and column leaching experiment.

2.2 *Elemental Quantification of Comminuted PCB Sample*

The selected metal content viz. Cu and Zn in the comminuted and homogenised PCBs sample was quantified in accordance with USEPA 3052 acid digestion method (1995). Comminuted PCB sample was placed in a Teflon vessel with 9 mL of nitric acid and 3 mL of hydrofluoric acid and heated up in a controlled environment of microwave digester (Ethos Easy, Milestone, Italy). Digested sample volume was madeup to 50 mL. Samples were then filtered by 0.22 μm filter paper. Suitable dilution was made to analyse the samples for metal content in an inductively coupled plasma mass spectrometer (ICP-MS) (7800, Agilent, USA). Metal quantification were done in multiplication of three and average values reported.

2.3 *Leaching of Selected Metals from PCBs of EoL-MPs Using Column Reactor by Applying Response Surface Methodology to Batch Study*

2.3.1 Optimisation Using Response Surface Methodology

To optimise experimental parameters and estimate the interaction effects of dynamic factors, response surface methodology (RSM), a mathematical tool was used. RSM was used to optimise and evaluate the effects of process parameters such as temperature, liquid–solid ratio, concentration of DTPA, pH of the solution and H_2O_2 in terms of molarity on the leaching efficiency. Five independent variables as mentioned in Table 1 are consecutively coded as A, B, C, D and E at three levels, i.e., −1, 0 and 1 with −1 and +1 being high and low, respectively. Model was developed in MiniTab 18 software to demonstrate the effect of each parameter and their interaction effects on the responses. The model acceptability and the validity of the

Table 1. Experimental independent factors with their low and high value

Factor	*Name*	*Low*	*High*
A	Temperature	20	50
B	pH	5	9
C	Molarity	0.3	0.7
D	L/S ratio	10	100
E	Mixing speed	150	450

Table 2. Dataset used for model development using RSM from Verma and Hait (2019)

Temperature	*pH*	*Molarity*	*L/S ratio*	*Mixing speed*	*Leaching efficiency*
20	7	0.3	50	450	33.2
20	7	0.5	50	450	45.8
20	7	0.7	50	450	22.4
20	5	0.5	50	450	46.6
20	7	0.5	50	450	45.1
20	9	0.5	50	450	68.2
20	9	0.5	100	450	62.8
20	9	0.5	10	450	16.4
20	9	0.5	50	450	63.9
35	9	0.5	50	450	75.2
50	9	0.5	50	450	85.6
50	9	0.5	50	300	67.7
50	9	0.5	50	150	47.7

optimisation phase were confirmed with previous study conducted by Verma and Hait (2019) and experimental data shown in Table 2 was used to find out the optimal conditions to scale up the process in the column reactor. Experimental data found from RSM model can be explained in the form of equation (1).

$$Y = b_0 + \sum_{i=1}^{k} b_i X_i + \sum_{i=1}^{k} b_{ii} X_i^2 + \sum_{i,j=1(i \neq j)}^{k} b_{ij} X_i X_j \tag{1}$$

where Y, b_0 and X_i and X_j are the expected outcome (leaching efficacy, %), constant coefficients and independent variables, respectively. Further, the terms b_i (i = 1, 2 and 3), b_{ii} and b_{ij} (i = 1, 2 and 3; j = 1, 2 and 3) are the linear, quadratic and interaction coefficients, respectively. While X_iX_j and X_i^2 represent the interaction and quadratic terms, respectively (Behera et al. 2018).

2.3.2 Column Leaching of Metals from End-of-Life Mobile Phone PCBs

A temperature controlled column reactor was designed to facilitate column leaching experiments. Other experimental parameters were optimised based on RSM model developed in previous section. The column experiments were conducted on optimal

Fig. 2. Methodology for extraction of metals using column reactor.

parameters found by employing RSM to find out the optimal flow rate. The column was operated in down-flow mode and was packed using glass beads and filtering nylon cloth. A column was placed vertically, filled with comminuted PCB sample and DTPA solution along with H_2O_2 were fed by a peristaltic pump from top to bottom (downward flow). The flow rate (mL/h) was experimentally optimised from 50, 100, 500 and 1000 in this study. The total experimental run was of 10 h and the samples were recycled throughout the experiment. Sample were drawn at every 30 min interval and analysed for the leached metal quantification. Collected sample were filtered through a 0.22 μm filter following the metal quantification using ICP-MS. Column leaching experimental methodology are graphically presented in Fig. 2 and a schematic of column reactor is presented in Fig. 3.

2.3.3 Statistical Analysis

To estimate the statistical parameters between the process parameters and the leaching efficiency of metals from waste PCB of EoL-MPs was assessed using analysis of variance (ANOVA). The statistical analyses were performed at the probability level $(P) < 0.05$. MiniTab 18 software was used for the regression analysis of experimental results, surface plots and contour plots at the optimised conditions. The significant value was checked by F-test. Further, R^2 coefficient was used to determine accuracy of fitted polynomial.

3 Results and Discussion

3.1 Metallic Content of EoL-MP PCB

The comminuted EoL-MP PCB was quantified for the selected metal content of Cu and Zn. Quantification studies results revealed that the Cu and Zn was 27.65% and 0.41% and were present in the EoL-MP PCB, respectively. Since the selected metal

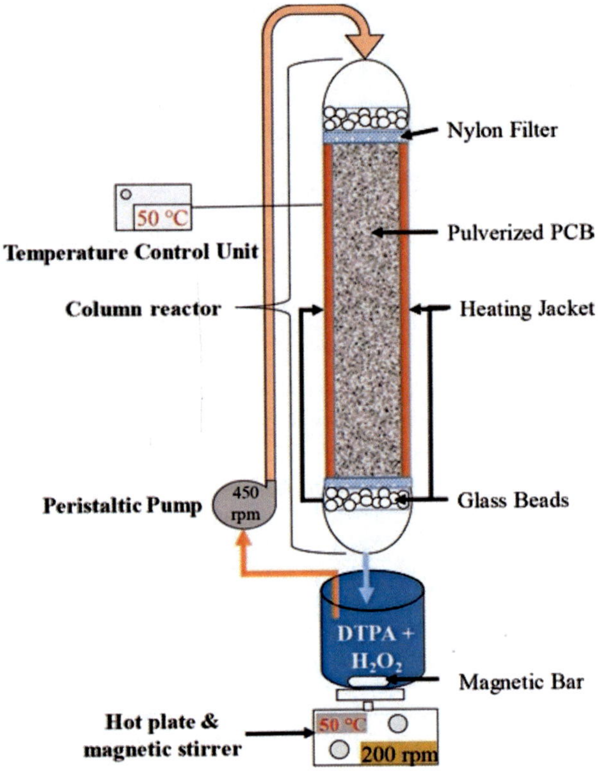

Fig. 3. Experimental setup for extraction of selected metals using column reactor.

presence was found to be in abundance, the comminuted EoL-MP PCB could be considered as a secondary metal reservior.

3.2 *Optimal Parameters for Column Leaching Using RSM*

The results obtained by RSM after modelling in MiniTab 18 software provided optimal parameters for the experimental study using column reactor. Final predicted equation for the leaching efficiency using DTPA in the uncoded form is shown in equation (2).

Table 3. Statistical factors obtained from the ANOVA for the models

S	R^2	*Adjusted* R^2
2.17830	0.9981	0.9888

$$\begin{aligned}\text{Leaching efficiency} = {} & -91.8 + 0.457\,\text{Temperature} - 33.20\,\text{pH} + 414.2 \\ & \text{Molarity} + 2.112\,\text{L/S ratio} \\ & + 0.1543\,\text{Mixing speed} + 0.0028\,\text{Temperature} * \\ & \text{Temperature} + 2.719\,\text{pH} * \text{pH} \\ & - 441.2\,\text{Molarity} * \text{Molarity} - 0.01451\,\text{L/S ratio} * \text{L/S ratio} \\ & - 0.000047\,\text{Mixing speed} * \text{Mixing speed}\end{aligned} \tag{2}$$

Using the equation, the leaching efficiency was maximised. Analysis of variance results revealed R^2 value of 0.9981, which shows the second order regression model is satisfactorily adjusted with the experimental results. The R^2 and adjusted R^2 values are shown in Table 3. Pareto chart of the standardised effects presented in Fig. 4 aids to visualise the most significant parameters in the experiment. The results obtained were found to be well fitted and in match with the previous experimental works conducted (Verma and Hait, 2019). The most optimum parameters with a leaching efficiency of around 94% are being shown in Fig. 5. Table 4 presents the analysis of variance data generated by equation (1) for metal leaching from waste PCB of EoL-MPs. The higher 'F' value and the lower 'P' value show statistical significance for the corresponding coefficient (Yi et al., 2010). The F-value of 106.51 shows that the model is significant. Further, the model terms are significant when the P values are <0.05 (Montgomery, 2017). In this case, A, B, D, E, BB, CC and DD are significant model terms.

The main part of the column leaching study was to determine the optimum leaching process parameters where maximum extraction of metals from EoL-MPs can be attained. Parameter optimisation was conducted by a numerical optimisation method. The response surface and contour plot at optimum operational parameters for maximum metal leaching were presented in Fig. 6(a) and (b), respectively. The optimum operational parameters for metal leaching process were 0.5 M DTPA: 50°C temperature: 1:75 solid to liquid ratio: 9 pH.

3.3 *Column Leaching of Selected Metals from EoL-MP PCB*

A column leaching study under continuous mode of operation was carried out to extract metals from pulverized particles of waste PCB of EoL-MPs. The process parameters of column leaching experiments were optimised using the RSM model based on the experimental data obtained from the shake flask leaching study (Verma and Hait, 2019). To experimentally optimise the lixiviant flow rate for metal

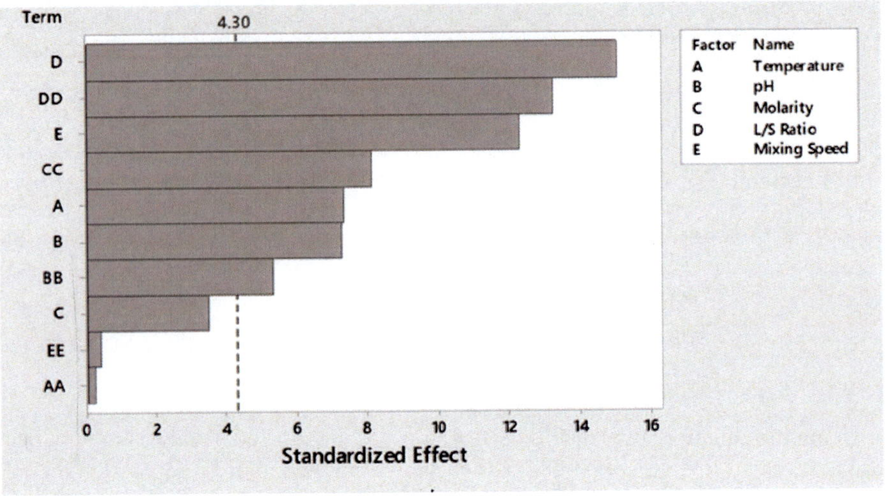

Fig. 4. Pareto chart of the standardised effects.

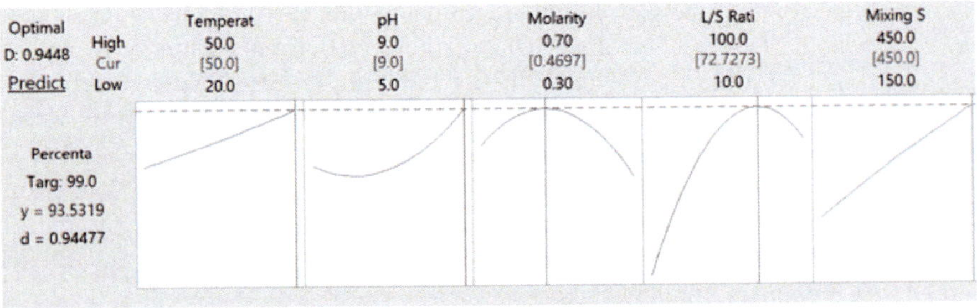

Fig. 5. Optimal parameters as obtained from the RSM model.

extraction, DTPA was introduced from the top of the column reactor using peristaltic pump at four different flowrates (mL/h) of 50, 100, 500 and 1000. Metal leaching is generally expected with the increase in lixiviant flow rate with recirculation as higher volume of lixiviant comes in contact with the PCB particles in the column reactor. However, the higher the contact time between lixiviant and PCB particles in column reactor the greater the amount of metal bind (Tuwati and Fan, 2015).

Maximum metal leaching has been found at 100 mL/h followed by 50 mL/h, 1000 mL/h and 500 mL/h. It is apparent from Fig. 7 that the maximum Cu leaching efficiency was increased drastically from 39.9%, to 97.0% in 10 h as the lixiviant flowrate increases from 50 mL/h to 100 mL/h. With further increase in flowrate Cu leaching efficiency was dropped to 27.1% and 23.6% at 1000 mL/h and 500 mL/h flowrate respectively in 10 h duration. From the column leaching experiment, the optimal flowrate was found to be 100 mL/h and it can be observed from Fig. 7. Further, Zn leaching was observed to be maximum at 100 mL/h followed by 50 mL/h flowrate with 98.9% and 48.9%, respectively. With the increase in the

Table 4. Summary of analysis of variance (ANOVA) for the models for metal leaching from waste PCB

Source	*Degree of freedom*	*Sum of square*	*Mean square*	*F-value*	*P-value*	*Remarks*
Model	10	5053.92	505.39	106.51	0.009	Significant
Linear	5	2165.41	433.08	91.27	0.011	Significant
Temperature (A)	1	254.80	254.80	53.70	0.018	Significant
pH (B)	1	252.20	252.20	53.15	0.018	Significant
Molarity (C)	1	58.32	58.32	12.29	0.073	
L/S ratio (D)	1	1076.48	1076.48	226.87	0.004	Significant
Mixing speed (E)	1	718.20	718.20	151.36	0.007	Significant
Square	5	1474.96	294.99	62.17	0.016	Significant
Temperature × temperature (A × A)	1	0.28	0.28	0.06	0.830	
pH × pH (B × B)	1	135.16	135.16	28.48	0.033	Significant
Molarity × molarity (C × C)	1	311.52	311.52	65.65	0.015	Significant
L/S ratio × L/S ratio (D × D)	1	837.44	837.44	176.49	0.006	Significant
Mixing speed × mixing speed (E × E)	1	0.74	0.74	0.15	0.732	
Error	2	9.49	4.75			
Total	12	5063.41				

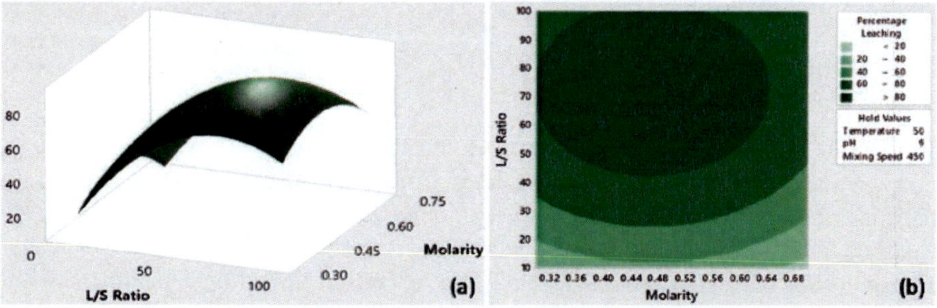

Fig. 6. Optimised operational parameters for maximum metal leaching (%) through plots (a) response surface and (b) contour plot.

lixiviant flowrate from 100 mL/h, Zn leaching has been found to be decreasing to 32.4% at and 30.6% at 500 mL/h and 1000 mL/h, respectively. Tuwati and Fan (2015) reported similar results that with increase in flowrate leaching of metals is also increasing in a column reactor running under down flow mode. It can be observed that column reactor facilitated more efficient metal extraction process from e-waste and process more than about ten times e-waste to the amount processed (4 g/L) earlier through batch studies.

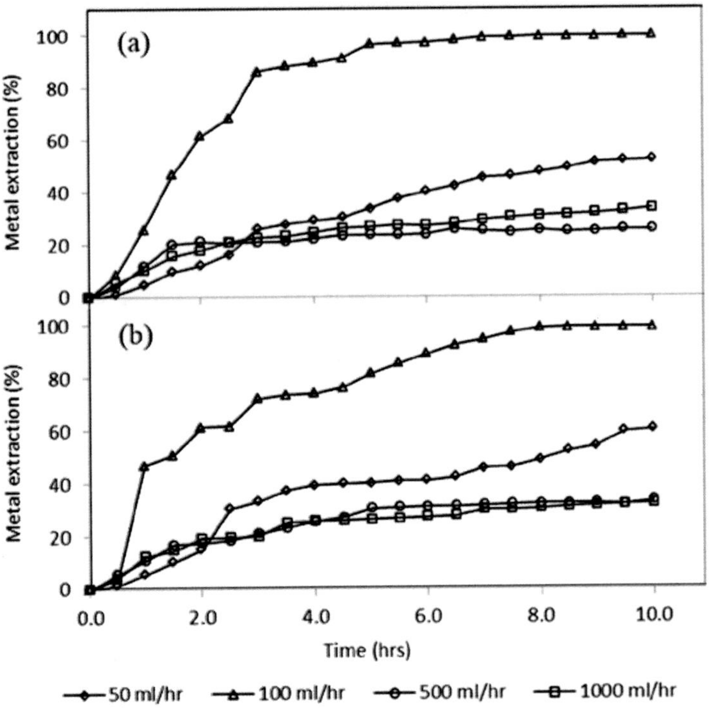

Fig. 7. Effect of flowrate on extraction efficiency of (a) Cu and (b) Zn from waste PCB of EoL-MP.

4 Conclusions

The results obtained from the present column leaching study revealed that metals viz., Cu and Zn can be efficiently extracted from EoL-MP under optimised conditions about 97% and 98.9%, respectively. Analysis of variance results revealed R^2 value of 0.9981, which shows the second order regression model is satisfactorily adjusted with the experimental results. The optimised operational parameters for metal leaching process using RSM were 0.5 M DTPA: 50°C temperature: 1:75 solid to liquid ratio: 9 pH: 0.9 M H_2O_2. Further from the column leaching study with varying flow rate from 50 to 1000 mL/h revealed the optimum flowrate of 100 mL/h for metal leaching using DTPA in combination with H_2O_2. Further, a step in the direction of scaling up the recycling process was taken. Column leaching experiment was performed under down flow (top to bottom) mode which might create a preferential path at high flow rate resulting in low leaching efficiency. Thus, a detailed study is warranted under different flow mode for performance evaluation of its effect on column leaching of metals from e-waste. In-addition, continuous stir tank reactor (CSTR) experiments could be performed to scale up the process.

References

Baldé, C.P., Forti, V., Gray, V., Kuehr, R. and Stegmann, P. (2017). The Global E-waste Monitor–2017, United Nations University (UNU), International Telecommunication Union (ITU) & International Solid Waste Association (ISWA), Bonn/Geneva/Vienna. *Electronic Version*, pp. 978–992.

Behera, S.K., Meena, H., Chakraborty, S. and Meikap, B.C. (2018). Application of response surface methodology (RSM) for optimization of leaching parameters for ash reduction from low-grade coal. *Int J Mining Sci Technol*, **28(4):** 621–629.

Beigl, P., Schneider, F. and Salhofer, S. (2012, February). Takeback systems for mobile phones: Review and recommendations. *In: Proceedings of the Institution of Civil Engineers-Waste and Resource Management*, **165(1):** 25-35. ICE Publishing.

Chauhan, G., Pant, K.K. and Nigam, K.D.P. (2015). Chelation technology: A promising green approach for resource management and waste minimization. *Environ Sci: Processes Impacts*, **17(1):** 12–40.

Deveci, H., Yazici, E., Aydin, U., Yazici, R. and Akcil, A. (2010, November). Extraction of copper from scrap TV boards by sulphuric acid leaching under oxidising conditions. *In: Proceedings of the Going Green-CARE INNOVATION 2010 Conference* (pp. 8–11).

E-waste (Management) Rules (2016). Ministry of Environment, Forest and climate Change, Government of India http://moef.gov.in/wp-content/uploads/2017/07/EWM-Rules-2016-english-23.03.2016.pdf. Accessed 2 July 2019.

Huang, K., Guo, J. and Xu, Z. (2009). Recycling of waste printed circuit boards: A review of current technologies and treatment status in China. *J Hazard Mater*, **164(2–3):** 399–408.

Jadhao, P., Chauhan, G., Pant, K.K. and Nigam, K.D.P. (2016). Greener approach for the extraction of copper metal from electronic waste. *Waste Manage*, **57:** 102–112.

Kasper, A.C., Berselli, G.B., Freitas, B.D., Tenório, J.A., Bernardes, A.M. and Veit, H.M. (2011). Printed wiring boards for mobile phones: Characterization and recycling of copper. *Waste Manage*, **31(12):** 2536–2545.

Lee, J.C., Song, H.T. and Yoo, J.M. (2007). Present status of the recycling of waste electrical and electronic equipment in Korea. *Resources Conserv Recycl*, **50(4):** 380–397.

Li, J., Lu, H., Guo, J., Xu, Z. and Zhou, Y. (2007). Recycle technology for recovering resources and products from waste printed circuit boards. *Environ Sci Technol*, **41(6):** 1995–2000.

Means, J.L., Kucak, T. and Crerar, D.A. (1980). Relative degradation rates of NTA, EDTA and DTPA and environmental implications. *Environ Pollut B, Chem Phys*, **1(1):** 45–60.

Mishra, D., Kim, D.J., Ahn, J.G. and Rhee, Y.H. (2005). Bioleaching: A microbial process of metal recovery; a review. *Met Mater Int*, **11(3):** 249–256.

Monteiro, M.R., Moreira, D.G., Chinelatto, M.A., Nascente, P.A. and Alcântara, N.G. (2007). Characterization and recycling of polymeric components present in cell phones. *J Polym Environ*, **15(3):** 195–199.

Montgomery, D.C. (2017). *Design and Analysis of Experiments*. John Wiley & Sons.

Osibanjo, O. and Nnorom, I.C. (2008). Material flows of mobile phones and accessories in Nigeria: Environmental implications and sound end-of-life management options. *Environ Impact Assess Rev*, **28(2–3):** 198–213.

Pastor, J., Aparicio, A.M., Gutierrez-Maroto, A. and Hernández, A.J. (2007). Effects of two chelating agents (EDTA and DTPA) on the autochthonous vegetation of a soil polluted with Cu, Zn and Cd. *Sci Total Enviro*, **378(1–2):** 114–118.

Priya, A. and Hait, S. (2017). Comparative assessment of metallurgical recovery of metals from electronic waste with special emphasis on bioleaching. *Environ Sci Pollut Res*, **24(8):** 6989–7008.

Robinson, B.H. (2009). E-waste: An assessment of global production and environmental impacts. *Sci Total Environ*, **408(2):** 183–191.

Tuwati, A. and Fan, M. (2015). Development of a contaminant leaching model for aquifer storage and recovery technology. *Cogent Environ Sci*, **1(1):** 1066177.

USEPA. (1995). *Microwave Assisted Acid Digestion of Siliceous and Organically based Matrices*. USEPA Method 3052, 3rd edn. United States Environmental Protection Agency, Washington, DC.

Verma, A. and Hait, S. (2019). Chelating extraction of metals from e-waste using diethylene triamine pentaacetic acid. *Process Saf Environ Protect*, **121:** 1–11.

Yang, H., Liu, J. and Yang, J. (2011). Leaching copper from shredded particles of waste printed circuit boards. *J Hazard Mater*, **187(1–3):** 393–400.

Yi, S., Su, Y., Qi, B., Su, Z. and Wan, Y. (2010). Application of response surface methodology and central composite rotatable design in optimizing the preparation conditions of vinyltriethoxysilane modified silicalite/polydimethylsiloxane hybrid pervaporation membranes. *Sep Purif Technol*, **71(2):** 252–262.

Chapter 23
Assessment of Sand Size on ECC Containing Waste Materials

Maninder Singh, Babita Saini, and H. D. Chalak

1 Introduction

Cement-based products are of great importance in the construction industry. Various types of cement products are available in the library of construction industry like mortar, normal concrete, high strength concrete fibre reinforced concrete, high performance fibre reinforced concrete (HPFRC), ultra-high performance fibre reinforced concrete (UHPFRC) etc. The most common/higher percentage of mortar and normal concrete are used for the construction purpose due to low cost and easily available ingredients. Due to poor durability, seismic effect, life span and brittle nature of these products lots of problem arise (Ramezanianpour et al., 2011; Gambhir, 2004; Belleghem et al., 2016; Abou-Zeid et al., 2001; Wong et al., 2012). However, many more cement-based products are available in the library of construction industry, which have better durability performance, seismic response, life span and different nature than normal concrete. Over the superb characteristics of some cement-based materials like HPFRC, UHPFRC, engineered cementitious composite (ECC) etc. (Woodson, 2012; Li, 1998; Singh et al., 2019a), the use of these materials is limited in construction industry due to high cost and huge consumption of cement. Numerous research investigation reported that the manufacturing process of cement create many types imbalances in the environmental ecosystem. On the other hand, industrial sector produced a large amount of solid waste products like glass waste, blast furnace slag (BFS), stone dust, rubber waste, palm oil fuel ash fly ash (FA), waste foundry sand, sewage sludge ash, plastic waste, rice husk ash,

M. Singh (✉) · B. Saini · H. D. Chalak
Civil Engineering Department, National Institute of Technology Kurukshetra, Kurukshetra, Haryana, India

A. S. Kalamdhad (ed.), *Integrated Approaches Towards Solid Waste Management*, https://doi.org/10.1007/978-3-030-70463-6_23

red mud, silica fume, jarosite, hyposludge etc., which also create lots of problems to the environment (Liew et al., 2017; Almeida et al., 2007; Singh et al., 2019b; Dos et al., 2017; Hu and Li, 2015). Therefore, due to environmental concern the recent advances have been made in construction industry by using the waste products as substitution of conventional products.

ECC is the novel class of cement-based products featuring ductile nature and tiny crack width. The unique characteristics of ECC like high tensile strain capacity, narrow crack width and excellent durability performance made it different from other kinds of materials. Literature reported that tensile strain capacity of ECC varies between 1% and 8% and crack width less than 100 μm (Singh et al., 2019c; Sahmaran et al., 2011; Li and Kanda, 1998; Lepech and Li, 2006). The design mechanism of ECC is based on the micromechanics of fibre bridging and matrix crack extension (Zhu et al., 2012; Pakravan et al., 2016). The cement, supplementary cementitious materials, silica sand and polymeric fibres are the most commonly used constituents in ECC. The bigger size of aggregates is not included in the design of ECC, because they affect the ductility performance of cementitious composites. Under tensile loading, the ECC behaviour depicts strain hardening after first crack rather than tension softening in conventional FRC (fibre reinforced concrete) (Li, 1998; Singh et al., 2019a; Zhu et al., 2012; Pakravan et al., 2016). The characteristics of ECC depicted that it is novel material; but, the higher consumption of cement and limited use of very fine aggregates enhanced the overall cost of the ECC matrix and limit the use of it on large scale structural applications. In the present study, effect of high amount of FA and BFS as supplementary cementitious materials and locally available river sand (RS) of different sizes as substitution of silica sand (SS) on performance of ECC has been studied to address the above-mentioned parameters. The utilisation of these products can save the environmental ecosystem and also provides better guidelines for future design.

2 Selection of Materials

The solid materials selected in this work were Portland cement (PC), FA and BFS as supplementary cementitious materials, silica sand (SS), locally available river sand (RS) of size 600 μm, 1.18 mm and 2.36 mm. The polypropylene (PP) and steel (SE) fibres were used as internal reinforcement in cementitious composite. To achieve the required flowability high range, water reducer admixture and water were used.

Mix Proportions

Total eight mixes were prepared to investigate the influence of various sizes of sand and waste products on performance of ECC. The proportions and designation of various mixes have been given in Table 1.

Table 1. Mix proportions of cementitious composite

		Ingredients							
Mix Id	*Size of sand*	*PC*	*FA*	*BFS*	*Sand*	*w/b*	*SP*	*PP fibre (%)*	*SE fibre (%)*
FA_SS	300 μm	1	2.5	–	0.8	0.27	0.60	1	1
FA_RS 1	600 μm	1	2.5	–	0.8	0.27	0.60	1	1
FA_RS 2	1.18 mm	1	2.5	–	0.8	0.27	0.60	1	1
FA_RS 3	2.36 mm	1	2.5	–	0.8	0.27	0.60	1	1
BFS_SS	300 μm	1	–	2.5	0.8	0.27	0.60	1	1
BFS_RS 1	600 μm	1	–	2.5	0.8	0.27	0.60	1	1
BFS_RS 2	1.18 mm	1	–	2.5	0.8	0.27	0.60	1	1
BFS_RS 3	2.36 mm	1	–	2.5	0.8	0.27	0.60	1	1

3 Casing and Testing Program

The mixing process of ECC was done by using power driven mortar mixer. The prepared fresh ECC mix was poured into the moulds of desired shape to harden the matrix. The cubical and prisms shaped specimens of size 70.6 mm × 70.6 mm ×70.6 mm and 500 mm × 100 mm × 100 mm respectively were casted to access the compressive and flexural behaviour of various mixes. The casted specimens were removed from the moulds after 24 h and then kept in water for required curing age.

3.1 Compressive Behaviour

To record the compression behaviour of various ECC mixes, the cubical shaped specimens were tested in the universal testing machine (UTM) as per IS 516:1959 and BS-EN-12390 part 3 specifications.

3.2 Flexural Response

To analyse the flexural response (strength and deflection capacity), the prism specimens were used and tested as per IS 516:1959 and BS-EN-12390 part 5 specifications by performing four point bending test.

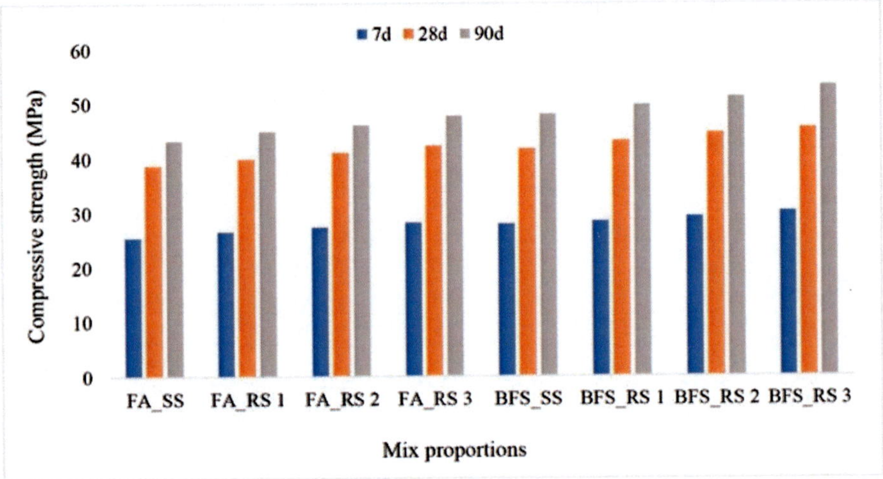

Fig. 1. Compressive strength of various ECC mixes.

4 Results and Discussion

4.1 Compressive Strength

The type of waste products and aggregate size adversely impact the compressive strength of ECC mixture as illustrated in Fig. 1. The compressive strength of BFS blended ECC mixes was higher than FA blended mixes with the usage of both type of sand. The compressive strength of BFS_SS mix proportion was higher by 7.37% as compared to FA_SS mix proportion after 28 days water curing. Similar observation was reported by Zhu et al. (2012) BFS as FA substitution enhanced the compressive strength of ECC matrix. The inclusion of RS as SS substitution improved the compressive strength with the usage of both type of waste products. The utilisation of RS improved the compressive strength of FA_SS and BFS_SS mix proportion up to 3.92% and 3.41% respectively. The experimental results revealed that increment in the size of RS improved the matrix toughness which resulted into compressive strength enhancement. The optimum compressive strength was observed in BFS_RS 3 mix proportion with the usage of 2.36 mm size of RS.

4.2 Flexural Strength

The type of waste products and aggregate size adversely impact the flexural response of different prism specimens. It has been noticed from Fig. 2 that the flexural strength of BFS blended ECC mixes was higher than FA blended mixes with the

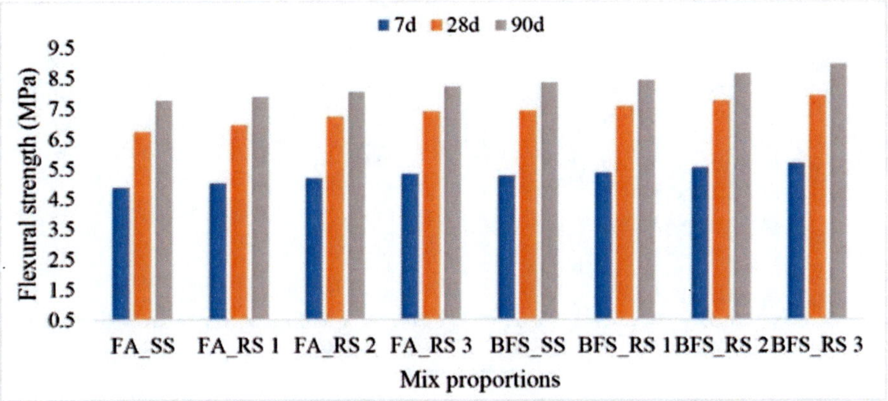

Fig. 2. Flexural strength of various ECC mixes.

usage of both type of sand. The flexural strength of BFS_SS mix proportion was higher by 10.26% as compared to FA_SS mix proportion after 28 days water curing. The inclusion of RS as substitution of SS improved the flexural strength with the usage of both type of sand. The utilisation of RS improved the flexural strength of FA_SS and BFS_SS mix proportion up to 3.27% and 2.03% respectively. The experimental results depicted that increment in the size of RS affect the flexural response of various mixes. The higher RS size enhanced the matrix toughness, which resulted into flexural strength enhancement. The best flexural strength was observed in BFS_RS 3 mix proportion with the usage of 2.36 mm size of RS.

4.3 *Mid Span Deflection*

The type of waste products, aggregate types and size strongly influence the mid span deflection of ECC mixes. It has been noticed from Fig. 3 that the deflection capacity of BFS incorporating ECC mixes was higher than FA mixes. The deflection capacity of BFS_SS mix proportion was higher up to 4.14% as compared to FA_SS mix proportion. The impact of aggregate types and size on deflection capacity was found vice versa of compressive and flexural strength. The utilisation of RS reduced the deflection capacity of FA_SS and BFS_SS mix up to 13.57% and 12.41% respectively. The maximum reduction was observed with the usage of 2.36 mm size aggregates. The observed results depicted that the unique bending performance of ECC was negatively affected with the increment in the size of aggregates. The optimum deflection capacity was observed in FA_SS mix proportion.

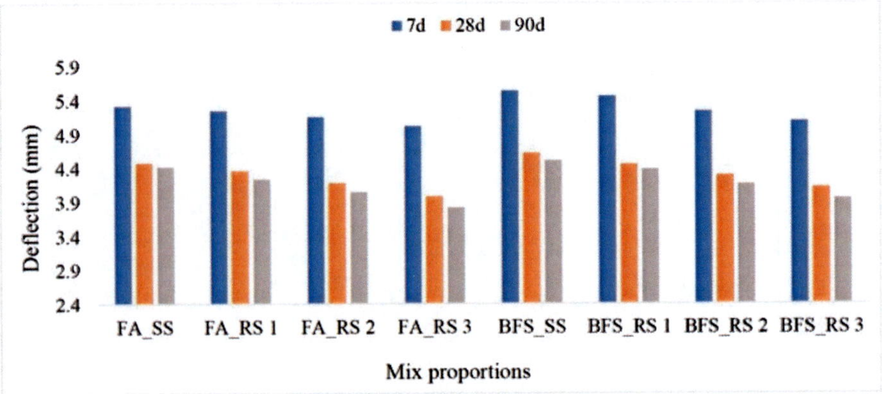

Fig. 3. Mid span deflection of various ECC mixes.

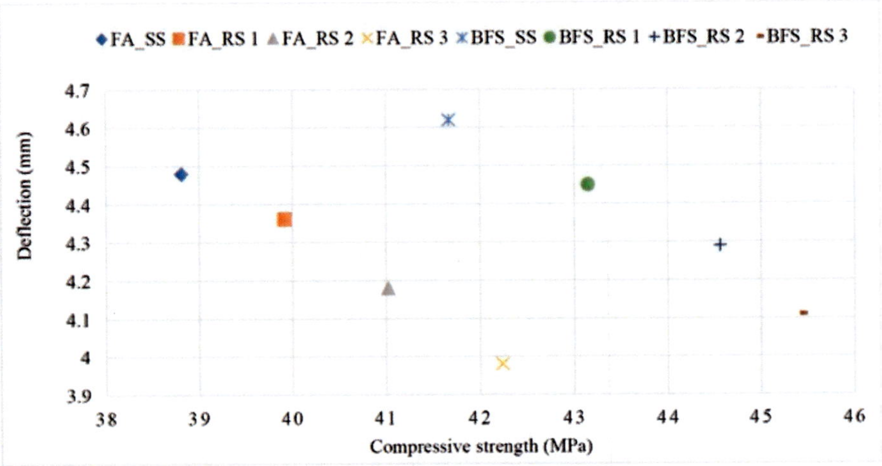

Fig. 4. Relationship between compressive strength and deflection capacity.

4.4 Relationship Strength Parameters and Deflection Capacity

Figures 4 and 5 depict the relationship between strength parameters and deflection capacity at 28 days water curing. From this relationship, it can be observed that the increment in the size of RS enhanced the strength parameters and as well as reduced the deflection capacity. The same trend was observed for 7 and 90 days water curing. The reported relationship confirms that the increment in the size of fine aggregates provide negative effect on ductile nature of ECC. The present relationship also revealed the superiority of BFS over FA containing ECC mixes.

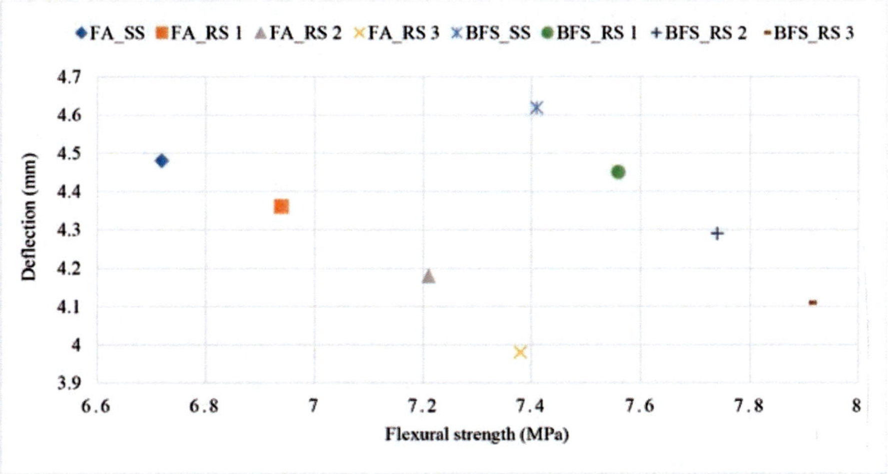

Fig. 5. Relationship between flexural strength and deflection capacity.

5 Conclusions

The present investigation describes the performance of ECC produced with different waste products and varying size of locally available river sand. The performance of ECC mixes was evaluated in terms of compressive, flexural strength and deflection capacity. The conclusions from the present study have been given as follows:

- The BFS utilised mix proportions depicted better performance than FA utilised mix proportions. The same trend was observed when the ECC was produced with RS in place of SS.
- The compressive strength, flexural strength and deflection capacity of BFS_SS mix proportion was higher by 7.37%, 10.26% and 3.12% as compared to FS_SS mix proportion after 28 days water curing.
- The strength parameters of RS containing mixes was higher than SS containing mixes; while, the deflection capacity reduced.
- Increase in size of RS (2.36 mm) improved the strength parameters of different ECC mixes; while reduced the deflection capacity for all curing ages.
- The relationships between strength parameters and deflection capacity confirm that the increment in the size of RS improved the strength parameters, while reduced the mid span deflection.
- In present investigation the successful utilisation of high-volume waste products solves the problem of disposal, water contamination nuisance, extraction of natural resources, toxicity and contribute in making the eco-friendly environmental ecosystem. The present research investigation helps in understanding the effect of waste products and aggregate size on performance of ECC.

Acknowledgement The authors feel obliged to the University Grants Commission, New Delhi for the financial assistance for research work.

References

Abou-Zeid, M., Fowler, D.W., Nawy, E.G., Allen, J.H., Halvorsen, G.T., Poston, R.W., et al. (2001). Control of cracking in concrete structures. *ACI Committee*, **224:** 12–16.

Almeida, N., Branco, F., Brito, J.D. and Santos, J.R. (2007). High-performance concrete with recycled stone slurry. *Cem Concr Res*, **37:** 210–220.

Belleghem, B.V., Montoya, R., Dewanckele, J., Steen, N.V.D., Graeve, D., Deconinck, J., Cnudde, V., Tittelboom, K.V. and Belie, N.D. (2016). Capillary water absorption in cracked and uncracked mortar - A comparison between experimental study and finite element analysis. *Constr Build Mater,* **110:** 154–162.

BS-EN-12390-3 (2009). Testing Hardened Concrete, Compressive Strength of Test Specimens. British Standard Institution.

BS-EN-12390-5 (2009). Testing Hardened Concrete, Flexural Strength of Test Specimens. British Standard Institution.

Dos Anjos, M.A.G., Sales, A.T.C. and Andrade, N. (2017). Blasted copper slag as fine aggregate in Portland cement concrete. *J Environ Manage*, **196:** 607–613.

Gambhir, M.L. (2004). Concrete Technology. The McGraw Hill companies.

Hu, C. and Li, Z. (2015). Property investigation of individual phases in cementitious composites containing silica fume and fly ash. *Cem Concr Composites*, **57:** 17–26.

IS 516-1959 (2006). Methods of Tests for Strength of Concrete (18th ed.). Bureau of Indian Standards, New Delhi.

Lepech, M. and Li, V.C. (2006). Long term durability performance of engineered cementitious composites. *Restor Build Monuments*, **12:** 119–132.

Li, V.C. (1998). Engineered cementitious composites - tailored composites through micromechanical modeling. Fiber reinforced concrete: Present and the future. *In:* Canadian Society for Civil Engineering, Montreal, pp. 64–97.

Li, V.C. and Kanda, T. (1998). Engineered cementitious composites for structural applications. *J Mater Civil Eng*, **10:** 66–69.

Liew, K.M., Sojobi, A.O. and Zhang, L.W. (2017). Green concrete: Prospects and challenges. *Constr Build Mater*, **156:** 1063–1095. doi: https://doi.org/10.1016/j.conbuildmat.2017.09.008.

Pakravan, H.R., Jamshidi, M. and Latifi, M. (2016) Study on fiber hybridization effect of engineered cementitious composite with low and high-modulus polymeric fibers. *Constr Build Mater*, **112:** 739–746.

Ramezanianpour, A.A., Pilvar, A., Mahdikhani, M. amd Moodi, F. (2011). Practical evaluation of relationship between concrete resistivity, water penetration, rapid chloride penetration and compressive strength. *Constr Build Mater*, **25:** 2472–2479.

Sahmaran, M., Ozbay, E., Lachemi, M., Li, V.C., Yucel, H.E. (2011). Effect of fly ash and PVA fiber on microstructural damage and residual properties of engineered cementitious composites exposed to high temperatures. *J Mater Civil Eng*, **23:** 1735–1745.

Singh, M., Saini, B. and Chalak, H.D. (2019a). Performance and composition analysis of engineered cementitious composite (ECC) – A review. *J Build Eng*, **26:** 100851. doi: https://doi.org/10.1016/j.jobe.2019.100851.

Singh, M., Saini, B. and Chalak, H.D. (2019b). Properties of engineered cementitious composites: A review. ICSWMD 2018, LNCE 21, pp. 1–11. doi: https://doi.org/10.1007/978-3-030-02707-0_54.

Singh, M., Saini, B. and Chalak, H.D. (2019c). Appraisal of hybrid fiber reinforced engineered cementitious composite. *In:* Proceedings of the 4th International Conference on Civil, Structural and Transportation Engineering (ICCSTE'19). Ottawa, Canada, June (2019). doi: https://doi.org/10.11159/iccste19.192.

Wong, H.S., Zimmerman, R.W. and Buenfeld. N.R. (2012). Estimating the permeability of cement pastes and mortars using image analysis and effective medium theory. *Cem Concr Res*, **42:** 476–483.

Woodson, R.D. (2012). Concrete materials. *In*: Concrete Portable Handbook. Elsevier Inc., Oxford, pp. 5–18.

Zhu, Y., Yang, Y. and Yao, Y. (2012). Use of slag to improve mechanical properties of engineered cementitious composites (ECCs) with high volumes of fly ash. *Constr Build Mater*, **36:** 1076–1081.

Chapter 24
Effect of Operating Parameters on Leachate Quantity and Quality Generated During Hydrolysis of Kitchen Waste

Rishi Gurjar and Manaswini Behera

1 Introduction

The global municipal solid waste (MSW) generation rate is 0.74 kg/capita/day, with an estimated annual waste generated to be 2.0 billion tonnes. It is poised to increase up to 3.4 billion tonnes by 2050 (Kaza et al., 2018). The composition of waste broadly comprises biodegradable and non-biodegradable fractions that depend on regional geography, economy, climate, lifestyle, social awareness and regulations. The biodegradable fraction majorly contributes between 40% and 50%, across the globe, dominated by kitchen waste, i.e., food, fruits and vegetables (Gerlagh et al., 1999; Futures, 2009; Yap and Nixon, 2015; Kaza et al., 2018). Fundamentally, these components are made up of carbohydrates, fats, proteins, lipids, inorganic salts and trace elements that led to a 40–70% bioenergy conversion efficiency (Ma et al., 2018). This high energy potential can be recovered by disintegrating waste into simpler compounds by anaerobic digestion (AD). The AD comprises hydrolysis, acidogenesis, acetogenesis and methanogenesis. It is considered as an effective technique to treat high moisture content waste (McKendry, 2002). Conventionally, the energy potential of organic waste has been used to obtain biogas and manure. The main component of biogas is methane (a GHG), posing a high risk to the environment. Also, the treatment of high moisture waste in thermal technologies reduces combustion efficiency drastically. Therefore, obtaining valuable products having a high commercial value such as ethanol, hydrogen, electricity and chemicals, broadens the waste to energy (WTE) concept.

The AD is required to operate under the acidogenic phase with the help of an intermediate step for achieving WTE. Popularly, a leach bed reactor (LBR) has been used to obtain methane from the AD of organic waste. Nevertheless, its ability to

R. Gurjar · M. Behera (✉)
School of Infrastructure, Indian Institute of Technology Bhubaneswar Argul, Bhubaneswar, Odisha, India
e-mail: manaswini@iitbbs.ac.in

A. S. Kalamdhad (ed.), *Integrated Approaches Towards Solid Waste Management*, https://doi.org/10.1007/978-3-030-70463-6_24

carry hydrolysed products in water and its subsequent replacement will promote acidogenic phase activity. Thus facilitating grounds to obtain other valuable by-products in a two-stage system, combining LBR and product-specific treatment technology, e.g., fermentation (hydrogen and ethanol) and microbial fuel cell (electricity). The LBR is a percolating dry anaerobic digestion column divided into three compartments, i.e., top, middle and bottom for gas collection, solid waste support and leachate collection, respectively. It carries out simultaneous liquefaction and acidification of high solid waste (>30%) and shows improved conversion efficiencies because of enhanced transport of volatile fatty acids (VFAs) by leachate generation, among reactors employing AD phenomenon (Mata-Alvarez et al., 2000; Salomoni et al., 2011). Leachate is an organic and nutrient-rich liquid medium formed by a collection of organic compounds such as acetic, butyric, propionic and valeric acid along with long-chain fatty acids termed as volatile fatty acids (VFAs). Their utilisation order, as suggested by Escapa et al. (2009), is as follows, acetic acid > butyric acid > propionic acid. These readily consumed VFAs are essential for any second-stage treatment technique to maximise product yield (Lee et al., 2008).

The generation of VFA-rich leachate hinges on several operating parameters of LBR, such as pH, temperature, solid content, waste particle size, carbon/nitrogen (C/N) ratio and leachate recirculation. A desirable pH for effective hydrolysis is proposed in the alkaline region between 7 and 9, but this range facilitates methanogens growth (Dahiya et al., 2015). Similarly, for the acidogenic phase, acidogenic bacteria can sustain between a broad range of 3–7 (Wu et al., 2006). The desired pH for acidogenic bacteria based on pK_a of acetic acid (4.75) should lie within 5.5–6.5 (Yu and Fang, 2002) and 4–6.5 (Speece, 1983). The initial pH of leachate depends on the type of waste, while low pH indicates organic acid generation. The low pH points to a higher hydrolysis rate of carbohydrates as compared to protein and lipids that promote alkalinity. Further, short solid retention time and low pH reduces the development of hydrogen consumers (methanogens) (Zahedi et al., 2013). The short retention time is governed by the organic loading rate (OLR) and make sure waste remains in the acidogenic phase.

Additionally, a C/N ratio also plays a vital role in the AD of waste. A higher ratio will develop deprived conditions as all life forms for survival require nitrogen; whereas, a lower ratio will lead to ammonia toxicity due to rapid protein hydrolysis (Chen et al., 2008). A C/N ratio between 20 and 30 is found satisfactory for efficient biological functioning (Khalid et al., 2011). Preferably, wastes not having the required ratio are treated by co-digestion, i.e., mixing of lower or higher nitrogen content waste to obtain the required ratio (Wu et al., 2010). Furthermore, the leachate recirculation serves as a substrate, regulates pH and acts as a medium between waste and microbes (Karim, 2012). Moreover, the smaller size particles improve the surface area for interaction with microbes. Therefore, the study aims to optimise operating conditions, i.e., pH (5, 6 and 7) and organic loading rate (5, 7.5 and 10 gVS/L·day) that will enhance and sustain acidogenic phase in an LBR during AD of KW.

2 Materials and Methods

2.1 Leach Bed Reactor Set-up

The LBR (7 cm × 7 cm × 30 cm) was fabricated with transparent acrylic sheets having a total volume of 1.47 L (effective volume = 1 L), provided with suitable inlets (top) and outlet (bottom). The inlets maintained desired OLR and collected/sparged gases while outlets collected daily leachate. A stirring mechanism (Geared Motor 12V DC, torque 0.2 N m; 60 rpm max) having a controller was mounted on top to mix grounded KW for two h at 30 rpm. The solid waste was placed on top of a gravel layer (1 cm thick) followed by stainless steel sieve (pore size 3 mm) and cotton cloth on a perforated acrylic plate (thickness, 0.5 cm) to establish a leachate collection system. The gravel layer was sandwiched between two SS mesh to avoid mixing in the solid waste bed. The schematic of the LBR is shown in Fig. 1.

Fig. 1. Schematic diagram of compact LBR.

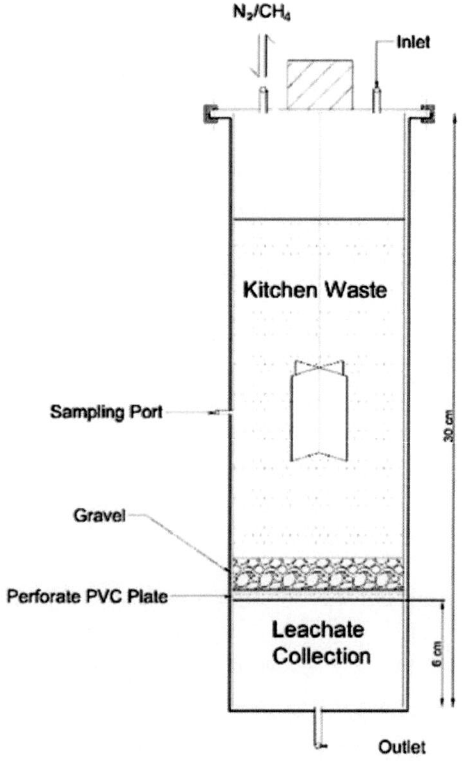

2.2 Analytical Procedure

The waste properties are unique to the source of its generation. Therefore, data must be collected for seven days in each season, i.e., summer, winter and rainy seasons (Municipal Solid Waste Management Manual, 2016). The various physicochemical parameters for solid waste viz. pH, chemical oxygen demand (COD), moisture content (MC), volatile solids (VS) and total solids (TS) were determined by Standard Methods (2005) and bulk density (ASTM E-1109-86, 2004), in a cubical wooden box (60 cm × 60 cm × 60 cm). The waste slurry was prepared by adding 1 g dry weight of waste in distilled water (50 mL) and later centrifuged for 15 min at 2500 revolution per minute (rpm), to determine pH and COD. The LBR's acidogenic leachate was analysed for pH, total chemical oxygen demand (TCOD), soluble COD (sCOD) as per Standard Methods (2005) and total volatile fatty acids (TVFA) as suggested by Anderson and Yang (1992) on alternate days. The LBR was buffered with the help of NaOH. The total and volatile solids present in the KW and inoculum were determined as suggested in Standard Methods (2005).

2.3 Experimental Setting

The KW (TS 32.37%, versus 90.30%) comprising cooked (rice, chapattis and curries) and pre-cooked food (fruits and vegetables with their peel) in 70:30 ratio (w/w) were first grounded to a thick paste with an electric mixer. This paste was mixed thoroughly with heat-treated inoculum obtained from the lake inside the Indian Institute of Technology - Bhubaneswar (IIT-BBS) campus. The inoculum (sieved at 1 mm) was heat-treated in the oven for 15 min at 100°C to suppress methanogenesis (Behera et al., 2010). At last, a 1L mix was prepared having inoculum (20%, v/v), water (300 mL) and grounded KW in LBR. The mix was allowed to acclimatise for three days with leachate recirculation and no replacement. Further, KW was added to LBR daily for maintaining the desired OLR (as per Table 1). After the acclimatisation period, leachate replacement of 50% was carried out by distilled water.

Table 1. The sequence for optimisation of the operating conditions forLBR treating KW

		Variable parameters	
S. no	*Run no.*	*Organic loading rate (g·VS/L·day)*	*pH*
1	1	7.5	6
2	2	5	5
3	3	5	7
4	4	10	5
5	5	7.5	6
6	6	10	7

3 Results and Discussions

3.1 Substrate Characterisation

The characteristics of the waste affect its AD in a leach bed reactor. One of the fundamental properties of waste is TS values, which influences loading and overall organic and nutrient production. Likewise, the VS present in waste contributes to its ease of degradability when subjected to AD. It ensures the high output of targeted parameters suitable for energy recovery in the second stage treatment technique. The waste characteristics are similar to other studies and are presented in Table 2 (Ten Brummeler, 2000; Aymerich et al., 2013). Additionally, the hydrolysis of waste is a rate-limiting factor in AD (Veeken et al., 2000) and several elements that contribute to its slow rate include moisture limitation, non-uniform shredding of waste, high bulk density and insufficient inoculum.

A high moisture content means more water present in contact with waste that stimulates waste hydrolysis. The hydrolysed products generated are easily solubilised in water and then transported with the help of leachate from one phase to another phase. However, the generation of leachate depends on waste's capacity to hold water, i.e., field capacity. It can be readily reached or exceeded by recirculating leachate or by adding water. The solubilisation, by leachate replacement with distilled/tap water, can be enhanced.

On the other hand, high bulk density indicates more amount of solid particles are in close contact with each other along with water, thus facilitating better contact (Barlaz et al., 1990). Furthermore, the inoculum improves the potential to hydrolyse waste with higher acid generation. Its addition ascertains a mixed culture of microorganisms is available. Nevertheless, the growth of specific organisms will depend on the operating conditions, as individual parameters promote/restrict the growth of various microbes. The hydrolysis of organic waste can be further expedited by decreasing waste size. It improves the wastes' surface area for better contact with microorganisms. The shredded waste is easily compacted and result in higher densities that can accelerate hydrolysis.

Table 2. Characterisation of the kitchen waste

Sl. No	*Parameters*	*Average Values*
1	pH	4.39 ± 0.10
2	Moisture content (%)	76.37 ± 4.17
3	Volatile solids (%)	92.48 ± 1.80
4	Total solids (%)	23.63 ± 4.17
5	Chemical oxygen demand (COD) (g/L)	42.32 ± 4.23
6	Bulk density (kg/m^3)	1069.92 ± 26.10

3.2 COD and TVFA Generation

The LBRs were subjected to various OLR and pH for a period of 21 days as per the sequence shown in Table 1. In all runs, the organic strength of leachate increased gradually with time, but the effect of different pH on the performance of LBR was contrasting. The higher OLR in comparison to lower OLR resulted in a high contribution to organics in leachate from the start; whereas, pH was vital to maintain the LBR under the acidogenic phase. Despite R2 (OLR 5 g·VS/L·day, pH 5) and R3 (OLR 5 g·VS/L·day, pH 7) having same OLR, the TVFA generation was at 4.01 g/L in R3 as compared to 5.40 g/L in R2. But maximum sCOD for R3 was higher at 48.79 g/L than 38.82 g/L, along with a higher TCOD of 53.21 g/L was produced by R3 in comparison to 47.10 g/L in R2. The average sCOD for R3 was 20% more than R2 (33.7 g/L), while the average TCOD was nearly similar for both runs at 50 g/L. A near-neutral pH resulted in better hydrolysis of the waste and simultaneous acidification in R3. The response of LBR for R2 and R3 is shown in Fig. 2(b) and (c), respectively. A similar trend in R4 (OLR 10 g·VS/L·day, pH 5) and R6 (OLR 10 g·VS/L·day, pH 7), is observed. The former showed marginally higher TVFA generation of 9.50 g/L than the latter having 8.91 g/L. The lower pH facilitated the growth of acidogens and resulted in higher waste acidification. Though the difference in TVFA generation is not significant, operating conditions had a stark difference in the hydrolysis of the KW as R6 produced higher hydrolysis, which resulted in a 43% TCOD increment from 64.42 g/L (R4) to 92.37 g/L. The obtained values of leachate properties in R4 and R6 are respectively illustrated in Fig. 2(d) and (f).

On the other hand, two identical runs R1 and R5 (OLR 7.5 g·VS/L·day, pH 6) backed up the narrative as observed in different runs. Both the runs provided a balance between hydrolysis and acidogenesis of the waste, as in both instances, TVFA and TCOD values remained almost unchanged. The acidification (TVFA/sCOD) and solubilisation efficiency remained close to 17% and 72%, for both runs, respectively. The R5 and R1 produced a maximum TVFA value greater than 9.99 g/L. However, a longer operational period for higher TVFA generation in R1 and R5 is desirable. The leachate characteristics obtained are presented in Fig. 2(a) and (e). Further, the pH values were under pH 7, thereby suggesting LBR's operation following an acidogenic phase. The pH values maintained during all the operational conditions are shown in Fig. 3.

Finally, all the runs produced a leachate volume of about 600–900 mL, which was consistent and steady throughout the study period. The R1 generated maximum leachate of 870 mL and a minimum of 653 mL by R6. The higher OLR in R6 might have resulted in packing the bed, thus hindering leachate generation or passage. The quantity of leachate obtained during the operational period for all the runs is shown in Fig. 4. The LBR performance shows KW as a high biodegradable material (Jagadabhi et al., 2011).

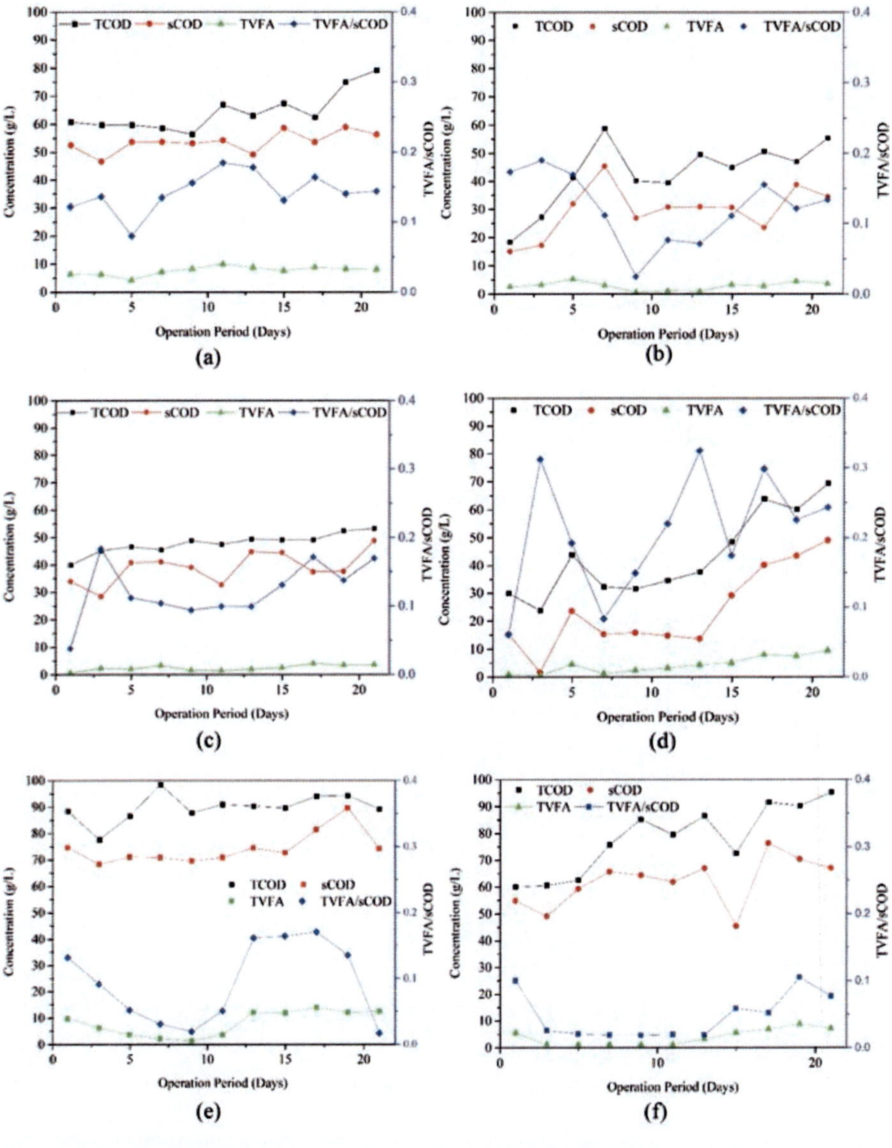

Fig. 2. COD, sCOD and TVFA concentrations for runs R1 (a), R2 (b), R3 (c), R4 (d), R5 (e) and R6 (f).

4 Conclusions

The study tried to attain operating conditions for carrying out acidogenesis of KW to be used for the recovery of valuable products instead of conventional by-products (biogas). In the present study, the LBR was subjected to various OLR and pH in a

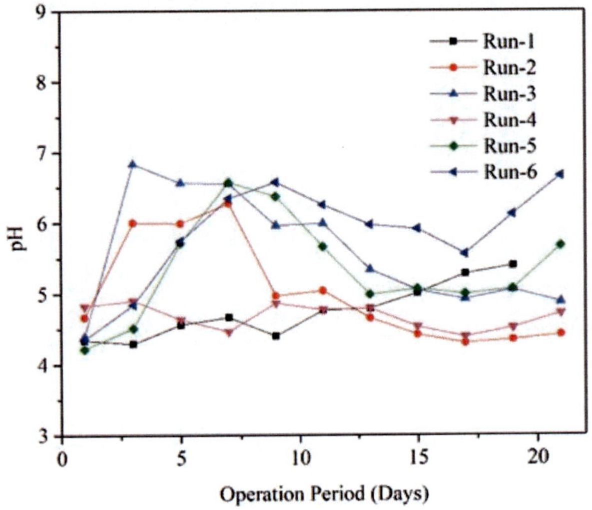

Fig. 3. pH values of leachate obtained from all the runs.

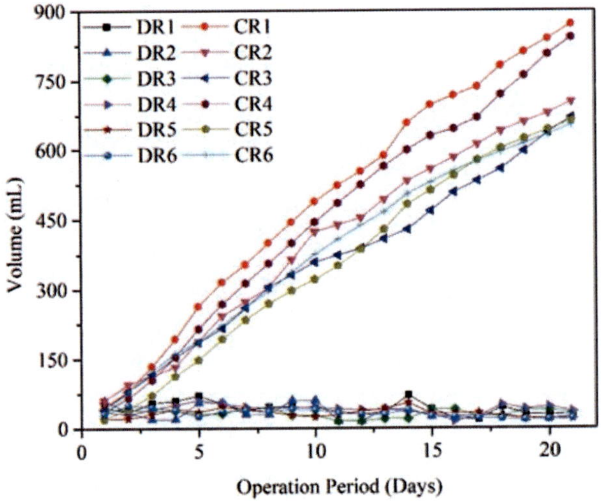

Fig. 4. Leachate quantity generated daily (D) and cumulative (C) for all the operating runs.

sequence of six runs to obtain VFA rich leachate. It was based on the premise that specific operating conditions can restrict LBR to only hydrolysis and acidogenesis steps of AD. From the study, it was observed that pH played an essential role in both phases. A lower pH encouraged the growth of acidogens, but decelerates hydrolysers growth, thereby reducing hydrolysis. Therefore, a pH promoting synergistic growth of both microbial consortia will provide an ideal scenario, in R1 and R5. These runs produced greater TVFA than 9.99 g/L, relative to other runs having higher OLR (R4 and R6). Contrastingly, at R3 a solubilisation efficiency of 92–74%

and 82% for R1 and R2 were obtained, respectively. Further, a higher TVFA/ sCOD ratio of 0.19 was obtained for R4 at pH 5 but resulted in weak hydrolysis. Additionally, a noticeable period of at least two weeks was required to generate organic acids with a steady leachate generation between 600 and 900 mL for all operational conditions. The study supplements that LBR can be operated in the acidogenic phase as a first stage treatment technique for a highly biodegradable KW to obtain VFA rich leachate for energy recovery.

In a quest to develop biodegradability of leachate, the presence of nutrients is essential, which requires characterisation. Moreover, optimising other parameters of LBR such as C/N and F/M ratio or use of electricity to enhance hydrolysis, i.e., electrohydrolysis, can improve the energy recovery potential of leachate.

Acknowledgements The grant (ECR/2016/002015) received from the Science and Engineering Research Board, Department of Science and Technology, Govt. of India, is duly acknowledged.

References

American Society for Testing and Materials (2004). Standard Test Method for Determining the Bulk Density of Solid Waste Fractions. ASTM E-1109-86.

Anderson, G.K. and Yang, G. (1992). Determination of bicarbonate and total volatile acid concentration in anaerobic digesters using a simple titration. *Water Environ Res*, **64(1):** 53–59.

Aymerich, E., Esteban-Gutiérrez, M. and Sancho, L. (2013). Analysis of the stability of high-solids anaerobic digestion of agro-industrial waste and sewage sludge. *Bioresour Technol*, **144:** 107–114.doi: https://doi.org/10.1016/j.biortech.2013.06.074.

Barlaz, M.A., Ham, R.K. and Schaefer, D.M. (1990). Methane production from municipal refuse: A review of enhancement techniques and microbial dynamics. *Crit Rev Environ Control*, doi: https://doi.org/10.1080/10643389009388384.

Behera, M., Jana, P.S. and Ghangrekar, M.M. (2010). Performance evaluation of low cost microbial fuel cell fabricated using earthen pot with biotic and abiotic cathode. *Bioresour Technol*, **101:** 1183–1189. doi: https://doi.org/10.1016/j.biortech.2009.07.089.

Chen, Y., Cheng, J.J. and Creamer, K.S. (2008). Inhibition of anaerobic digestion process: A review. *Bioresour Technol*, **99(10):** 4044–4064. doi: https://doi.org/10.1016/j.biortech.2007.01.057.

Dahiya, S., Sarkar, O., Swamy, Y.V. and Venkata Mohan, S. (2015). Acidogenic fermentation of food waste for volatile fatty acid production with co-generation of biohydrogen. *Bioresour Technol*, **182:** 103–113. doi: https://doi.org/10.1016/j.biortech.2015.01.007.

Escapa, A., Manuel, M.F., Morán, A., Gómez, X., Guiot, S.R. and Tartakovsky, B. (2009). Hydrogen production from glycerol in a membraneless microbial electrolysis cell. *Energy Fuels*, **23(9):** 4612–4618. doi: https://doi.org/10.1021/ef900357y.

Futures, R. (2009). Municipal waste composition: A review of municipal waste component analyses. Department for Environment Food and Rural Affairs. London, UK. http://randd.defra.gov.uk/Default.aspx?Module=More&Location=None&ProjectID=1513.3

Gerlagh, R., Beukering, P.V., Verma, M., Yadav, P.P. and Pandey, P. (1999). Integrated Modelling of Solid Waste in India. International Institute for Environment and Development. https://www.researchgate.net/profile/Pjh_Van_Beukering/publication/228913748_Integrated_modelling_of_solid_waste_in_India/links/00b7d51efe5524656d000000.pdf.

Jagadabhi, P.S., Kaparaju, P. and Rintala, J. (2011). Two-stage anaerobic digestion of tomato, cucumber, common reed and grass silage in leach-bed reactors and upflow anaerobic sludge blanket reactors. *Bioresour Technol*, **102(7):** 4726–4733. doi: https://doi.org/10.1016/j.biortech.2011.01.052.

Karim, A.H.A. (2012). Evaluation of a trickle flow leach bed reactor for anaerobic digestion of high solids cattle manure. M. Tech Thesis, Colorado State University, USA.

Kaza, S., Yao, L., Bhada-Tata, P. and Van Woerden, F. (2018). What a waste 2.0: A global snapshot of solid waste management to 2050. World Bank Publications. https://openknowledge.worldbank.org/handle/10986/30317.

Khalid, A., Arshad, M., Anjum, M., Mahmood, T. and Dawson, L. (2011). The anaerobic digestion of solid organic waste. *Waste Manag*, **31(8):** 1737–1744. doi: https://doi.org/10.1016/j.wasman.2011.03.021.

Lee, Z.K., Li, S.L., Lin, J.S., Wang, Y.H., Kuo, P.C. and Cheng, S.S. (2008). Effect of pH in fermentation of vegetable kitchen wastes on hydrogen production under a thermophilic condition. *Int J Hydrogen Energy*, **33(19):** 5234–5241. doi: https://doi.org/10.1016/j.ijhydene.2008.05.006.

Ma, C., Liu, J., Ye, M., Zou, L., Qian, G. and Li, Y.Y. (2018). Towards utmost bioenergy conversion efficiency of food waste: Pretreatment, co-digestion, and reactor type. *Renew Sustain Energy Rev*, **90(October 2017):** 700–709. doi: https://doi.org/10.1016/j.rser.2018.03.110.

Mata-Alvarez, J., Mace, S. and Llabres, P. (2000). Anaerobic digestion of organic solid wastes: An overview of research achievements and perspectives. *Bioresour Technol*, **74 (1):** 3–16. doi: https://doi.org/10.1016/S0960-8524(00)00023-7.

McKendry, P. (2002). Energy production from biomass (part 3): Gasification technologies. *Bioresour Technol*, **83(1):** 55–63. http://linkinghub.elsevier.com/retrieve/pii/S0960852401001201%0Afile:///Files/ce/cef45de9-df33-49f6-811b 8af68619b1bc.pdf%0Apapers3://publication/doi/10.1016/S0960-8524(01)00120-1

Municipal Solid Waste Management Manual (MSWMM) (Part I and II) (2016). Central Public Health and Environmental Engineering Organisation (CPHEEO). Ministry of Urban Development, New Delhi, India. http://cpheeo.gov.in/cms/manual-on-municipal-solid-waste-management-2016.php.

Salomoni, C., Caputo, A., Bonoli, M., Francioso, O., Rodriguez-Estrada, M.T. and Palenzona, D. (2011). Enhanced methane production in a two-phase anaerobic digestion plant, after CO_2 capture and addition to organic wastes. *Bioresour Technol*, **102(11):** 6443–6448. doi: https://doi.org/10.1016/j.biortech.2011.03.079.

Speece, R.E. (1983). Anaerobic biotechnology for industrial wastewater treatment. *Environ Sci Technol*, **17(9):** 416A-427A. doi: https://doi.org/10.1021/es00115a001.

Standard Methods (2005). *Standard Methods for Examination of Water and Wastewater*, 21st ed. American Public Health Association.

Ten Brummeler, E. (2000). Full scale experience with the BIOCEL process. *Wat Sci Tech*, **41(3):** 299–304. doi: https://doi.org/10.2166/wst.2000.0084.

Veeken, A., Kalyuzhnyi, S., Scharff, H. and Hamelers, B. (2000). Effect of pH and VFA on hydrolysis of organic solid waste. *J Environ Eng*, **126:** 1076–1081. doi: https://doi.org/10.1061/(ASCE)0733-9372(2000)126:12(1076).

Wu, M.C., Sun, K.W. and Zhang, Y. (2006). Influence of temperature fluctuation on thermophilic anaerobic digestion of municipal organic solid waste. *J Zhejiang Univ Sci B*, **7(3):** 180–185. doi: https://doi.org/10.1631/jzus.2006.B0180.

Wu, X., Yao, W. and Zhu, J. (2010). Biogas and CH_4 productivity by co-digesting swine manure with three crop residues as an external carbon source. *Am Soc Agric Biol Eng Annu Int Meet 2010*, ASABE 2010 **2(11):** 1091–1104. doi: https://doi.org/10.1016/j.biortech.2010.01.052.

Yap, H.Y. and Nixon, J.D. (2015). A multi-criteria analysis of options for energy recovery from municipal solid waste in India and the UK. *Waste Manag*, **46:** 265–271. doi: https://doi.org/10.1016/j.wasman.2015.08.002.

Yu, H.G. and Fang, H.H. (2002). Acidogenesis of dairy wastewater at various pH levels. *Wat Sci Tech*, **45(10):** 201–206. doi: https://doi.org/10.2166/wst.2002.0331.

Zahedi, S., Sales, D., Romero, L.I. and Solera, R. (2013). Hydrogen production from the organic fraction of municipal solid waste in anaerobic thermophilic acidogenesis: Influence of organic loading rate and microbial content of the solid waste. *Bioresour Technol*, **129:** 85–91. doi: https://doi.org/10.1016/j.biortech.2012.11.003.

Chapter 25
Comparative Performance Evaluation of Toxicity Assessment Tests on Waste Li-Ion Laptop Batteries

Aparna Sharma, Arvind K. Nema, and Saif Khan

1 Introduction

Li-ion or lithium-ion batteries (also called LIBs) in the modern time are used everywhere be it the automobile sector, IT sector or anywhere else. As we are heading towards development, the advancements in technology have led to an increase in electric and electronic appliances, which would in turn result in an increased demand for batteries. Hence, the amount of battery waste is also rising day by day. Globally, the amount of e-waste, which is disposed off per year is around 20–50 million tonnes, of which batteries are a major part. Although there is no definite official data on the generation and disposal of battery waste in India, the amount can be estimated indirectly using the approximate e-waste data, which is based on independent studies conducted by the NGOs or government agencies. According to the Battery University, the global battery market is about 50 billion US dollars out of which approximately 5.5 billion US dollars is for rechargeable batteries. The annual growth rate was about 6% through 2006 which was estimated to increase steadily after that. As per a UN report, in India, there is a prediction of 500% rise in the waste from discarded computers by 2020, and rise in the waste from discarded mobile phones is estimated to be about 18 times than 2007. In spite of the advancements in technology, the waste disposal and management practices are still poor, outdated and sometimes illegal.

Besides, they have a negative effect on environment. The improper disposal of waste batteries can result in toxic metal contamination. They can pollute ground and surface water, soil, air, flora, fauna and also humans. For instance, Guiyu, Hong

A. Sharma (✉) · S. Khan
Department of Civil Engineering, IIT Delhi, New Delhi, India

A. K. Nema
Environmental Engineering Division, Department of Civil Engineering, IIT Delhi, New Delhi, India

A. S. Kalamdhad (ed.), *Integrated Approaches Towards Solid Waste Management*, https://doi.org/10.1007/978-3-030-70463-6_25

Kong is facing acute water shortages as the surface and well waters have become undrinkable due to high lead content. A person lost his life and six were hospitalised due to cobalt-60 radiation tragedy at Mayapuri, Delhi. Similarly, in Mundka, Delhi, a fire broke out due to improper handling of e-waste risking the life of workers. According to the Blacksmith Institute, the poisonous lead which gets released during the recycling process affects more than 12 million people in the developing world, stunting children's physical and mental growth. Such incidents and the continuously rising quantities of waste emphasise on the urgent need to tackle this problem in developing countries like India.

Several risks are related to the use and disposal of batteries. Researchers have stated a lot about the harmful effects of heavy metals. Human nervous system, blood, reproductive system and kidneys and brain development in children is affected by lead. Cadmium exposure may affect kidneys and bones. Antimony is a toxic compound causing dermatitis, is also stated as a possible carcinogen. Mercury affects respiratory system and can cause skin disorders and long-lasting damage to brain. Carcinogenic metals like chromium affects the DNA and can cause asthmatic bronchitis, and beryllium can result in lung cancer. Heart, spleen and liver are damaged by barium which also causes muscle weakness (Rajya Sabha Report on E-waste, 2011).

Also, according to the GreenIT report, the main four battery related concerns are lead mining, effects on manufacturing workers' health, effects on building occupants in case of a fire and disposal at the end of useful life. The effects of chemical contamination due to batteries are based upon many factors like the chemical that one comes in contact with, the concentration of the element in the environment and the exposure duration (Nriagu and Simmons, 1990). There are two ways in which the toxins can enter the environment, through leachate itself or the gases produced by the leachate. If the batteries come into contact with a liquid, the resulting leachate makes it easier for the toxic substances present in them to enter the environment through soil. From there, they will enter the ground and surface water systems and in turn will pollute the entire ecosystem. Another way for toxins to enter the environment is through the gases that are produced by leachate, which will mix with air and can contaminate the atmosphere. Because of these, if batteries are dumped into landfills, they can be hazardous for the ecosystem. And once humans are contaminated by one or more of the toxic chemicals, it can result in serious health problems including neurological and kidney damage, birth defects, deafness, vision problems and cancer. These are the effects which we know until now, but there is still much more to be explored about the other threats that these toxic chemicals can pose to humans.

Several researchers in the past have worked on the different types of batteries. Dillon (1994) in his work had explored the existing lithium-based battery technology, the development in batteries through the years and their applications. Using the information on estimated lithium battery consumption, a risk assessment of waste lithium battery disposal was also performed. He explained the environment related policies of the United States, Europe, and industry as well. van der Kuijp et al. (2013) studied the health hazards of China's lead-acid battery industry. It was clear from the study that to prevent millions of children in China from getting poisoned further, it's urgent for to take immediate regulatory action. As per a USEPA study, the batteries with nickel and cobalt cathodes and solvent-based electrode processing are most likely to cause serious environmental impacts, including global warming, resource depletion, eco-toxicity and impacts on human health. The results also suggest that the use stage of a battery contributes more to the life cycle impacts, especially in the case of battery use with more carbon-intensive grids. Also, according to Sivaramanan (2013), the low income for recycling workers compels them to work more and more without thinking about their health.

Hence, it can be concluded that proper disposal and management of hazardous waste is a world-wide problem. Ineffective hazardous waste management affect local communities almost in all countries; the most obvious effect being the illegal transboundary trade, mostly from industrialised/developed countries (Marsili et al., 2009). Due to this, the middle- and low-income countries get affected principally. The burden of diseases due to waste-related exposures increases. It has also been reported to be the chief cause of soil and groundwater contamination (Fazzo et al., 2017).

This has created an urgent need to evaluate the environmental impact of waste disposal/dumping, which further led to the development of several leaching procedures. During a leaching test, the material to be tested is brought into contact with a liquid known as leachant under a set of reference test conditions, the resultant liquor known as leachate is then analysed. The purpose of these tests is to gather information about the concentration and release of the constituent material from waste under the conditions closely resembling the actual disposal site conditions. The data from these tests can also be used to classify the materials or waste products according to the criteria or by comparing their concentration to the relevant limits (Lewin, 1996).

During the recent years, the concentration of trace metal mobile fractions has become important as compared to the total contents of the environmental media in risk assessment as well as minimization studies of contaminated sites. In this context, a lot of metal speciation, leaching and extraction methods have been developed (Centioli et al., 2008).

Various leaching methods are available worldwide that have been designed to address some specific aspects of leaching. A range of objectives for these tests include:

(a) Classification of waste as hazardous/non-hazardous
(b) Evaluation of process modification in waste generation/treatment
(c) Quality control in waste treatment
(d) Designing leachate treatment system
(e) Risk assessment
(f) Disposal issues of waste
(g) Waste reuse

Most of the leaching tests are focused on determination of inorganic contaminants rather than organic ones. Every leaching test has been designed to serve a specific purpose. Literature data suggests that most of the tests fail to simulate field leaching conditions that will be developed over the period of time. Even the use of aggressive leaching tests (such as TCLP) may not be protective because of the increased leachability of some constituents under high pH conditions. The solubility of most metals is highly sensitive to field pH conditions. If the site-specific conditions are unknown, the use of any leaching test may not provide useful results. Therefore, assessment of performance of a leaching test is application specific. Till date, no single laboratory leaching test has been able to evaluate the leaching behaviour of a wide variety of material in a broad range of management scenario (Tiwari et al., 2015). However, when used within a proper framework, leaching test can provide useful information for environmental decision making. Most widely followed leaching tests across the world are batch leaching tests such as TCLP, SPLP, EN 12457/1-4 and column leaching tests NEN 7341, CEN/TS 14405. Threshold limits for leaching of various contaminants have been defined in various regulatory norms and management practices followed worldwide.

2 Materials and Methods

The waste lithium ion batteries were collected from different vendors in Delhi, India. These batteries were dismantled and the components were reduced to the sizes required for different tests. Five leaching tests have been carried out in this study, which are briefly explained below. After leaching, the collected leachates are analysed for determining four metals, nickel (Ni), iron (Fe), copper (Cu) and lead (Pb) using atomic absorption spectrophotometer (AAS). The parameters of the tests are summarised in Table 1.

Table 1. Parameters of the tests performed

	Parameters					
Test name	*L:S per step*	*Sample mass required*	*Size reduction*	*Extraction fluid*	*Duration/ contact time*	*Agitation*
NEN 7341	50:1 (v/m)	8 g	<125 mm	Deionised water & acidified water (pH 4)	6 h (3 h per step)	Horizontal, 30 ± 2 rpm
DIN 38414-S4	10:1	20 g	<4 mm	Distilled water	24 h	Horizontal, 30 ± 2 rpm
JLT-13	10:1	10 g	0.5–5 mm	Distilled water	6 h	Horizontal, 200 rpm
GB5086.1-1997	10:1	75 g	<5 mm	Distilled water (pH 5.8–6.3)	18 h	End-to-end, 30 ± 2 rpm
NIEA R201.12C	20:1	100 g	<9.5 mm	Acetic acid	18 ± 2 h	End-to-end, 30 ± 2 rpm

2.1 *JLT-13 (Used in Tables and Figures as JLT)*

In a glass flask, 10 g sample was added to 100 mL of distilled water, to reach an L/S of 10:1. The flask is then closed with a cap and is kept in an orbital shaker for 6 h for horizontal shaking at the speed of 200 oscillations per minute. After shaking, the leachate thus obtained was filtered through Whatman filter paper grade 40 (pore size: 8 μm). The filter pore size that was specified in Japanese leaching test No. 13 (JLT-13) is 1 μm, but in this case grade 40 Whatman filter paper was used for direct comparisons with other leaching test results in this study. The leachate is then sent for metal analysis. (Environment Agency of Japan, 1973; Kim et al., 2003).

2.2 *GB5086.1-1997 (Used in Tables and Figures as GB)*

In this method, 750 mL deionised water was added to 75 g of dried sample having particle size less than 5 mm in a 1000 mL capped bottle, i.e., the L/S ratio was kept 10:1. This bottle was rolled in a rolling agitator at a speed of 30 ± 2 rpm for 18 h continuously, then left for 30 min to settle down. The leachate thus obtained is filtered using grade 40 Whatman filter paper and preserved for metal analysis (CEPA, 1997; Jiang et al., 2007).

2.3 *NEN 7341 (Used in Tables and Figures as NEN)*

This test is performed at the pH values of 4 and 7. For the test, about 5 g of sample (less than 125 mm size) is added to distilled water at L/S ratio of 50:1 and extracted twice in succession using an orbital shaker. The first stage of the test includes horizontal shaking for 3 h at 30 ± 2 rpm at constant pH 7. In the second stage, the material is again shaken horizontally for 3 h at constant pH 4, or a lower pH if the sample drives the pH lower. The leachates are filtered using Whatman filter paper grade 40 and then analysed (Netherlands Standardization Institute, 1993).

2.4 *DIN 38414 S4 (Used in Tables and Figures as DIN)*

It is an agitated extraction test in which 20 g sample is mixed with distilled water maintaining a liquid-to-solid ratio of 10:1. The extraction slurry is horizontally agitated for 24 h. The mixture is then filtered on Whatman filter paper grade 40 and the resulting leachate is analysed for its metal components. (German Institute of Standardization, 1984).

2.5 NIEA R201.12C (Used in Tables and Figures as NIEA)

In this test, 100 g of sample (size < 9.5 mm) is put into the glass flask containing acetic acid (pH 2) with L/S ratio of 20 and agitated end-to-end for a duration of 18 ± 2 h. The collected leachate is filtered and sent for metal analysis (TEPA, 2002).

3 Results and Discussion

The results of the metal analysis of the samples, which was performed using an atomic absorption spectrophotometer, can be seen in Table 2. The average values of the contents of the four metals (Ni, Fe, Cu and Pb) has been shown graphically in Fig. 1.

As it can be seen from the results of the metal analysis (Table 2 and Fig. 1), all the tests except for NIEA R201.12C gave metal content values that are comparable to each other. This was due to the fact that only NIEA R201.12C used acetic acid as the leaching medium. So, it can be concluded that the metal leaching is better in case of using an acid as a leaching solution. Also, the values of NEN 7341 step 1 and step 2 are a clear indication of the fact that all of the metals were leached during initial leaching, and not the prolonged one. Apart from that, it is also seen that the time duration of the leaching is not affecting the amount of metal leached to a great extent. Hence, it can be suggested that amongst the five tests that were compared, NIEA R201.12C gave better results than the other four.

Table 2. Metal analysis results of the leaching tests on waste Li-ion battery samples

Tests	*Ni*	*Fe*	*Cu*	*Pb*	*Tests*	*Ni*	*Fe*	*Cu*	*Pb*
GB 1	1.627	0.259	13.23	0.459	**JLT 1**	0	0.071	0	0.013
	1.855	0.259	11.64	0.397		0	0.071	0	0.029
	1.906	0.259	12.96	0.428		0	0.071	0	0.013
GB 2	1.931	0.259	25.64	0.521	**JLT 2**	0	0.099	0	0.013
	2.033	0.259	27.22	0.521		0	0.071	0	0
	2.083	0.259	25.06	0.521		0	0.071	0	0
NIEA 1	24.620	37.33	34.58	1.136	**DIN 1**	0	0.156	0	0
	24.560	37.52	34.49	1.189		0	0.156	0	0
	24.530	37.35	36.27	1.121		0	0.071	0	0
NIEA 2	24.930	36.70	37.19	0.793	**DIN 2**	0	0.128	0.035	0
	24.620	36.56	37.79	0.793		0	0.099	0.054	0
	24.400	37.27	38.60	0.762		0	0.099	0.072	0
NEN 1 S1	2.350	2.053	4.009	0.169	**NEN 2 S1**	0.757	0.864	2.894	0
	2.069	1.911	4.215	0.138		0.913	0.751	2.472	0
	2.444	1.855	6.437	0.107		1.038	0.552	2.762	0
NEN 1 S2	0.351	0.071	0	0.029	**NEN 2 S2**	0	0.071	0	0
	0.288	0.071	0	0.044		0	0.071	0	0
	0.195	0.071	0	0.060		0	0.071	0	0

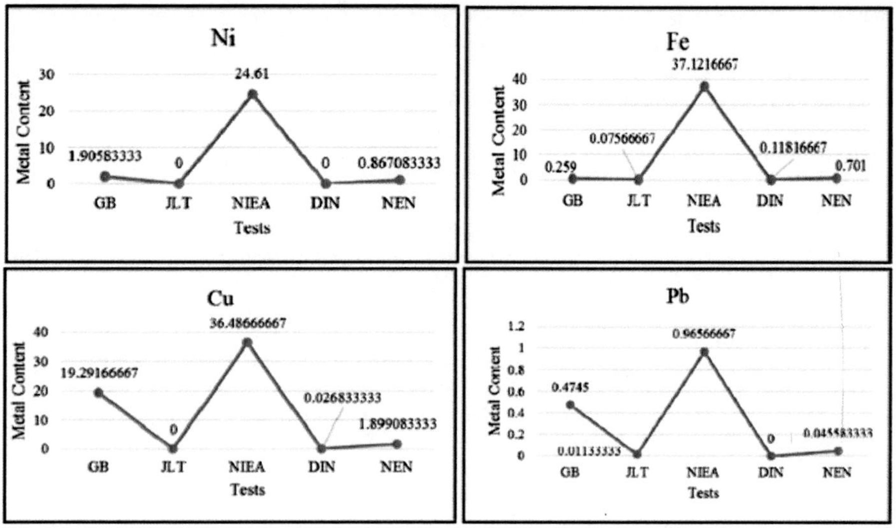

Fig. 1. Average values of the metal contents in the leachate samples.

4 Conclusions

On the basis of the current study, the conclusions that can be drawn are that acid leaching gave better metal leaching values and prolonged leaching is not quite necessary for determining the four metals that were studied here. NIEA R201.12C was found to be the better leaching test in comparison with the other four tests. But all of the tests were found to be time consuming. Also, because leaching behaviour can change dramatically over time and under different leaching conditions, more than one test should be used to provide information necessary for characterising the potential leaching behaviour. Hence, it is further suggested to carry out more of the available leaching tests and to develop better leaching tests in order to verify the results obtained here.

References

Centioli, D., Comans, R.N., Gaudino, S., Galas, C. and Belli, M. (2008). Leaching tests: Useful tools for the risk assessment of contaminated sediments. *Annali dell'Istituto superiore di sanita*, **44(3):** 252–257.

CEPA, Chinese Environment Protection Association (1997). Test method standard for leaching toxicity of solid wastes—Horizontal vibration extraction procedure (GB5086.1-1997). Chinese National Standard Press, Beijing. p. 1 (in Chinese).

Dillon, C. (1994). The ecological impact of batteries. Thesis report. Carnegie Mellon University.

DIN 38414. (1984). Determination of leachability by water (S4) German standard methods for the examination of water, waste water and sludge. Sludge and sediments. DIN 38414 Part 4, October.

Environment Agency of Japan. (1973). Japanese Environment Agency Notification No. 13 (in Japanese).

Fazzo, L., Minichilli, F., Santoro, M., Ceccarini, A., Della Seta, M., Bianchi, F., Comba, P. and Martuzzi, M. (2017). Hazardous waste and health impact: A systematic review of the scientific literature. *Environ Health*, **16(1):** 107.

GreenIT report. Batteries for IT Systems in Buildings Environmental Issues.

Jiang, J. G., Xin, X. U., Jun, W. A. N. G., Yang, S. J. and Zhang, Y. (2007). Investigation of basic properties of fly ash from urban waste incinerators in China. *J Environ Sci*, **19(4):** 458–463.

Kim, S.Y., Matsuto, T. and Tanaka, N. (2003). Evaluation of pre-treatment methods for landfill disposal of residues from municipal solid waste incineration. *Waste Manage Res*, **21(5):** 416–423.

Lewin, K. (1996). Leaching tests for waste compliance and characterisation: Recent practical experiences. *Sci Total Environ*, **178(1–3):** 85–94.

Marsili, D., Fazzo, L. and Comba, P. (2009). Health risks from hazardous waste disposal: The need for international scientific cooperation. *Eur. J. Oncol*, **14(3):** 151–159.

NEN 7341 (1993). Determination of leaching characteristics of inorganic components from granular (waste) materials. NNI, Delft (Netherlands). Formerly NVN 2508.

Nriagu, J.O. and Simmons, M.S. (1990). *Food Contamination from Environmental Sources* (Vol. 23). John Wiley.

Secretariat, R.S. (2011). E-waste in India. India Research Unit (Larrdis), Rajya Sabha Secretariat, New Delhi.

Sivaramanan, S. (2013). E-waste management, disposal and its impacts on the environment. *Universal J Environ Res Technol*, **3(5):** 531–537

TEPA, 2002. NIEA R201.12C. (2002). Toxicity Characteristic Leaching Procedure (TCLP) for Industrial Waste, Taipei, Taiwan.

Tiwari, M.K., Bajpai, S., Dewangan, U.K. and Tamrakar, R.K. (2015). Suitability of leaching test methods for fly ash and slag: A review. *J Radiat Res Appl Sci*, **8(4):** 523–537. doi: https://doi.org/10.1016/j.jrras.2015.06.003.

Van der Kuijp, T.J., Huang, L. and Cherry, C.R. (2013). Health hazards of China's lead-acid battery industry: A review of its market drivers, production processes, and health impacts. *Environ Health*, **12(1):** 61.

Chapter 26
Recycling Construction and Demolition Waste: Potential Applications and the Indian Scenario

Yuvraj Singh and Harvinder Singh

1 Introduction

Rapidly increasing urbanisation and the high pace of development has brought with itself an attention-seeking and a very challenging issue of managing C&DW. Haphazard development in past, improper designing, insufficient strength, revised codal guidelines, damage due to seismic forces or other deteriorating factors are amongst few reasons under which demolition of a structure may be opted for. Demolition of a structure followed by re-construction gives rise to demolition and construction waste, respectively. Managing or disposing of this waste has become a vital and mandatory challenge for all stakeholders. Disposing C&DW into landfills may be thought of a convenient way but besides leading to huge space consumption; it leads to land degradation and thus impacts the environment. According to 40 CFR section 258.2, the C&D debris received by landfills constitutes of roadwork and excavated material; waste from demolition, construction, renovations and site clearances (Electronic Code of Federal Regulations, 2019). Also, the manufacturing of construction materials like cement, aggregates, etc. contributes to environmental pollution and thus it becomes important to plan its use judiciously. Wherever possible, demolition activities should be avoided and in case of strength in-deficiency, strengthening techniques like the use of fibre-reinforced polymer composites may be opted for and when unavoidable, the produced construction and demolition waste should be carefully managed by recycling it. Reducing and recycling C&DW can largely conserve landfill space and can thus prove to be instrumental in diminishing the impact of producing and using new construction materials on the environment. It also results in providing employment and can reduce the overall cost of the building project through avoided purchase or disposal cost (EPA, US, 2016).

Y. Singh (✉) · H. Singh
Department of Civil Engineering, Guru Nanak Dev Engineering College Ludhiana, I.K. Gujral Punjab Technical University, Kapurthala, Punjab, India

A. S. Kalamdhad (ed.), *Integrated Approaches Towards Solid Waste Management*, https://doi.org/10.1007/978-3-030-70463-6_26

With the augmenting population and the declining reserves of non-renewable resources, the crucial and the vital challenge to mankind is to keep our planet liveable for the coming generations. With this, it becomes mandatory to turn the economy from linear to circular. And concrete being versatile and hence a widely used construction material has to become a part of this change (Potier, 2019). Keeping in view, the importance and urgency of dealing with C&DW, an overview of the current regulations, practices, and initiatives regarding recycling the C&DW in and outside India are presented in this paper. Furthermore, the paper also presents an insight into the ongoing research on transforming construction waste into a resource, its related challenges, and potential applications.

2 Global Scenario

Around the globe, the government and the municipalities have been framing legislation intending to promote the recycling of C&DW. Realising the urgency of managing this waste, the British Standards Concrete Committee planned to incorporate provisions for recycled aggregates (RA) into its concrete standard (BS8500-2). And in order to introduce this provision, rigorous research was desirable to answer the related queries. In an effort towards the same, a research carried out at the University of Dundee proved instrumental in providing insight on the queries of British Standards Institution (BSI) and as a result a complete specification for the application and products of coarse recycled concrete aggregates (RCA) and a partial specification for other types of coarse RA were introduced in 2002. Realising that the only way of promoting the use of recycled C&DW is by making it competitive with the primary materials, governments of the UK and the Netherlands took measures in this direction and emerged successfully. One vital and key initiative towards achieving this target was the initiation of landfill taxes. Provided that the fly-tipping involving illegal dumping is minimised, this measure can be instrumental and effective in declining the volume of C&DW in India too (Harrison, 2019). Likewise, a waste framework directive was revised in Europe where each member of the union is bound to include a requirement of reuse or recycle 70% of the C&DW by 2020 (BIO Intelligence Service, 2011). In order to accomplish the target of 70%, the technique varies across the continents since each member state of the European Union must include the requirement into their respective legislation (Jeffrey, 2011). Efforts are being made around the globe by various researchers and governing bodies to contribute towards managing C&DW. Furthermore, in order to reduce the burden on landfills, many practitioners are moving from full demolition to the concept of deconstruction.

3 Indian Scenario

Keeping in view the requirement of good and planned infrastructure in India, a major portion of the annual budget is allocated towards infrastructure development by the planning commission during the past couple of decades. Moreover, a rapid increase in the population and augmenting infrastructure development has led to a shortage and hence increased the cost of building materials than ever before. Besides this, the debris or C&DW produced from demolished structures are being largely disposed of in landfills leading to environmental degradation and reduced availability of such places, especially in urban areas. This makes it inevitable to cope with C&DW by recycling and reusing it in order to address these issues. It has been reported that CPCB estimated that out of the total solid waste produced in India; the construction industry is responsible for contributing more than 25% of it (Singh and Sharma, 2007).

In India, infrastructure development, and large housing and industrial construction projects are ongoing at a high pace. Reconstruction of existing buildings, development of industrial corridors and economic zones augments the magnitude of C&DW. To address this issue, stringent regulations are being formulated by the municipalities but the implementation of the same is lacking widely. In most of the cases, the contractors dump/dispose of C&DW to the empty or self-owned land, low-lying areas for a cost, or more commonly, the C&DW is dumped along the roads or other public areas in an unauthorised manner along (BMPTC, 2018). While the C&DW from individual households finds its way into nearby empty plots or municipal bins making the municipal waste heavy and degrading its quality for treatments such as composting or energy recovery. On the other hand, the activities regarding C&DW management, processing and re-use have got energised after issuance of C&DW Management Rules, 2016 by Ministry of Environment, Forest and Climate Change (MoEF&CC).

4 Challenges

Ram et al. (2019) reported that the estimated amount of C&DW produced in India is between 150 and 716 million tonnes/year. Recycling C&DW is the potential solution aiming sustainability and is hence opted worldwide. As far as the awareness among Indian contractors and other potential stakeholders is concerned, a survey carried out related to the hindrance in the application of C&DW reports that 70% of the respondents are unaware of the various recycling methodologies whereas remaining 30% respondents have even indicated unawareness of recycling possibilities (Singh and Sharma, 2007). Another major lacking area or the hurdle towards handling C&DW in India is the availability of just a few recycling facilities. Amongst various barriers towards the establishment of recycling facilities in India, lack of strict regulations, inadequate incentives, unawareness related to recycling methodologies and missing guidelines for confident acceptance and application have been accounted in the literature. Furthermore, the influence of human factors including attitude and the industry norms are among another major factor leading to a hindrance to widespread acceptance. Another vital concern in regard to the operation of the processing plants involving crushing and screening which accompanies the generation of noise and dust cannot be overlooked. Therefore, it is crucial to select and examine the location of the plant, environmental conditions of the surrounding area and related legal requirements with utmost care, and take inevitable counter measures to mitigate the associated effects (Hansen, 1986). Challenges obstructing the wide and confident acceptance and application of C&DW should, therefore, be addressed in order to achieve sustainable solutions towards managing C&DW.

5 Current Indian Regulations and Guidelines

MoEF&CC, BMTPC, CPCB, CPWD and BIS/IRC issued C&DW management rules 2016, Guidelines for utilisation of C&DW in construction of dwelling units and related infrastructure, Guidelines on environment management of C&DW, Guidelines regarding use of C&DW and standards/guidelines respectively. These guidelines facilitate understanding, comprehension and appropriate action by all stakeholders. All materials that are produced as waste during demolition, repair, re-modelling, retrofitting or new construction are categorised as C&DW. Nothing is excluded. The owner of the civil structure is usually considered as the generator of C&DW or the contractor in case the construction of the structure is contracted and the contract states very clearly that the handling and management of the malba/debris, i.e. C&DW generated would be the responsibility of the contractor. The C&DW generator is responsible for the proper management of C&DW and the waste generation site.

Guidelines for proper management of C&DW: Appropriate management/ handling of C&DW and its site include restricting littering of debris or any type of pollution, i.e. fugitive dust and noise. It involves separation and storage of the reusable items such as bricks, doors, windows, cupboards, kitchen & bathroom fittings, other fittings, wooden items, glass, wall panels, roof slabs, other structural elements, etc. without damaging supplying to re-users/second-hand markets. The recyclable items like electrical wires, metals, glass, plastics, paper boards, gypsum boards, etc. must be collected separately and supplied to re-melters. Concrete debris must be kept separately and be processed/supplied to processing plant without mixing with masonry, soil and other debris. MSW, toxic waste, electronic waste and other hazardous waste, etc., should not be mixed with the C&DW and must be disposed off separately carefully as prescribed in SWM rules. C&DW should be segregated in different streams such as concrete, soil, bricks and mortar and other streams stated above for supply to second raw materials market/ re-melters/re-users/processors. The C&DW generated at the site must be evacuated at the required frequency to avoid mix up and spilling over to neighbouring areas. The safety of equipment and manpower must be ensured. Rules and regulations of local agencies regarding permissions required and other aspects must be followed.

6 Research Findings

During the past couple of decades, keeping the alarming need of moving towards sustainability in mind, many researchers have been working towards turning waste into a resource by incorporating it directly or indirectly into a useful material. Among many such materials, recycling C&DW is vital and has been researched globally. It has been reported that recycled fine and coarse aggregates find applications for non-structural purposes like flooring, filling, etc., and has not more than 10–20% less strength than conventional concrete (Singh and Singh 2019). Besides non-structural applications and other downstream products, recycled C&DW has been researched by many researchers for use in the making of structural concrete as well.

Hansen (1986) collated the research carried on RCA and RA concrete from about 4 decades before then. On reviewing the extensive experiments and field tests carried out during that period, it was concluded that aggregates obtained from recycled concrete find applications as a drainage material, shoulders, bituminous concrete, stabilised base courses, lean mix, sub-bases and also as new concrete pavements. It has also been reported that due to lower density and augmented water absorption and high variation range of RA as compared to conventional aggregates, it becomes vital to monitor these properties prior to actual application and incorporate the effect of the same in the design mix of RA concrete. Singh and Sharma (2007) in their investigation on the use of recycled aggregated reported that RA possesses comparatively inferior mechanical and physical properties and augmented water absorption in comparison to that of conventional aggregates and hence these factors need to be incorporated in the design. Moreover, as stated by other researchers it has been confirmed that the compressive strength also reduces to about 15–20% and this variation is largely influenced by the source of concrete from which aggregates are obtained. Keeping in view the potential applications of RA concrete, more research work and projects were proposed in order to modify the design codes and for clear cut specifications and procedure for use of RA concrete (Singh and Sharma 2007). Yehia et al. (2015) also confirmed the variation of the properties of concrete incorporated with RA that were obtained from different sources. It has been emphasised that the issues or deficiencies that may be encountered in the obtained RA concrete can be eliminated or avoided by careful proportioning. In another experimental interrogation on RA concrete, the enhancement of properties of RA concrete on incorporating bacteria was evaluated (Sahoo et al., 2016). Taffese (2018) on investigating the suitability and potential of RCA for the making of concrete confirmed that RCA obtained manually by crushing the concrete can be employed as a replacement to normal aggregates for producing the concrete up to 10–20% depending upon the application and design considerations.

Bheel et al. (2018) reported that the use of RA alone should be avoided due to a detrimental effect on the strength and durability parameters. Furthermore, it has been proposed that RA in combination with other additives or supplementary materials like rice husk ash should be added up to an optimum percentage in order to compensate for the reduced strength. Besides this, the use of plasticisers is also recommended in order to achieve the desired workability (Bheel et al., 2018).

Bravo et al. (2019) reported that the source and the quality of C&DW has a major effect on the characteristics of recycled concrete and this effect was observed to be more pronounced in case of replacement of natural fines with that of recycled ones. In order to address this issue, it has been proposed that wherever possible, selective demolition should be opted for. This practice would allow obtaining the material from C&DW with the minimum amount of contamination and hence will enable it's more confident and effective use in construction or in making of other downstream products (Silva et al., 2014). Furthermore, as evident from the extensive experimental investigation reported in the literature, the properties of RA concrete varies and depends upon several factors including source, size of aggregates, percentage replacement, etc., therefore, it becomes vital to investigate the composition as well as the properties of the obtained RA or other recycled materials prior to its use for the project under consideration. This will enable a more effective and confident application of recycled materials. Therefore, careful examination including processing and categorising of the RA or other recycling materials is a mandatory step towards considering C&DW fit and effective for application in construction as per codal specifications.

7 Concept of Deconstruction

Deconstruction may be defined as the engineered demolition involving scrutiny of all types of materials and elements in a building under consideration before demolition takes place. Full demolition involves turning the entire building into debris resulting in mixing all types of material, structural elements, electrical and sanitary fitting together and therefore leaves no scope of re-use and makes the process of recycling altogether a lot more challenging if not impossible. On the other hand, planned or systematic way of demolition referred as engineering demolition or deconstruction provides an opportunity to extract the usable including wooden/steel doors and windows and their frames, sanitary and plumbing fittings (including sinks, taps, etc.), wooden floorings, outer wall/floor panels, cupboards, etc.) from the building leading to decreased demolition waste besides economic and social benefits. Not only this, even structural elements and pre-cast members may also be re-used if opted for deconstruction. Deconstruction can thus salvage enormously the materials and components for reusing besides reducing the volume of debris generated (BMPTC, 2018).

8 Potential Applications

The only sustainable approach towards managing waste is to convert it into a resource. Though every effort should be made to minimise C&DW by considering options like retrofitting or strengthening, opting for deconstruction over full demolition, etc., once generated, it becomes mandatory to manage or recycle the C&DW in order to prevent environmental degradation and to lessen the burden on landfills.

Keeping in view the performance, environmental viability and the cost, the recycled aggregates (RA) can be considered as the potential surrogated for natural aggregates (Hackenhaar et al., 2019). The produce of recycled C&DW can be incorporated directly or as downstream products. The direct product of recycling of C&DW includes fine aggregates, RA or recycled concrete aggregates (RCA) of varying sizes obtained from mixed waste and concrete waste respectively, or as manufactured soil. Manufactured soil finds its application as a substitute for excavated soil and is quite suitable for landfills and landscaping and can prevent the environment from degrading low-lying areas. Whereas, fine and coarse aggregates obtained as a product of recycled C&DW are a ready raw material for RMC plants, construction sites and road construction (BMPTC, 2018).

Besides the direct produce, recycled C&DW finds applications in the manufacturing of several downstream products which includes bricks, blocks, tiles, hollow bricks, wall tiles, pavers, kerbstones, park benches, drain covers, planters, precast compound wall, fence post, tree guards, tree pit covers, manhole covers, underground cable covers, precast boundary wall panels and poles, etc. In order to promote the application of the produce of recycled C&DW, the marketing of the products should not be restricted only to the building construction projects. It has also been reported that RCA, being an engineered material can be effectively incorporated into highway applications in several ways and can prove to be highly beneficial economically, environmentally and socially (Snyder et al., 2018). As a result, RCA finds its successful and common application in pavement base and fill applications due to several reasons including the stable trait of typically angular and rough-textured particles; augmented stability due to secondary cementation; relative inertness or insensitivity of the material towards materials encountered in the pavement environment; economic benefits due to alleviated hauling costs and tipping fees for disposal; environmental sustainability due to resource conservation and alleviation in the processing and hauling energy; and the excellent performance potential (Snyder et al., 2018). Therefore, the recycled products of C&DW can also be marketed to road projects including flyovers and allied structures, railway bridges, metro projects; reclamation of low-lying areas, plinth areas and landscaping in the form of manufactures soil. Furthermore, in order to promote and augment confidence of users on the recycled products, the details of products including quality, environmental impact and supportive government policy should be advertised to the potential clients. Moreover, a complete track of the forthcoming construction projects of the government should be kept and the use of the produce of recycled C&DW must be included in the detailed project report and tender preparation stages itself. Besides this, technical training and consumer awareness seminars must be held on regular bases for widespread acceptance and implementation. Regular monitoring by the government body is mandatory and the construction projects that do not incorporate the use of C&DW produce as per the advisory of local authorities should be brought to the notice of the concerned and a suitable action should be taken (BMPTC, 2018). Snyder et al. (2018) highlighted the importance of engaging all key stakeholders of a project under consideration in order to maximise the benefits of recycling. It has also been recommended that the benefits resulting from the recycling should be publicised in order to promote recycling in future projects (DETR, 2000).

9 Concluding Remarks

- It has been established that recycling C&DW is as essential as the construction of new infrastructure. Furthermore, besides the prime aim of reducing the environmental degradation and burden on landfills, recycling C&DW can lead to comparatively lesser demand for new construction materials which further promotes sustainability.
- It has been found that Indian legislation has incorporated the much-needed target of 100% recycling of C&DW. But, this target is not possible without dedicated involvement and implementation of all stakeholders.
- From ongoing research, it has been found that, if used in proper proportions and with an appropriate technique, recycled C&DW can confidently be used along with appropriate additives in various applications.
- It has been established that opting for deconstruction over full demolition is undoubtedly one of the vital steps towards reducing C&DW.
- Recycled construction and demolition waste find several applications including direct and downstream products. Thus, turning this waste into a resource can lead to enormous economic as well as environmental benefits.

Acknowledgements The authors are grateful for the support and assistance provided by Guru Nanak Dev Engineering College, Ludhiana; I,K, Gujral Punjab Technical University, Kapurthala; and TEQIP III (*a Government of India Project assisted by the World Bank*).

References

Bheel, N. et al. (2018). Experimental study on recycled concrete aggregates with rice husk ash as partial cement replacement. *Civil Eng J*, **4 (10):** 2305–2314.

BIO Intelligence Service (2011). Service contract on management of construction and demolition waste – SR1. Final report for the European Commission.

BMTPC (2018). Utilisation of recycled produce of construction & demolition waste – A ready reckoner. Building Materials and Technology Promotion Council, Ministry of Housing and Urban Affairs, Government of India.

Bravo, M., Brito, J. and Evangelist, L. (2019). The effect of the quality of construction and demolition waste on recycled aggregates concrete properties. *Indian Concrete J*, **09:** September.

DETR (2000). Controlling the environmental effects of recycled and secondary aggregated production: Good practice guidance. Department of the Environment Transport, and the Regions. Crown Publishing, London, England

Electronic Code of Federal Regulations (2019). Title 40: Protection of Environment. Part 258. Criteria for Municipal Solid Waste Landfills.

Environmental Protection Agency, Unites States (2016). Industrial and Construction and Demolition landfills.

Hansen, T.C. (1986). Recycled aggregates and recycled aggregate concrete second state-of-the-art report developments 1945–1989. *Mater Struct*, **19:** 201–246.

Harrison, T. (2019). Background to the BSI provisions for RCA in concrete. *Indian Concrete J*, **09:** September.

Hackenhaar, I.C., et al. (2019). Life cycle assessment applied to construction and demolition waste treatment: Proposal of a Brazilian scenario. *IOP Conf Ser: Earth Environ Sci*, **323:** 012054.

Jeffrey, C. (2011). Construction and demolition waste recycling – A literature review. Dalhousie University's Office of Sustainability, Nova Scotia, Canada.

Potier, J.M. (2019). Point of view: Recybeton. *Indian Concrete J*, **09:** September.

Ram, V.G., Prajapati, R. and Poulose, M.K. (2019). Challenges and prospects of demolition waste recycling in India. *Indian Concrete J*, **09:** September.

Sahoo, K.K., et al. (2016). Enhancement of properties of recycled coarse aggregate concrete using bacteria. *Int J Smart Nano Mater*, **7** (1)**:** 22–38.

Silva, R.V., Brito, J., Dhir, R.K. (2014). Properties and composition of recycled aggregates from construction and demolition waste suitable for concrete production. *Constr Build Mater*, **65:** 201–217.

Singh, H. and Singh, Y. (2019). Applications of recycled and waste materials in infrastructure projects. *In:* Singh, H., et al. (eds.), ICSWD 2018, LNCE 21, pp. 569–585.

Singh, S.K. and Sharma, P.C. (2007). Use of recycled aggregates in concrete – A paradigm shift. NBMCW, October.

Snyder, M.B., et al. (2018). Recycling concrete pavement materials: A practitioner's reference. National Concrete Pavement Technology Center, Iowa State University, Ames, IA.

Taffese, W.Z. (2018). Suitability investigation of recycled concrete aggregates for concrete production: An experimental case study. Advances in Civil Engineering (Hindawi), 2018.

Yehia, S., et al. (2015). Strength and durability evaluation of recycled aggregate concrete. *Int J Concrete Struct Mater*. **9:** 219–239.

Chapter 27
Biochemical Methane Potential and Kinetics of *Parthenium hysterophorous* with Different Food to Microorganisms (F/M) Ratios

Biswanath Saha, Visva Bharati Barua, Gautam Vrindra, Ajay S. Kalamdhad, and Meena Khwairakpam

1 Introduction

The rapid growth of terrestrial weeds is a worldwide predicament, because of its harmful effect and competitiveness with agricultural land with their natural resources like sunlight, nutrients and soil.

Anaerobic digestion of animal waste got a great attraction because of oil crises (Gunaseelan, 1986). *P. hysterophorous* can be managed by the different ways like compost, biogas (Gunaseelan, 1987), bioethanol (Ghosh and Hallenbeck, 2012). Anaerobic digestion of *P. hysterophorous* is an effective way of management of this terrestrial weed, where biogas is the end product, which is a source of renewable energy. Renewable energy produced from *P. hysterophorous* can be an alternative for fossil fuels, as its rapid burning is one of the dilemmas and this toxic weed is abundant. In the last few decades' anaerobic digestions got the attention due to its sustainable conversion of waste to energy and also the waste can be managed. Anaerobic digestion reduced the volume of the organic matter around 30–70% (Elliott and Mahmood, 2007). Also, the sludge produced after the digestion can be used as a fertilizer. Biogas produced from *P. hysterophorous* can be converted into vehicle fuel or electricity (Espinosa-Solares et al., 2009).

B. Saha (✉) · G. Vrindra · M. Khwairakpam
Centre for Rural Technology, Indian Institute of Technology, Guwahati, Assam, India

V. B. Barua
Centre for the Environment, Indian Institute of Technology, Guwahati, Assam, India

A. S. Kalamdhad
Department of Civil Engineering, Indian Institute of Technology Guwahati, Guwahati, Assam, India

A. S. Kalamdhad (ed.), *Integrated Approaches Towards Solid Waste Management*, https://doi.org/10.1007/978-3-030-70463-6_27

Table 1. Characterization of *P. hysterophorous* and cow dung

Parameter	*P. hysterophorous*	*Cow dung*
pH	6.76 ± 0.4	7.63 ± 0.5
sCOD (mg/L)	7329 ± 205	2343 ± 185
TS (%)	13.5 ± 0.4	19.8 ± 0.2
VS (%)	81.33 ± 3.0	79 ± 0.4
VFA (%)	345 ± 4.0	320 ± 3.0
MC (%)	83.6 ± 3.0	81.2 ± 2.0
TKN	2.54 ± 0.02	3.3 ± 0.05
TOC	48.22 ± 2.0	44 ± 2.0

This study focuses on the sustainable conversion of this noxious weed to biogas. Concentration and composition of substrate play a vital role in anaerobic digestion (Zhou et al., 2011). When the substrate concentration is increased, it increases the efficiency of anaerobic digestion but once the substrate amount is excess there is a chance of production of excess amount of ammonia, nitrogen and volatile fatty acid (VFA), which hamper the anaerobic digestion process (Sánchez et al., 2002; Fernández et al., 2008; Kun et al., 2015). Therefore, a proper F/M ratio should be evaluated and maintained to run the efficient anaerobic digestion process. Different F/M ratio was studied from 0.5 to 3. Batch reactor study was performed with working capacity of 15 L. This study shows *P. hysterophorous* have the potential to produce biogas, which can be used for different purposes. Though there are lots of studies on BMP of different lignocellulosic biomass, there is no literature on optimising the best F/M ratio for *P. hysterophorous* (Table 1).

2 Material and Methods

2.1 Substrate and Inoculum

P. hysterophorous was collected from the premises of Indian Institute of Technology, Guwahati (IITG) and fresh cow dung was collected from a neighbouring village, Amingaon. BMP ratios studied were 0.5, 1, 1.5, 2, 2.5, 3 and control. The whole plant including root was taken out, then cut into an average size of 1 cm uniformly and ground properly. Smaller particle size has a higher potential for biogas production (Sharma et al., 1988). If the particle size is larger, it reduces gas production and also affects the microbial activity. Grinding is also required to reduce the volume of the reactor, without effecting the biogas production (Gollakota and Meher, 1988; Moorhead and Nordstedt, 1993). Then the BMP ratio was calculated based on Volatile solids (VS), the cow dung was mixed properly with *P. hysterophorous* in different ratios according to Table 2.

Table 2. Various amount of *P. hysterophorous* and cow manure was used for different F/M ratio based on VS

F/M ratio	*P. hysterophorous (g)*	*Cow manure (g)*
Control	–	50
0.5	31.01	50
1.0	62.02	50
1.5	93.03	50
2	124.04	50
2.5	155.05	50
3	186.06	50

2.2 BMP Test

The information of bio-methanation of the specific substrate is obtained from BMP test. Studying of BMP set up is a remarkable method to determine the methane production potential under the different optimum conditions (Hussain and Dubey, 2015). BMP has been examined in a batch scale to conclude the maximum amount of biogas yield per gram of volatile solid (VS) contained in the substrate within the anaerobic setup. The BMP set up is run as described by Owen et al. (1979). 1000 mL of reagent bottle is used as a BMP reactor and rubber cork is used for closing the bottles. Twenty-one bottles are used for different F/M ratio, 18 bottles are used for different F/M ratio and in 3 bottles only cow dung was fed, which was considered as the control. Micro- and macronutrients are equally distributed to all the reactors for the microorganisms to grow. Daily methane production was measured by using 1.5 N of NaOH. Quite a few environmental conditions should be maintained inside the reactor like pH, stirring intensity, etc. (Browne and Murphy, 2013). Temperature 30–38 °C is good for the breakdown of the organic substrate. To maintain anaerobic condition nitrogen gas is purged.

2.3 Analysis of Sample

Different parameters were studied such as VS, total solid (TS), moisture content (MC), and sCOD. The standard protocol has been followed according to APHA (2005). For pH measurement, 10 g of *P. hysterophorous* was taken in a conical flask along with deionised water, mixed for 2 h at 150 rpm in a horizontal shaker, then pH was checked (DiLalo and Albertson, 1961). pH titration method is used for analyses of the VFA. Total Kjeldahl nitrogen was measured as described by Bremner (1965).

3 Result and Discussion

3.1 *Initial Characterisation of Parthenium hysterophorous and Inoculums*

P. hysterophorous shows decent VS and MC which is beneficial for the anaerobic digestion. Volatile solid and moisture content of *P. hysterophorous* was found as 81.33% VS and 83.6 % MC respectively. Table 1 shows the initial characterisation of *P. hysterophorous* and cow dung used as the inoculum. The composition of lignin content in *P. hysterophorous* is around 15.34%, hemicellulose 28% and cellulose 47%. Soluble COD is found to be high; pH is near to 7, which is also suitable for anaerobic digestion. Volatile fatty acid come under the optimum range, however, higher VFA reduced the pH, which affect the microorganism which leads to hampering the digestion process (Tada et al., 2005).

3.2 *BMP Test*

The BMP test was performed with different F/M ratio i.e., 0.5–3 for 50 days and the ambient temperature varies between 28 and 35 °C. Higher temperature is suitable for the methane production but it can affect the microorganism growth, that can be troubled for methane production (Chae et al., 2008). In anaerobic digestion F/M ratio is the most significant parameter (Chudoba et al., 1992). There number of bacteria's play a key role to break down the organic matter to methane. One type of microorganism uses the product of other microorganisms, as a result, the growth of one partner is improved. There are a number of microorganisms involved to carry out the specific reaction throughout the anaerobic digestion. 1.5 N NaOH is used to calculate methane gas production as shown in Fig. 1. Amount of methane produced is measured as the amount of NaOH collected whereas the CO_2 is absorbed through NaOH and produced sodium bicarbonate, follow the principle of water displacement method.

3.2.1 Biogas Production

The rate of methane production is proportional to the breakdown of organic substrate (Esposito et al., 2012). There are various factors like inoculum, VS, VFA, sCOD, pH and temperature, which affect methane production. Biogas yield was measured daily; it was measured by the liquid displacement method by the amount of NaOH displaced (Fig. 2a). In ratio 2, it shows higher methane production up to 140 mL in 28 days followed by the ratio 1.5 (99 mL), highest cumulative biogas is observed in ratio 2 with 3319 mL followed by ratio 2.5, which shows 2903 mL. Among all the F/M ratio 0.5 and 3 shows the lowest amount of biogas production 2400 mL

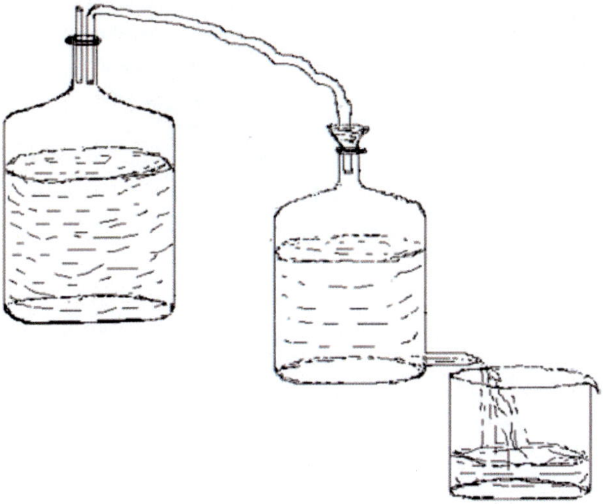

Fig 1. BMP Setup.

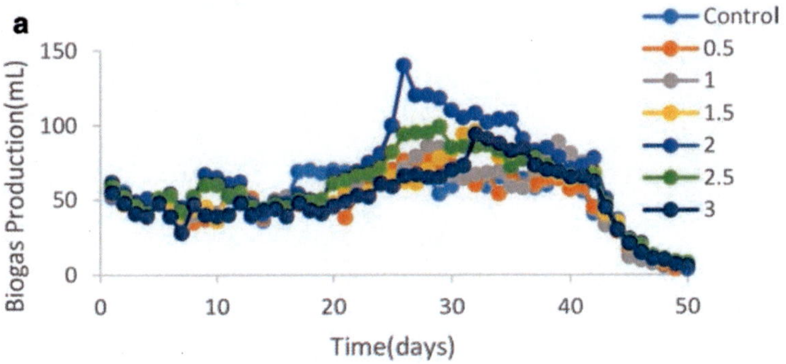

Fig. 2a. Biogas production.

and 2527 mL respectively as shown in the Fig. 2b. Biogas production increased from 0.5 to 2, then it started to decrease because of imbalance in the F/M ratio. For *P. hysterophorous*, the best order of F/M ratio is 2 > 2.5 > 1.5 > 1 > 3 > 0.5. Methane production was observed to increase from F/M ratio 0.5 to 2 after that it started decreasing at ratio 2.5 and 3. Initially, it takes time to produce biogas because of the presence of lignocellulose in the substrate. Inoculum concentration is higher compared to the substrate in the lower F/M ratio. As a result, biogas production is less in the lower ratio. Similarly, in the higher ratios, the substrate concentration is higher compared to the inoculum which may lead to increase the inhibitor like ammonia, nitrate so the proper degradation does not take place inside the reactor. Microorganisms require balanced nutrient ratio for their efficient growth. A similar study was observed by Barua and Kalamdhad (2017).

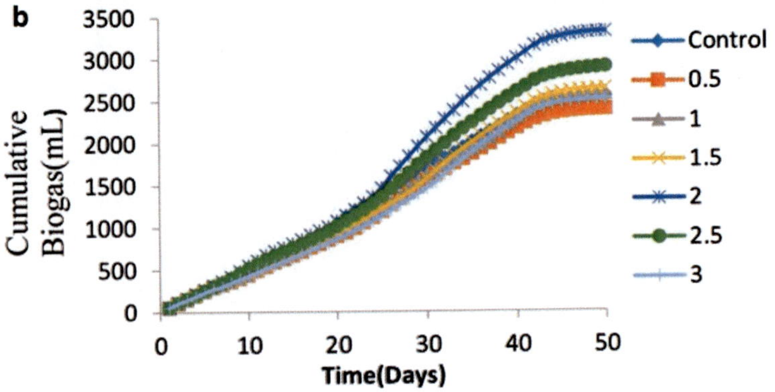

Fig. 2b. Cumulative Biogas production.

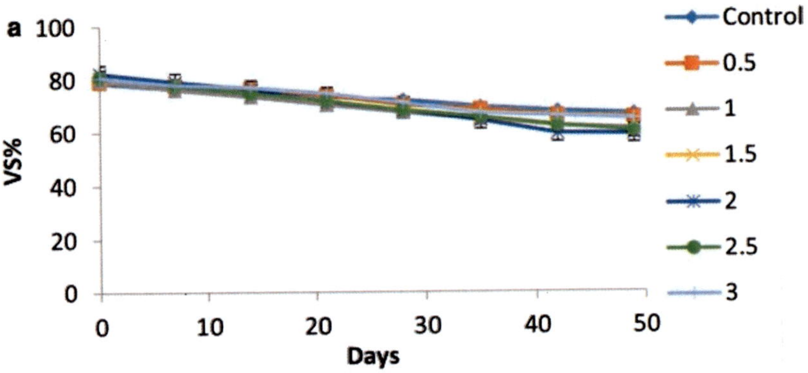

Fig. 3a. Volatile solid.

3.2.2 Volatile Solid

In anaerobic digestion, higher the VS reduction higher is the degradation and at the same time higher the probabilities of biogas production (Dhamodharan et al., 2015). In the BMP studies, highest VS reduction is observed in ratio 2 (about 28.40%), followed by ratio 2.5 with 26.03% and cumulative biogas production shows 3319 mL and 2903 mL respectively. Volatile solid reduction increases from 0.5 to 2. The ratio greater than two (>2) started decreasing, since degradation increased with increasing the ratio up to 2 due to that biogas yield was also less in lower ratio and also higher than ratio 2. The reduction order followed by *P. hysterophorous* is 2 > 2.5 > 1.5 > 1 > 3 > 0.5. In this study, it can be correlated that higher the VS reduction higher is the production of biogas (Figs. 2a and 3a).

3.2.3 Effect of sCOD

A complex molecule is converted into simpler molecules with the help of bacteria, which leads to increased sCOD. The amount of hydrolysis and solubilisation can correlate with sCOD removal efficiency. In the present BMP study, highest sCOD was obtained as 14780 mg/L in 28 days and it got reduced to 5720, which is about 61.29% in F/M of ratio 2. For sCOD, the reduction order followed by *P. hysterophorous* is 2 > 2.5 > 1.5 > 1 > 3 > 0.5. The sCOD increased up to 28 days and later on it decreased because of the acids produced by the microbial activity. Similar results were found by Moukazis et al. (2017). With an increase in the sCOD, methane production is also increased (Passos et al., 2015; Barua and Kalamdhad, 2016a, b). Highest methane production was shown in 28 days in ratio 2. Soluble COD can be correlated with methane production. As the sCOD increased VFA also increased. Similar results were found by Barua and Kalamdhad (2016a, b) as shown in Fig. 3b and 3c.

3.2.4 VFA

As the digestion started sugar was converted into acid which leads to a rise of VFA. For VFA concentration highest amount was found in F/M ratio 2 in 28 days followed by the 1.5 ratios. 674 ± 5 mg/L of VFA was obtained from ratio 2 followed by F/M ratio 1.5 around 665 ± 7. Methane production gets hampered once the VFA concentration crosses 13,000 mg/L (Viéitez and Ghosh, 1999). It was observed that VFA increased from 28 to 35 days in ratios 0.5 to 3, after that it started decreasing. The same trend was found in all the ratios which are due to the beginning of

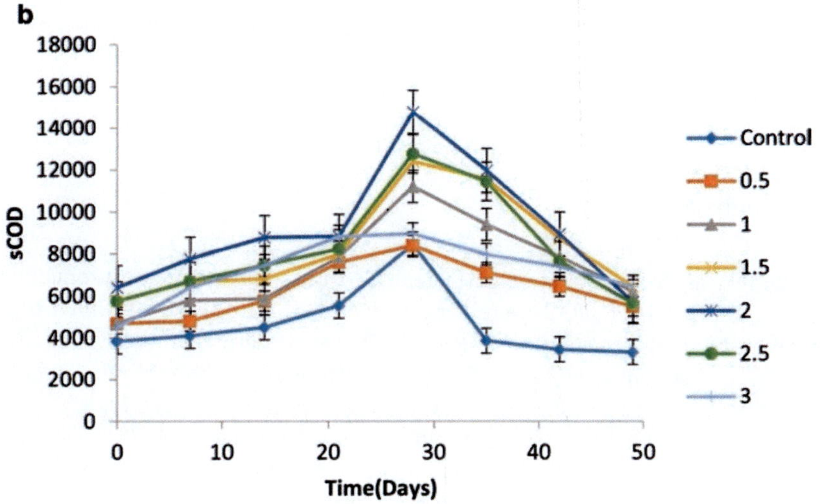

Fig. 3b. sCOD.

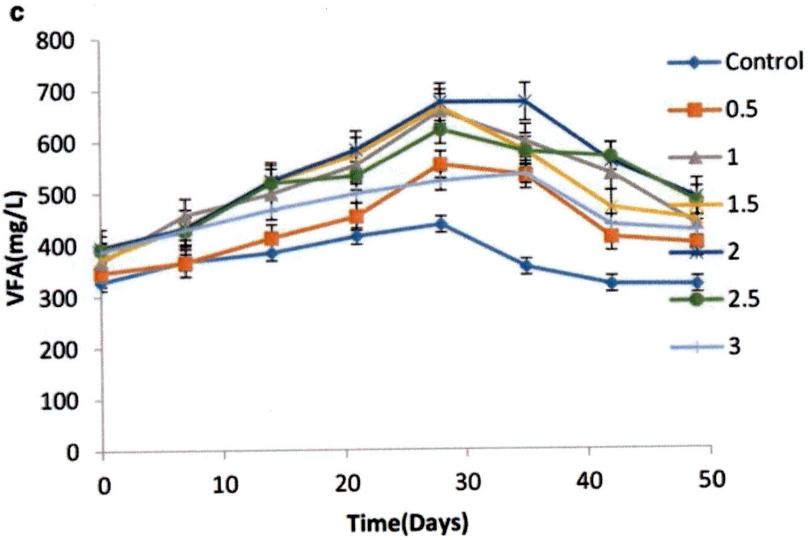

Fig. 3c. Volatile Fatty acid.

methanogenic phases. Same outcomes were observed by Barua and Kalamdhad (2017) in water hyacinth substrate. During anaerobic digestion the amount of VFA, acetic acid should come under 2000 mg/g. Methanogenic bacterial activity gets affected when the pH value drops to 6.0–6.5 (Speece, 1983; Lim et al., 2009). It was observed that above pH 5, more than 75% of biogas production was achieved (Jain and Mattiasson, 1998). pH started to decrease after three days, which was maintained between 6.5 and 7.5 using sodium bicarbonate (Owen et al., 1979). Production of acid leads to a drop in pH which affects the activity of microbes (Ahring, 1995). Therefore, if the VFA production in the reactor passed beyond a limit it inhabits the microbial activity, which leads to affect acitogenesis stage in the anaerobic digestion process.

4 Conclusion

This study was done to find out the ideal F/M ratio for maximum production of biogas. Different F/M ratios were observed from 0.5 to 3, out of which highest methane production potential was seen in F/M ratio 2, around 140 ± 3 mL in 28th days. Highest amount sCOD and VS reduction was achieved in F/M ratio 2. Amount of inoculum and substrate calculated according to F/M ratio 2 feed for 15.5 L scale reactor where cumulative methane yield was observed around 4978 L ± 3 in 26th days. So, ratio 2 shows the best combination of F/M ratios. *P. hysterophorous* can be converted into energy in the form of biogas. Further pretreatment is required for the reduction of the hydrolysis period and faster energy recovery.

Acknowledgements The author is grateful to Dr. Visva Bharati Barua, Centre for Rural Technology and Centre for the Environment, Indian Institute of Technology, Guwahati for laboratory support.

References

Ahring, B.K. (1995). Methanogenesis in thermophilic biogas reactors. *A. Van Leeuw*, **67:** 91– 102.

Albertson, O.E. (1961). Volatile acids by direct titration. *J. Water Pollut. Control Fed*, 356–365.

APHA (2005). Standard Methods for the Examination of Water and Wastewater, 20th ed. APHA American Public Health Association.

Barua, V.B. and Kalamdhad A. (2016a). Effect of various type of pretreatment technique on hydrolysis, compositional analysis and characterization of Water Hyacinth. *Biosour Technol*, **227 (2017):** 147–154.

Barua, V.B. and Kalamdhad, S.A. (2016b). Water hyacinth to bio gas: A review. *Poll Res*, **35 (3):** 491–501.

Barua, V.B. and Kalamdhad, S.A. (2017). Biochemical methane potential test of untreated and hot air oven pretreated water hyacinth: A comparative study. *J Cleaner Prod*, **166 (2017):** 273–284.

Bremner, J.M. (1965). Inorganic forms of nitrogen. *In:* Black, C.A. (Ed.), *Methods of Soil Analysis: Chemical and Microbiological Properties*. Vol. 2. American Society of Agronomy, Madison, Wisconsin, pp. 1179–237.

Browne, J.D. and Murphy, J.D. (2013). Assessment of the resource associated with biomethane from food waste. *Appl. Energy*, **104:** 170–177.

Chae, J., Jang, A., Yim, S.K. and Kim, S. (2008). The effects of digestion temperature and temperature shock on the biogas yields from the mesophilic anaerobic digestion of swine manure. *Bioresour Technol*, **99:** 1–6.

Chudoba, P., Capdeville, B. and Chudoba, J. (1992). Explanation of biological meaning of So/Xo ratio in batch cultivation. *Water Sci Technol*, **26:** 743–751.

Dhamodharan, K., Kumar, V., and Kalamdhad, A.S. (2015). Effect of different livestock dungs as inoculum on food waste anaerobic digestion and its kinetics. *Bioresour Technol*, **180:** 237–241.

Elliott, A. and Mahmood, T. (2007). Pretreatment technologies for advancing anaerobic digestion of pulp and paper bio-treatment residues. *Water Res*, **41:** 4273–4286.

Espinosa-Solares, T., Valle-Guadarrama, S., Bombardiere, J., Domaschko, M. and Easter, M. (2009). Effect of heating strategy on power consumption and performance of a pilot plant anaerobic digester. *Appl Biochem Biotechnol*, **156:** 35–44.

Esposito, G., Frunzo, L., Liotta, F., Panico, A. and Pirozzi, F. (2012). Bio-methane potential tests to measure the biogas production from the digestion and co-digestion of complex organic substrates. *Open Environ Eng J*, **5:** 1–8.

Fernández, J., Pérez, M. and Romero, L.I. (2008). Effect of substrate concentration on dry mesophilic anaerobic digestion of organic fraction of municipal solid waste (OFMSW). *Bioresour Technol*, **99 (14):** 6075–6080.

Ghosh, D. and Hallenbeck, P.C. (2012). Advance Bioethanol Production. Springer, New York, pp. 165–181.

Gollakota, K.G. and Meher, K.K. (1988). Effect of particle size, temperature, loading rate and stirring on biogas production from castor cake. *Biol Wastes*, **24:** 243–249.

Gunaseelan, V.N. (1986). *Parthenium* as an additive with cattle manure in biogas. *Biol. Wastes*, **21:** 195–202.

Gunaseelan, V.N. (1987). *Parthenium* as an additive with cattle manure in biogas. *Biol. Wastes*, **21:** 195–202.

Hussain, A. and Dubey, S.K. (2015). Specific methanogenic activity test for anaerobic degradation of influents. *Appl Water Sci*, **7:** 535–542. doi: https://doi.org/10.1007/s13201-015-0305-z.

Jain, S.R. and Mattiasson, B. (1998). Acclimatization of methanogenic consortia for low pH biomethanation process. *Biotech Lett*, **20 (8):** 771–775.

Kun, L., RongHou, L. and Chen, S. (2015). Comparison of anaerobic digestion characteristics and kinetics of four livestock manures with different substrate concentrations. *Bioresour Technol*, **198:** 133–140.

Lim, S., Kim, S. and Fox, P. (2009). Biological nutrient and organic removal from meat packing wastewater with a unique sequence of suspended growth and fixed- film reactors. *Water Sci Technol*, **60:** 3189–3197.

Moorhead, K.K. and Nordstedt, R.A. (1993). Batch anaerobic digestion of water hyacinth: Effects of particle size, plant nitrogen content and inoculum volume. *Bioresour Technol*, **44 (1):** 71–76.

Owen, W., Stuckey, D., Healy, J., Young, L. and McCarty, P. (1979). Bioassay for monitoring biochemical methane potential and anaerobic toxicity. *Water Res*, **13:** 485e492.

Passos, F., Carretero, J. and Ferrer, I. (2015). Comparing pretreatment methods for improving microalgae anaerobic digestion: Thermal, hydrothermal, microwave and ultrasound. *Chem Eng J*, **279:** 667–672. doi: https://doi.org/10.1016/j.cej.2015.05.06.

Sánchez, E., Borja, R., Weiland, P., Travieso, L. and Martin, A. (2002). Effect of substrate con centration and temperature on the anaerobic digestion of piggery waste in a tropical climate. *Process Biochem*, **37 (5):** 483–489.

Sharma, S.K., Mishra, I.M., Sharma, M.P. and Saini, J.S. (1988). Effect of particle size on biogas generation from biomass residues. *Biomass*, **17:** 251–263.

Speece, R.E. (1983). Anaerobic biotechnology for industrial wastewater treatment. *Environ Sci Technol*, **17:** 416Ae427A.

Tada, C., Yang, Y., Hanaoka, T., Sonoda, A., Ooi, K. and Sawayama, S. (2005). Effect of natural zeolite on methane production for anaerobic digestion of ammonium rich organic sludge. *Bioresour Technol*, **96 (4):** 459–64.

Viéitez, E.R. and Ghosh, S. (1999). Biogasification of solid wastes by two-phase anaerobic fermentation. *Biomass Bioenerg*, **16:** 299–309.

Zhou, Y., Zhang, Z., Nakamoto, T., Li, Y. and Yang, Y. (2011). Influence of substrate-to-inoculum ratio on the batch anaerobic digestion of bean curd refuse-okara under mesophilic conditions. *Biomass Bioenergy*, **35 (7):** 3251–3256.

Chapter 28
Removal of Lead and Copper by Using Bentonite as an Adsorbent

Saswati Ray, Anil Kumar Mishra, and Ajay S. Kalamdhad

1 Introduction

Due to the rapid growth in population, the generation of waste has been increasing many folds. Waste from various industries like smelting, painting mining, electroplating, petrochemical, etc. has been discharged into the environment, which can have a severe impact on health. Heavy metals present in the waste are potentially very noxious and non-biodegradable. Exposure to heavy metals can cause serious health issues for human beings. The permissible limit for drinking water is 2 mg/L and 0.05 mg/L for Cu^{2+} and Pb^{2+}, respectively, as per the WHO standard (Baylan and Meriçboyu, 2016).

The technique implemented for the discarding of waste by several nations is landfilling. A part of a landfill liner system is a liner material which is placed below the landfill system for preventing the percolation of pollutant in the surrounding environment. Bentonite is suitable as a barrier material in a waste containment system because of its high Cation exchange capacity (CEC), high montmorillonite mineral content, excellent sorption capability, low permeability and high specific surface area (SSA) (Dutta and Mishra 2015; Ray et al., 2018; Ray et al., 2019).

In the recent past, numerous techniques like membrane processes coagulation, ion exchange, flotation have been employed to eliminate heavy metals. The above techniques are highly expensive. Adsorption is one of the common and effective techniques which uses harmless and relatively low-cost material for the elimination of metal ions (Owolabi and Unuabonah, 2010; Azad et al., 2015). Bentonite is a locally available, low-priced natural resource generated from volcanic ash and can be used as an adsorbent (Pawar et al., 2016). In literature, numerous works are reported by various researchers on the removal of heavy metal by using clay mineral. The sorption properties of bentonite were examined by Kubilay et al. (2007) in

S. Ray (✉) · A. K. Mishra · A. S. Kalamdhad
Department of Civil Engineering, Indian Institute of Technology, Guwahati, Assam, India
e-mail: r.saswati@iitg.ac.in

A. S. Kalamdhad (ed.), *Integrated Approaches Towards Solid Waste Management*, https://doi.org/10.1007/978-3-030-70463-6_28

the presence of Cu^{2+}, Zn^{2+} and Co^{2+} at various pH and concentration (15–70 mg/L) and concluded the sequence of sorptivity at 20 °C was $Zn^{2+} > Cu^{2+} > Co^{2+}$ at pH 3.0, 5.0 and 7.0 and $Zn^{2+} > Co^{2+} > Cu^{2+}$ at pH 9.0. A study reported by Freitas et al. (2017) on examination of sorption capacity on bentonite in the existence of binary solution of Cu^{2+} and Ag concluded that Cu^{2+} possessed better sorption capacity as compared to Ag. Likewise, several works have been conducted by various researchers on the elimination of various metals like Zn^{2+} (Kaya and Ören, 2005; Olu-Owolabi and Unuabonah, 2010; Araujo et al., 2013), Cd^{2+} (Kozar et al., 1992; Pradas et al., 1994), Ni^{2+} (Liu and Zhou, 2010; Futalan et al., 2011; Taha et al., 2017) by using bentonite as an adsorbent. The investigation aims to study the effect of pH, initial concentration, kinetic, and agitation speed on bentonite in the presence of Cu^{2+} and Pb^{2+}.

In the current study, Cu^{2+} and Pb^{2+} were carefully chosen since it is generally found in leachate and can cause a serious threat to the ecosystem. The investigation will be beneficial for engineers for designing a barrier in landfills that would be able to hold back contaminant percolation.

2 Materials and Methods

2.1 Characterisation of Bentonite

Bentonite samples were obtained from Barmer district, Rajasthan, India, in powder form. Table 1 depicts the characterisation of bentonite used in the current study. Ammonium acetate process is given by Chapman (1965) and Pratt (1965) was followed for analysing the Cation exchange capacity (CEC) of the soil. By using the procedure given by Cerato and Lutenegger (2002) specific surface area (SSA) (340.4 m^2/g) of the bentonite was determined. As per ASTM D854-92 (1994) and ASTM D5890 (2006), specific gravity and free swell of the soil were analysed. The liquid and plastic limit of the soil was examined as per ASTM D4318 (2010). Optimum moisture content (OMC), as well as maximum dry density (MDD), was

Table 1. Characterisation of bentonites

Properties	*Bentonite*
CEC (meq/100 g)	36.2
Specific surface area (SSA) (m^2/g)	340.4
Free swelling (mL/2 g)	20.0
pH	9.2
Montmorillonite content (%)	64
Liquid limit (%)	305.0
Plastic limit (%)	39.0
Specific gravity	2.7
Optimum moisture content (OMC) (%)	33.5
Maximum dry density (MDD) g/cm^3	1.30

obtained by following ASTM D698 (2012). To analyse the minerals like montmorillonite, illite and kaolinite present in the soil, X-ray diffraction (XRD) test was performed. The chemical composition of the bentonite clay was analysed by conducting an X-ray fluorescence (XRF) experiment (Table 2).

2.2 *Reagents and Solutions*

Preparation of stock solutions was done by dissolving the required amount of copper nitrate [$Cu(NO_3)_2{\cdot}3H_2O$] and lead nitrate [$Pb(NO3)_2$] in deionised DI water. The desired concentrations of Cu^{2+} and Pb^{2+} were prepared from the stock solution. In the present investigation, reagents used are of analytical grade.

2.3 *pH Study*

In 100 mL of Cu^{2+} and Pb^{2+} solution of 500 mg/L, 5 g of soil is added at pH ranges from 2.0 to 8.0. For adjusting to the desired pH of the mixture, 0.1 M diluted HNO_3 and NaOH solutions were used. The solution was agitated in a rotary shaker at 150 rpm for 24 h. The solution was centrifuged at 3000 rpm for 10 min and filtered using filter paper. The concentration of the supernatant was analysed using atomic absorption spectrophotometer (AAS).

2.4 *Batch Sorption Study*

Batch sorption study was conducted as per ASTM (2008). A series of the batch study was done in triplicates to analyse the retention of Cu^{2+} and Pb^{2+}. In 100 mL of Cu^{2+} and Pb^{2+} solution of a chosen concentration (100, 250, 500, 750 and 1000),

Table 2. Chemical composition of bentonite of the major element (expressed as weight percent)

Major and minor chemical oxides	*Bentonite*
SiO_2	50.70
Al_2O_3	19.09
Fe_2O_3	16.29
MnO	0.313
MgO	1.06
CaO	1.10
Na_2O	3.59
K_2O	1.52
TiO_2	2.13
P_2O_5	0.21

5 g of clay was added in a stoppered conical flask. By the addition of 0.1 M diluted HNO_3 and NaOH solution the pH was adjusted to 5 ± 0.1. In the rotary shaker, the solution was shaken for 24 h for room temperature at 150 rpm. After shaking the solution for 24 h, it was centrifuged for 10 min at 3000 rpm. The supernatant obtained was filtered using Whatman 42 filter paper. The equilibrium concentration of both metals was checked using AAS. The number of metal ions adsorbed and the following equations computed the removal percentage (R_p)

$$q_e = \frac{(C_i - C_e)v}{m} \tag{1}$$

$$R_p = \frac{C_i - C_e}{C_i} \times 100 \tag{2}$$

where q_e (mg/g) is the equilibrium concentration of metal ions adsorption on the bentonite, C_i (mg/L) is the initial metal ion concentration, C_e (mg/L) the aqueous metal concentration in equilibrium solution, m (g) the quantity of soil used, and V (L) the volume of metal solution.

2.5 *Kinetic Study of Metals on Bentonite*

In batch mode, the kinetic study of Cu^{2+} and Pb^{2+} was carried out on bentonite. In a stoppered conical flask, 5 g of soil was dissolved in 100 mL of 500 mg/L of Cu^{2+} and Pb^{2+} solution. The solutions were standardised to 5 ± 0.1 by adding drop by drop 0.1 M NaOH and 0.1 M HNO_3. The solutions were agitated for 5, 10, 20, 40, 60, 80, 120, 160 and 200 min at 150 rpm at room temperature using shaker. After extracting the solutions from the shaker at the desired time, the solutions were centrifuged. The supernatants obtained were analysed in AAS for residual heavy metal concentration.

3 Results and Discussion

3.1 *Influence of pH on Bentonite*

The pH of the solution highly influences the adsorption process. Fig. 1 shows that with the rise in pH from 2 to 3, the removal percentage of Cu^{2+} and Pb^{2+} increased from 30.68 to 42.07 % and 65.12 to 76.84%. The removal percentage of both the metals is low due to the competition of H^+ ions and metal ions for the adsorption site. With the rise in pH from 4 to 5, the removal percentage increases gradually.

Fig. 1. pH study of Pb^{2+} and Cu^{2+}.

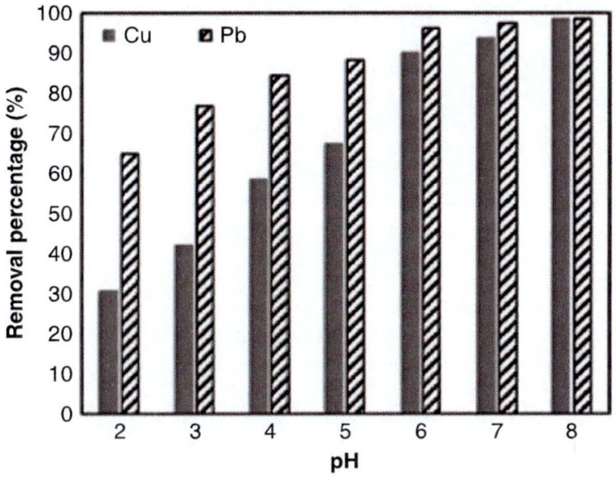

At pH 5, the removal was found to be 67.41 and 88.29% for Cu^{2+} and Pb^{2+}. At pH 4–5, the basic process that governs the process is ion exchange and adsorption. From pH 6 to 8, the removal percentage of Cu^{2+} and Pb^{2+} abruptly increases from 90.18 to 98.53% and 96.21 to 98.64% respectively for Cu^{2+} and Pb^{2+}. This is due to the formation of a hydroxyl compound (Kaya and Ören, 2005). Therefore, to evade precipitation of metals, pH 5 was carefully selected for further investigation.

3.2 *Influence of Initial Concentration of Cu^{2+} and Pb^{2+}*

Figure 2 depicts the plot between percentage removal of Cu^{2+} and Pb^{2+} and initial metal concentration. Figure 2 shows that with the rise in metal concentration from 100 to 1000 mg/L, the removal of both the metals decreased. The removal of Cu^{2+} and Pb^{2+} decreased from 89.85 to 67.43% and 97.37 to 88.46%, with the rise in concentration from 100 to 1000 mg/L. A huge number of sorption sites are available at a lower concentration. With the rise in the concentration of metal ions, the sorption sites of the bentonite get occupied due to adsorption. As a result, the removal of metal ions declined.

Figure 3 shows the relationship between the amount of metals adsorbed and initial metal concentration. Figure 3 depicts that with the rise in concentration from 100 to 1000 mg/L, the adsorption capacity of both the metal increased. At 1000 mg/L, the adsorption capacity of bentonite was found to be 13.49 and 17.63 mg/g for Cu^{2+} and Pb^{2+}, respectively. For Pb^{2+}, the metal uptake was higher as compared to Cu^{2+}. The hydrated ionic radius of Pb^{2+} (4.01 Å) is lower than Cu^{2+} (4.19 Å), causing decreasing in adsorption of Cu^{2+} as compared to Pb^{2+} (Baylan and Meriçboyu, 2016).

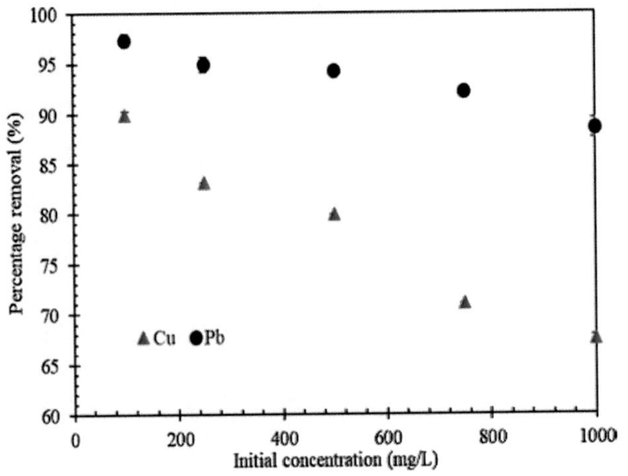

Fig. 2. Effect of percentage removal of Cu^{2+} and Pb^{2+}.

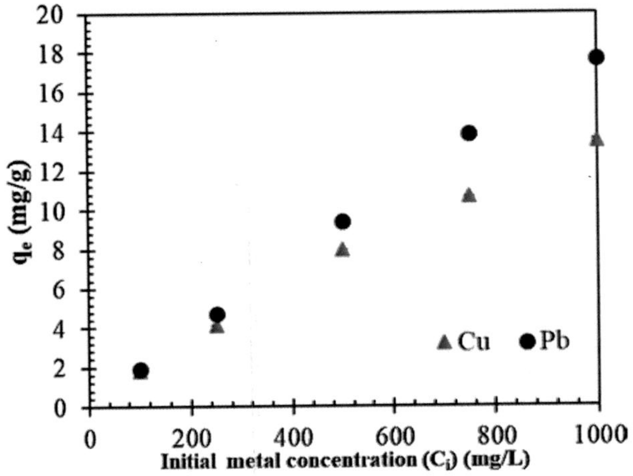

Fig. 3. Effect of initial heavy metal concentration.

3.3 *Influence of Contact Time in the Presence of Cu^{2+} and Pb^{2+}*

Adsorption Kinetic study was performed on bentonite. 5 g bentonite was added to 100 mL of 500 mg/L of metal solution at room temperature. It can be seen from Fig. 4 that the percentage removal of both the metals increased with the increase in shaking time. The removal of Cu^{2+} was 63.57% by 5 min and reached equilibrium within 120 min. However, for Pb^{2+}, 81.28% removal was achieved in 5 min and reached equilibrium within 10 min.

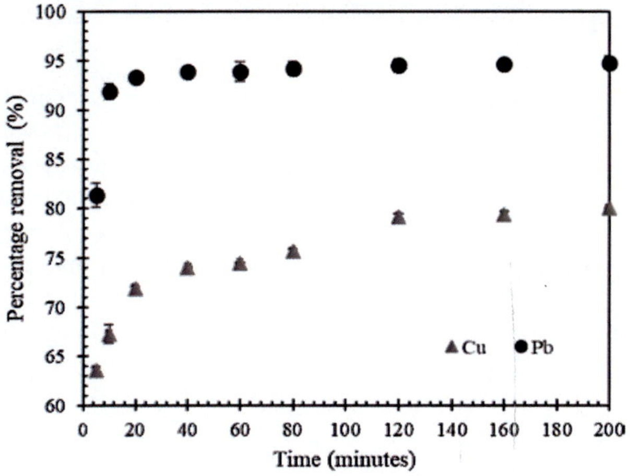

Fig. 4. Effect of contact time in the presence of heavy metals.

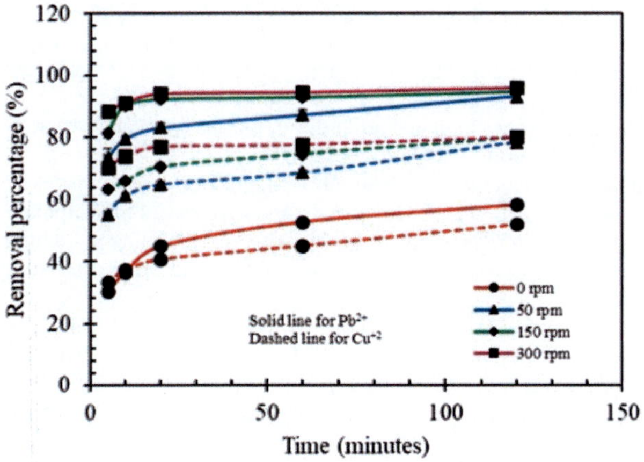

Fig. 5. Effect of agitation speed in the presence of heavy metals.

3.4 Influence of Agitation Speed in the Presence of Cu^{2+} and Pb^{2+}

Figure 5 shows the effect of agitation speed (0, 50, 150 and 300 rpm) on metal uptake by bentonite. Figure 5 indicates that with the rise in agitation speed, the uptake of both the metal increased. At 0 rpm (without agitation), 51.91 and 58.31% removal was obtained for Cu^{2+} and Pb^{2+} whereas, the highest removal was 79.99 and 95.52% at 300 rpm, respectively. With the increase in agitation, speed metal comes in contact with a large no of sorption sites. Therefore, metal uptake rises with the rise in agitation speed (Inglezakis et al., 2007).

4 Conclusion

The investigation was conducted to determine the removal of Cu^{2+} and Pb^{2+} by using bentonite as an adsorbent. The pH study revealed that for the removal of Cu^{2+} and Pb^{2+}, pH 5 was found to be effective. 13.49 and 17.63 mg/g of sorption capacity of bentonite was obtained in the presence of 1000 mg/L of Cu^{2+} and Pb^{2+}. Removal percentage declines with the rise in the concentration of both the metals. Pb^{2+} shows higher removal as compared to Cu^{2+}. The study also shows that the equilibrium reaches within 120 and 10 min for Cu^{2+} and Pb^{2+} on the bentonite surface. The removal of both the metal is significantly influenced by the agitation speed.

References

Araujo, A.L.P.D., Gimenes, M.L., Barros, M.A.S.D.D. and Silva, M.G.C.D. (2013). A kinetic and equilibrium study of zinc removal by Brazilian bentonite clay. *Mater Res*, **16(1):** 128–136.

ASTM. (1994). Standard test method for specific gravity of soils. ASTM D854-92. ASTM, West Conshohocken, PA.

ASTM. (2008). Standard test method for 24-h batch-type measurement of contaminant sorption by soils and sediments. ASTM D4646.ASTM, West Conshohocken, PA.

ASTM. (2010). Standard test methods for liquid limit, plastic limit, and plasticity index of soils. ASTM D4318.ASTM, West Conshohocken, PA.

ASTM. (2012). Standard test methods for laboratory compaction characteristics of soil using standard effort. D 698, American Society for Testing and Materials, Philadelphia.

ASTM D. (2006). Standard test method for swell index of clay mineral component of geosynthetic clay liners, 5890-06. American Society for Testing and Materials, West Conshohocken, Pennsylvania.

Azad, F.N., Ghaedi, M., Dashtian, K., Montazerozohori, M., Hajati, S. and Alipanahpour, E. (2015). Preparation and characterization of MWCNTs functionalized by N-(3-nitrobenzylidene)-N′-trimethoxysilylpropyl-ethane-1,2-diamine for the removal of aluminum (III) ions via complexation with eriochrome cyanine R: spectrophotometric detection and optimization. *RSC Adv*, **5(75):** 61060–61069.

Baylan, N. and Meriçboyu, A.E. (2016). Adsorption of lead and copper on bentonite and grapeseed activated carbon in single-and binary-ion systems. *Sep Sci Technol*, **51(14):** 2360–2368. doi: https://doi.org/10.1080/01496395.2016.1212888.

Cerato, A.B. and Lutenegger, A.J. (2002). Determination of surface area of fine-grained soils by the ethylene glycol monoethyl ether (EGME) method. *Geotech Test J.*, **25(3):** 315–321.

Chapman, H.D. (1965). Cation-exchange capacity. 1. Methods of soil analysis. Part 2. Chemical and microbiological properties., pp. 891–901.

Dutta, J. and Mishra, A.K. (2015). A study on the influence of inorganic salts on the behaviour of compacted bentonites. *Appl Clay Sci*, **116:** 85–92.

Freitas, E.D., Carmo, A.C.R., Neto, A.A. and Vieira, M.G.A. (2017). Binary adsorption of silver and copper on Verde-lodo bentonite: Kinetic and equilibrium study. *Appl Clay Sci*, **137:** 69–76.

Futalan, C.M., Kan, C.C., Dalida, M.L., Hsien, K.J., Pascua, C. and Wan, M.W. (2011). Comparative and competitive adsorption of copper, lead, and nickel using chitosan immobilized on bentonite. *Carbohydrate Polym*, **83(2):** 528–536.

Inglezakis, V.J., Stylianou, M.A., Gkantzou, D. and Loizidou, M.D. (2007). Removal of Pb (II) from aqueous solutions by using clinoptilolite and bentonite as adsorbents. *Desalination*, **210(1):** 248–256.

Kaya, A. and Ören, A.H. (2005). Adsorption of zinc from aqueous solutions to bentonite. *J Hazard Mater*, **125(1–3):** 183–189.

Kozar, S., Bilinski, H., Branica, M. and Schwuger, M.J. (1992). Adsorption of Cd (II) and Pb (II) on bentonite under estuarine and seawater conditions. *Sci Total Environ*, **121,** 203–216.

Kubilay, Ş., Gürkan, R., Savran, A. and Şahan, T. (2007). Removal of Cu (II), Zn (II) and Co (II) ions from aqueous solutions by adsorption onto natural bentonite. *Adsorption*, **13(1):** 41–51.

Liu, Z.R. and Zhou, S.Q. (2010). Adsorption of copper and nickel on Na-bentonite. *Process Saf Environ Protect*, **88(1):** 62–66.

Olu-Owolabi, B.I. and Unuabonah, E.I. (2010). Kinetic and thermodynamics of the removal of Zn^{2+} and Cu^{2+} from aqueous solution by sulphate and phosphate-modified Bentonite clay. *J Hazard Mater*, **184(1–3):** 731–738.

Pawar, R.R., Bajaj, H.C. and Lee, S.M. (2016). Activated bentonite as a low-cost adsorbent for the removal of Cu (II) and Pb (II) from aqueous solutions: Batch and column studies. *J Ind Eng Chem*, **34:** 213–223.

Pradas, E.G., Sánchez, M.V., Cruz, F.C., Viciana, M.S. and Pérez, M.F. (1994). Adsorption of cadmium and zinc from aqueous solution on natural and activated bentonite. *J Chem Technol Biotechnol*, **59(3):** 289–295.

Pratt, P.F. (1965). Sodium. Methods of Soil Analysis. Part 2. Chemical and Microbiological Properties., pp. 1031–1034.

Ray, S., Chowdhury, B.R., Mishra, A.K. and Kalamdhad, A.S. (2018). Impact of heavy metals on consolidation properties of bentonite. *In*: The International Congress on Environmental Geotechnics. Springer, Singapore, pp. 567–574.

Ray, S., Mishra, A.K. and Kalamdhad, A. (2019). Influence of various concentration of lead on consolidation parameters of bentonite. *Int J Geotech Eng*, 1–7.

Taha, A.A., Ahmed, A.M., Abdel Rahman, H.H., Abouzeid, F.M. and Abdel Maksoud, M.O. (2017). Removal of nickel ions by adsorption on nano-bentonite: Equilibrium, kinetics, and thermodynamics. *J Dispers Sci Technol*, **38(5):** 757–767.

Chapter 29
Impact of Precipitation on Biodegradation of Fresh Municipal Solid Waste in Anaerobic Simulated Reactor

Chejarla Venkatesh Reddy, Shekhar D. Rao, and Ajay S. Kalamdhad

1 Introduction

The generation of MSW in urban areas has increased with rapid urbanization and population growth. Leachate, a liquid origin from MSW, has been seen as a significant waste product that affects environmental resources such as surface and groundwater, public health and hygiene. The liquid generated due to the biodegradation process and precipitation travels downwards under self-weight (Naveen et al., 2017a). The leachate composition depends upon the landfill age, buried solid waste material. Decomposition of waste material through the biochemical process and other parameters like time, temperature and moisture content (Fatta et al., 1999). Unscientific management of solid waste management in India has led to the generation of a high amount of leachate. The contamination levels were high in surface and groundwater due to surface runoff and percolation of rainwater where the landfills do not have proper lining system, collection system of leachate and treatment facilities. In developing countries like India, there are not many studies on characteristics of leachate to design leachate treatment facilities (Naveen et al., 2017a). The solid waste in the landfills is converted organic and inorganic compounds in liquid and gaseous states through the degradation process. The degradation process is mainly divided into five stages: (1) aerobic phase, (2) acidogenic phase, (3) acetogenic phase, (4) methanogenic phase and (5) maturation phase (Gujer and Zehnder, 1983). The pollution from landfills is associated with waste composition, while in India, organic waste (wet waste) has a greater share (>50%) of the total composition (Joshi and Ahmed, 2016). These problems constantly related with high loads of organic pollutants like BOD_5, COD, solids, cations and anions like sulphate, chloride, iron, zinc and heavy metals including Fe, Cd, Cu and Hg (Christensen et al., 2001).

C. V. Reddy (✉) · S. D. Rao · A. S. Kalamdhad
Department of Civil Engineering, Indian Institute of Technology Guwahati, Guwahati, Assam, India
e-mail: Chejarla@iitg.ac.in

A. S. Kalamdhad (ed.), *Integrated Approaches Towards Solid Waste Management*, https://doi.org/10.1007/978-3-030-70463-6_29

Recently, important features of waste management studies have been the study of the process, characteristics, and patterns of landfill pollution. Such types of studies have used landfill simulation reactors (LSRs) to get a more realistic picture of waste degradation and leaching (Ejlertsson et al., 2003; Ledakowicz and Kaczorek, 2004; Sponza and Ağdağ, 2004). During the simulating phases, organic matter of waste is broken down in to small and converted to the gaseous and liquid phase. This process allows for the study of the emissions. Most of the researches have explored the recirculation of leachate which is a core functionality of bioreactor. Researchers have found the advantages of recirculation of leachate to stabilize the waste (Pohland, 1975; Reinhart and Grosh, 1998). In simulated anaerobic reactors, the effect of recirculation of leachate and rate of recirculation on anaerobic treatment of MSW has been investigated. Recirculation of leachate has positive effects on waste stabilization (Ledakowicz and Kaczorek, 2004; Reinhart and Grosh, 1998; Sponza and Ağdağ, 2004) and addition of precipitation to the reactor was significant for the decomposition of waste as to dilute and flush out large concentrations of TVA and establish favourable methanogenic conditions (Petchsri et al., 2006). There are no studies in India, observing the changes in the quality of leachate under the biodegradation process with the addition of moisture content only (rainfall). Such conditions simulate the real-life landfill cells subject only to rainfall and degradation under anaerobic conditions, which flush out leachate. The research paper aims to characterize the physical and biological parameters of simulated anaerobic landfill leachate. The study investigates on reactors operated with and without rainfall to understand the behaviour of landfill for 39 weeks. To determine the effect of precipitation on processes of degradation and leaching, and to obtain appropriate data for India's specific waste composition as well as to establish the pattern of degradation of the various parameters in the landfill leachate, this study is useful for understanding the change in the quality of leachate with and without rainfall. The quality of leachate in two different scenarios: (1) mainly through waste biodegradation; (2) addition of water (rainfall) have been tested for application of proper treatment of leachate for prevention of adverse environmental effects.

2 Materials and Methods

2.1 Collection of MSW Material

Seventy three percent of wet waste (i.e., food) and 27% of dry waste (i.e., paper, plastic wood and metals) were the waste components considered in this study and composition of waste data obtained from CPCB, 2017 report. Collection of waste components from individual sources instead of a waste transfer station or landfill. For example, paper waste being absorbent quickly absorbs the moisture and nutrients from food waste. To avoid such mixtures, waste components were obtained from individual waste sources. Food waste was obtained from the fancy bazar market mostly contained fruits, vegetables and kitchen waste from IITG hostels. A mixture of paper, plastic, wood and metals are collected form recycling bins and local stores. Individual components are mixed and fed into the reactor. Before filling the waste into reactor initial characterization of waste will be carried out (physical characterization, i.e. specific density, moisture content, pH and biological characterization, i.e., volatile solids (VS), total organic carbon (TOC).

2.2 Initial Characterization of Waste

The moisture content will be determined by drying samples at 105 °C using hot air oven for 24 h (Tiquia and Tam, 1998). Volatile solids (VS) content (weight loss on ignition at 550 °C) will be determined using muffle furnace (Kalamdhad and Kazmi, 2008). pH value will be measured by mechanical shaking of samples for 2 h with distilled water with waste to water extraction of 1:10 (w/v) (Standard, 2006).

2.3 Setup of Landfill Simulation Reactor

The laboratory-scale landfill simulation reactor (LSR) under anaerobic condition was fabricated to study the behaviour of degradation of MSW in the landfill. Two landfill simulators (LSR) were fabricated with Mild Steel sheets of gauge 2 mm. Reactor1 (R1) operated without rainfall and Reactor2 (R2) with rainfall. The single setup has dimensions of 1.1 m × 1 m × 1 m as shown in Fig. 1. Headspace between the top surface of the waste and the top lid was kept 10 cm, to accommodate

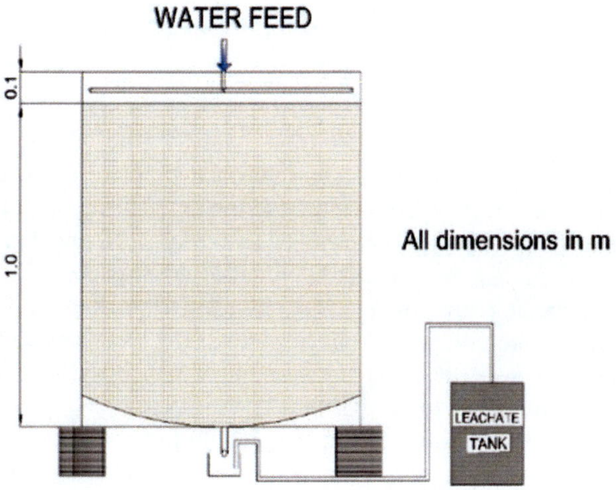

Fig. 1. Setup of landfill simulator.

showerhead to feed the simulated rainwater as tap water. The showerhead was used only in R2 to feed the rainwater.

Unshredded mixed waste (organic and inorganic) was fed into the reactor. The weighted average rainfall according to area received by India was 1020.8 mm in 2018; based on the information of rainfall and dimensions of the reactor the amount of water to be added was calculated. 19.8 L of water was added to the simulator reactor through the showerhead, which ensures even distribution of the rainwater across the surface.

2.4 *Characterization of Leachate*

Raw leachate was collected from the bottom of the reactor. pH and EC of the leachate samples were measured using a pH meter and conductivity meter respectively. Biological oxygen demand (BOD_5) by titration method and COD was measured by the dichromate method (Association et al., 1920).

2.5 *Reactor Monitoring*

The reactors were monitored once a week. Every week leachate samples were collected and stored at 4 °´C then used for further analysis.

3 Results and Discussions

3.1 *Initial Characterization of MSW (Large-scale Reactor Study)*

Individual fresh waste components like food, paper, textile, plastic, and inorganic waste were collected from different sources since they were not allowed to be mixed, and waste was mixed before feeding to the reactor. The moisture content of individual waste components was determined and results presented in Table 1. Food waste had maximum moisture content (73%), followed by paper (9%). Textile waste did not have much moisture since it was collected from local tailors. Similarly, paper and plastic waste collected from recycling bins were not expected to have much moisture. However, the moisture content of paper waste (9%) and plastic waste (5%) can be attributed to these sources.

The fresh waste components were mixed according to wet (73%) and dry (27%). The highest ratio of wet to dry waste composition observed in India was chosen and selected parameters of mixed composition are presented in Table 2. The paper and

Table 1. The moisture content of individual waste components

Components	*Composition (%)*	*Moisture content (%)*
Food waste	73	85
Paper	10	9
Plastic	10	5
Wood waste	5	35
Dirt and textiles	2	12

Table 2. Initially selected parameters for the characterisation of mixed waste

Parameters	*Mean ± Stdev*
Moisture content (%)	79.6 ± 1.41
pH	5.64 ± 0.26
EC (mS/cm)	5.82 ± 1.22
Density (kg/m^3)	676.24 ± 196.69
VS (%)	56.22 ± 1.15
TOC (%)	31.23 ± 0.64
*Wet (kg)	292
*Dry (kg)	108

textiles have the maximum absorption capacity of moisture. The inorganic waste contained metal, dirt, plastic.

The moisture content of fresh waste observed in the study was compared with those reported in the literature in Table 3. It was observed that the values of moisture content obtained in this study were close to the values found in earlier studies.

3.2 Comparison of Leachate Parameters with and without Rainfall Addition

3.2.1 pH

pH is one of the important parameters, which indicates biodegradation of waste (microbial activity) inside the reactor. R1 operated without rainfall and R2 operated with rainfall. Figure 2 shows the trend of pH as a function of time for the two reactors R1 and R2. Initial weeks of pH was in the acidic phase due to production of volatile organic acids, resulting in accumulation of carboxylic acids and pH drops in both the reactors. Young leachates of pH ranged from 5.41 to 8.66 in R1 and 4.91 to 5.35 in R2. pH in R1 has increased with time as acidogenic bacteria were dominated by methanogenic bacteria. While pH values of R2 increased very gradually to 5.35 such an increase might be attributed to the flushing out of VFA which is discussed in a later section as the leachate is collected. While the lower rate of increase as compared to R1 might be due to build-up of carbon dioxide gas and its

Table 3. Comparison of moisture content in this study with the previous study

Waste components	*Moisture content (% wet wt)*	*Author*	*Moisture content(% wet wt) found in this study*
Food waste			
Mixed food waste	68.3	(Cho et al., 1995)	85
Boiled rice	65		
Cooked meat	47		
Fresh cabbage	95		
Food waste	50–80		
Paper waste			
Paper	4–10	(Tchobanoglous et al., 1993)	9
Cardboard	4–8		
Yard waste			
Yard waste	30–80	(Tchobanoglous et al., 1993)	
Wood	15–40		35
Textile waste			
Textiles	6–15	(Tchobanoglous et al., 1993)	12

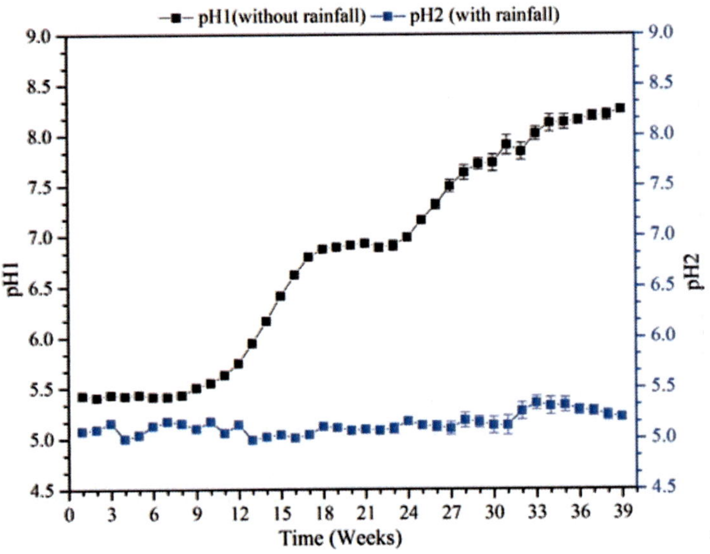

Fig. 2. Trend of pH as a function of time.

dissolution in leachate. Evolved CO_2, due to the degradation process, exerts high partial pressure on leachate; it dissolves, forming carbonic acid, which decreases the pH(HYPERLINK \l "_ENREF_13" \o "Khattabi, 2002 #370"Khattabi et al., 2002a). The pH values below 7 are usually softer waters and the acidity is due to carbonic, humic, fulvic and other organic acids (Mahapatra et al., 2011b; Mahapatra et al., 2011a)

3.3 *Electrical Conductivity*

The variation of electrical conductivity (EC) is illustrated in Fig. 3. EC is influenced by the amount of dissolved organic and inorganic matter present in the leachate used to indicate the degree of salinity and mineral content. The high values of EC can be seen in R1 as represented in EC1 ranged from 15.59 to 38.31 mS/cm due to production of carbonates and bicarbonates due to high biological activity (Wang and Pelkonen, 2009). The values were found to be higher than the reported values during the acidogenic phase (1.6 to 17 mS/cm) (Reinhart and Grosh, 1998). This could be due to the presence of the high amount of inorganic salts like sodium and chloride in young leachate at initial weeks of biodegradation of waste (Karthik, 2012). Moreover, pH greater than 7 can carry a greater load of dissolved substances which cause an increase in EC (Naveen et al., 2017b). In R2, EC2 values ranged from 5.05 to 16.28 mS/cm. EC values were found to be low compared to without rainfall R1

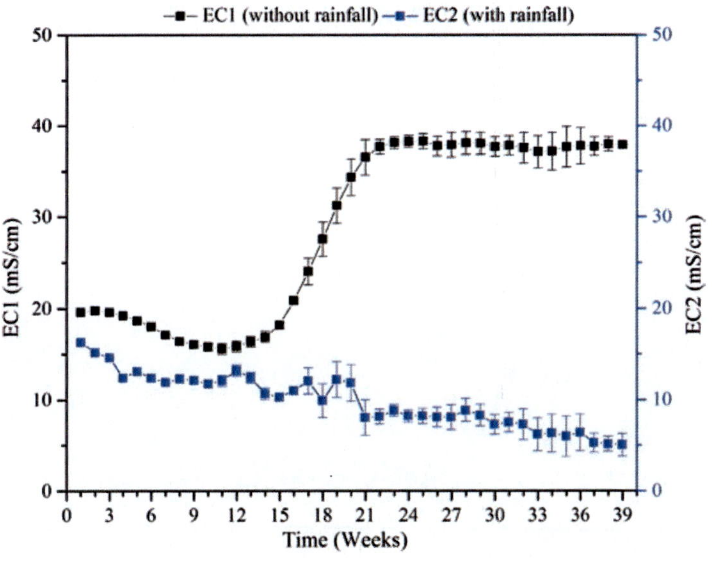

Fig. 3. Trend of EC as a function of time.

because of dilution due to water being dominant and added water had very low EC leaching results in decrease of pH and EC.

3.4 Volatile Fatty Acids (VFA)

Variation of volatile fatty acids is presented in Fig. 4. In R1, VFA values ranged from 3600 to 37500 mg/L due to the degradation of organic waste to form volatile acids i.e., acetic, propionic and butyric (Robinson and Maris, 1983). As pH increases, there is a decrease in VFA because of phase change from acidogenic to methanogenic (bacterial activity). In R2, VFA ranged from 3750 to 23125 mg/L. Initial weeks of VFA concentrations are low compared to R1 which might be due to the water addition. Weeks 6–14 paralleled with greater degradation hence more production of VFA. The acidogenic phase has led to a build-up of VFAs and decomposition of waste is limited. The addition of water could positively affect waste degradation because pH increases, and VFA production decreased with time. Multiple peaks in both reactors could be due to degradation of wastes at different times. As literature reported that VFA is toxic to methanogenic activity as pH should be maintained at 6–8 and VFA should be 10,000 mg/L is a favourable condition to microorganisms (Petchsri et al., 2006). High VFA content is expected, as acid production is enhanced if the moisture of the solid waste is high (Wang and Pelkonen, 2009). Lower values are observed as the organic content in the waste is lower or leached out as observed in a study conducted with old mined waste (Khattabi et al., 2002b).

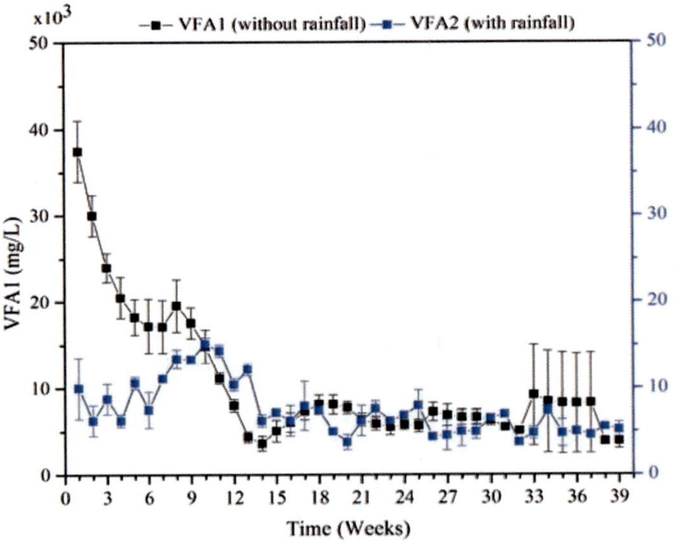

Fig. 4. Trend of VFA as a function of time.

3.5 Biochemical Oxygen Demand (BOD_5)

Fresh MSW generate leachates which are known to have characteristically high organic loads (El-Fadel et al., 2002; Swati et al., 2007; Tatsi and Zouboulis, 2002; Warith, 2002). BOD_5 represents microbial activity and organic pollution in leachate (Service et al., 1977). Figure 5 shows the degradation pattern of BOD_5 with time. In R1, BOD_5 ranged from 13800 to 50250 mg/L as a high amount of organic matter present in the landfill indicates a high concentration of BOD_5 in initial weeks of degradation of waste which diminished over time. In the acidogenic phase (pH at 5), BOD_5 values were higher and as pH increases, BOD_5 values were found to be decreased due to break down of organic matter and leach out the high concentrations. A similar trend can be seen in R2, the values ranged from 6180 to 41,100 mg/L as found to be higher at initial weeks later decreased compared to without rainfall reactor since adding water resulting in dilution of water and organics were leached out from the solid waste. Higher concentrations of BOD_5 in the reactor without rainfall, due to the absence of water for dilution and leaching of hydrolytic and acidogenic materials, resulted in a buildup of high organic intensity and a decrease in the decomposition of waste (Petchsri et al., 2006).

3.6 Chemical Oxygen Demand (COD)

COD represents microbial activity and organic pollution in leachate (Service et al., 1977). Figure 6 shows the degradation pattern of COD with time. In R1, COD ranged from 20,432 to 92,160 mg/L as a high amount of organic matter was present

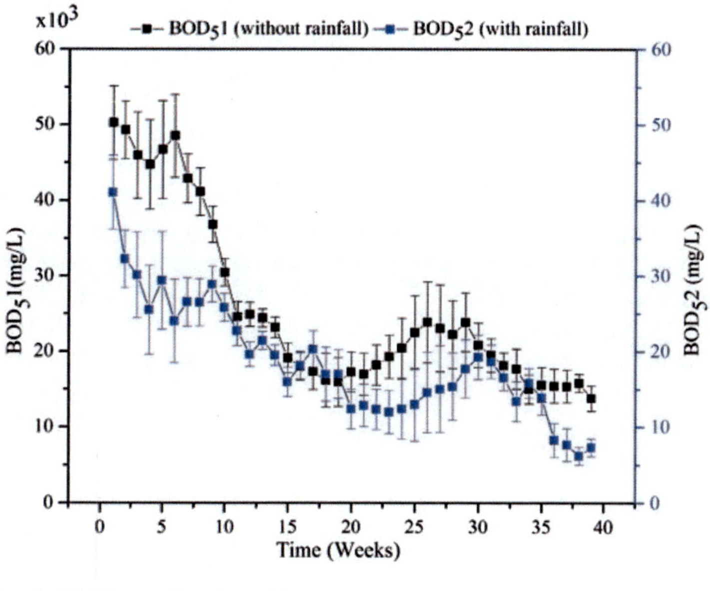

Fig. 5. Trend of BOD_5 as a function of time.

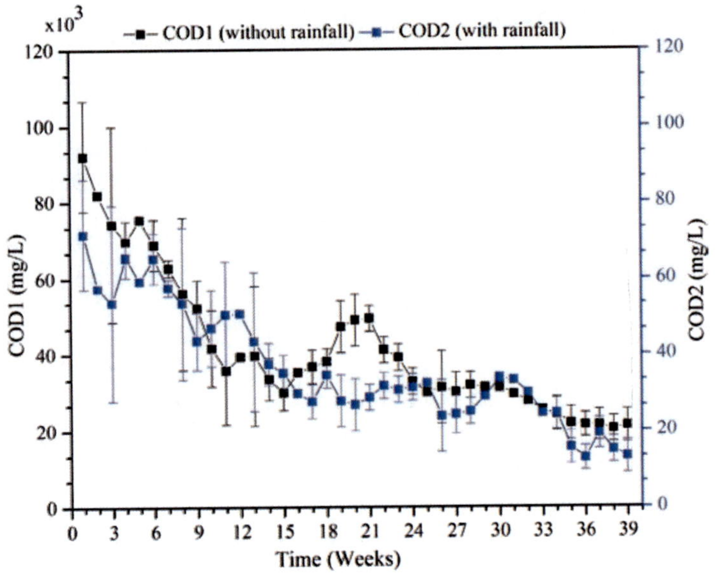

Fig. 6. Trend of COD as a function of time.

in the landfill and waste was hydrolysed to organic and inorganic constituents while organics get converted to simpler compounds or gases and inert materials settle out in leachate and bottom of the reactor. In the acidogenic phase (pH at 5), COD values were higher as pH increases. COD values decreased due to the break down of organic matter and leaching out. A similar trend can be seen in R2 where reactor operated with the addition of tap water at the initial phase of leachate. The values ranged from 71,630 to 12,781 mg/L as found to be higher at initial weeks later decreased as compared to without rainfall reactor since the effect of adding water resulting in dilution of water and leachout. As VFA contribute to COD, the increased VFA concentrations seen in weeks 6–9 paralleled with the corresponding spike in COD values. Higher concentrations of VFA and COD in the reactor without rainfall, due to the absence of water for dilution and leaching of hydrolytic and acidogenic materials, resulted in a buildup of high organic intensity and a decrease in waste decomposition (Petchsri et al., 2006). The main reason for this phenomena—by soluble microbial product (SMP) contributes to COD in the effluent (Shoeybi and Salvacion, 2012). SMP is a product of bacterial metabolism. Various researchers established that generation of SMP happens in the anaerobic degradation of organic waste such as proteins and polysaccharides (Knox and Goddard, 2013).

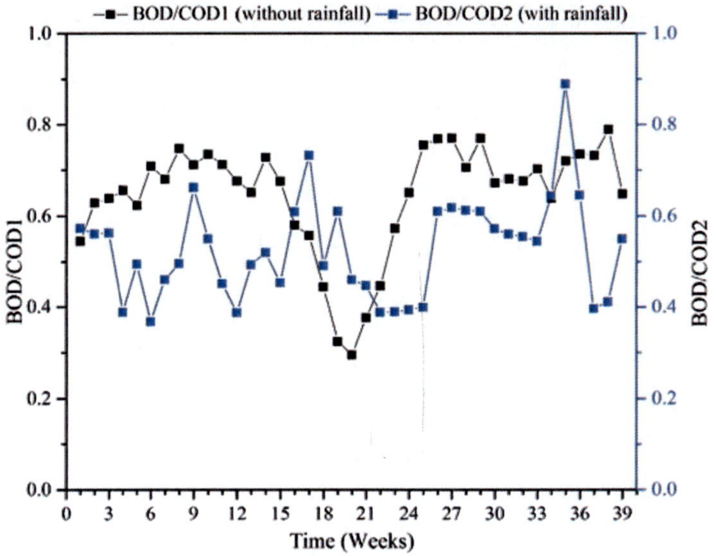

Fig. 7. Trend of BOD_5/COD ratio as a function of time.

3.7 *BOD_5/COD Ratio*

BOD_5/COD ratio is presented in Fig. 7. Average BOD/COD ratio of R1 was 0.64 and R2 was 0.52 respectively. The results showed that both reactors were in the acidogenic phase. Unstabilised Young leachate has a BOD_5/COD ratio of 0.5 or more (Swati et al., 2011). The phase of waste stabilization in landfill concerning BOD_5/COD ratio also indicates biochemically degradable organic matter to total organic matter. It also represents the degradation of organic matter in a landfill. R1 and R2 graphs of BOD_5 and COD were found to be a decline in concentrations are constantly observed as organic matter removed by the washout effect and degradation process (Adhikari and Khanal, 2015). In R2 operated with the addition of rainfall, where leachate is degraded biologically and chemically oxidize organic matter. This is proven that decrease in BOD_5 and COD shows a decrease in organic load of leachate. Both Landfill simulators indicate landfill is in the acidogenic phase and still high amount of organic matter is yet to be degraded. Lower BOD_5/COD of R2 indicates better degradation and removal of organic matter as compared to R1.

4 Conclusions

Key conclusions have been made from experimental results, the impact of with and without rainfall on the degradation of waste are as follows:

(1) Reactor2 operated with rainfall has a positive effect on the acceleration of biological activity as proven that by a significant reduction in BOD_5 and COD but leachate pH increase is lower due to carbon dioxide gas dissolution in leachate/water to form carbonic acid due to higher partial pressure exerted by CO_2 as compared to Reactor1 operated without rainfall, which has a high concentration of leachate as seen from the experiment.
(2) These experimental results explained that high VFA was observed in R1 in the initial phase due to absence of water dilution and leaching of hydrolytic products. The acidogenic phase has led to the build-up of organic strength and decomposition of waste is limited. R2 operated with rainfall showed that addition of water could be a positive effect on the degradation of waste because leachate pH increased and VFA decreased with time is significant for providing favourable conditions.
(3) Addition of rainfall (R2) has positive effects on the landfill than reactor operated without rainfall (R1). pH value of R2 has negatively impacted due to addition of rainfall since generated carbon dioxide was dissolved in water/leachate to form carbonic acid causing the lower rate of increase of pH value but other parameters show a decline in concentrations as compared to Reactor1 operated without rainfall.
(4) Greater degradation of both BOD_5 and COD was observed in R2 (85% and 82% respectively) as compared to 73% and 78% reduction of BOD_5 and COD in R1.
(5) In a broadway, it was observed that the reduction of BOD_5 and COD trends for the same reactor were similar (in both cases considered separately) which indicated pronounced biological and chemical degradation.

References

Adhikari, B. and Khanal, S.N. (2015). Qualitative study of landfill leachate from different ages of landfill sites of various countries including Nepal. *J Environ Sci Toxicol Food Technol*, **9:** 2319–2399.

Association, A.P.H., Association, A.W.W., Federation, W.P.C. and Federation, W.E. (1920). Standard methods for the examination of water and wastewater. American Public Health Association.

Cho, J.K., Park, S.C. and Chang, H.N. (1995). Biochemical methane potential and solid state anaerobic digestion of Korean food wastes. *Bioresource Technology*, **52:** 245–253.

Christensen, T.H., Kjeldsen, P., Bjerg, P.L., Jensen, D.L., Christensen, J.B., Baun, A., Albrechtsen, H.-J. and Heron, G. (2001). Biogeochemistry of landfill leachate plumes. *Appl. Geochem.*, **16:** 659–718.

Ejlertsson, J., Karlsson, A., Lagerkvist, A., Hjertberg, T. and Svensson, B.H. (2003). Effects of co-disposal of wastes containing organic pollutants with municipal solid waste—a landfill simulation reactor study. *Adv Environ Res*, **7:** 949–960.

El-Fadel, M., Bou-Zeid, E., Chahine, W. and Alayli, B.J.W.M. (2002). Temporal variation of leachate quality from pre-sorted and baled municipal solid waste with high organic and moisture content. **22:** 269–282.

Fatta, D., Papadopoulos, A. and Loizidou, M. (1999). A study on the landfill leachate and its impact on the groundwater quality of the greater area. *Environ Geochem Health*, **21:** 175–190.

Gujer, W. and Zehnder, A.J. (1983). Conversion processes in anaerobic digestion. *Water Sci Technol*, **15:** 127–167.

Joshi, R. and Ahmed, S. (2016). Status and challenges of municipal solid waste management in India: A review. *Cogent Environ Sci*, **2:** 1139434.

Kalamdhad, A.S. and Kazmi, A. (2008). Mixed organic waste composting using rotary drum composter. *Int J Environ Waste management*, **2:** 24–36.

Karthik, O. (2012). Pilot Scale Simulation of Landfill Bioreactor and Controlled Dumping of Fresh and Partially Stabilized Municipal Solid Waste in a Tropical Developing Country.

Khattabi, H., Aleya, L. and Mania, J. (2002a). Changes in the quality of landfill leachates from recent and aged municipal solid waste. *Waste Manage Res*, **20:** 357–364.

Khattabi, H., Aleya, L. and Mania, J.J.W.M. (2002b). Changes in the quality of landfill leachates from recent and aged municipal solid waste. **20:** 357–364.

Knox, K. and Goddard, A. (2013). Non-degradable COD in landfill leachate – A review: occurrence, chemical characteristics, environmental properties and treatment, Fourteenth International Waste Management and Landfill Symposium, Sardinia, Italy.

Ledakowicz, S. and Kaczorek, K. (2004). Laboratory simulation of anaerobic digestion of municipal solid waste. *J Environ Sci Health A*, **39:** 859–871.

Mahapatra, D.M., Chanakya, H. and Ramachandra, T. (2011a). Assessment of treatment capabilities of Varthur lake, Bangalore, India. **14:** 84.

Mahapatra, D.M., Chanakya, H. and Ramachandra, T. (2011b). C:N ratio of sediments in a sewage fed urban lake. **5:** 86–92.

Naveen, B., Mahapatra, D.M., Sitharam, T., Sivapullaiah, P. and Ramachandra, T. (2017a). Physico-chemical and biological characterization of urban municipal landfill leachate. *Environ. Pollut.*, **220:** 1–12.

Naveen, B., Mahapatra, D.M., Sitharam, T., Sivapullaiah, P. and Ramachandra, T.J.E.P. (2017b). Physico-chemical and biological characterization of urban municipal landfill leachate. *Environ Pollut*, **220:** 1–12.

Petchsri, P., Towprayoon, S., Chaiprasert, P. and Nopharatana, A. (2006). The effect of precipitation on municipal solid waste decomposition and methane production in simulated landfill bioreactor with leachate recirculation. *Songklanakarin J Sci Technol*, **28:** 615–631.

Pohland, F., 1975. Sanitary landfill stabilization with leachate recycle and residual treatment, Report for EPA grant No. R-801397, USEPA National Environmental Research Center, Cincinnati, OH.

Reinhart, D.R. and Grosh, C.J. (1998). Analysis of Florida MSW landfill leachate quality, Analysis of Florida MSW landfill leachate quality. University of Central Florida/CEED.

Robinson, H.D. and Maris, P.J. (1983). The treatment of leachates from domestic wastes in landfills—I: aerobic biological treatment of a medium-strength leachate. *Water Res*, **17:** 1537–1548.

Service, U.S.N.T.I., Chian, E.S. and DeWalle, T.B. (1977). Evaluation of Leachate Treatment. Vol. 1: Characterization of Leachate. US Environmental Protection Agency, Office of Research and Development.

Shoeybi, M. and Salvacion, J.L. (2012). Landfill leachate degradation in tropical maritime climate; an experimental laboratory scale study. International Conference on Environment Science and Engieering, pp. 103–108.

Sponza, D.T. and Ağdağ, O.N. (2004). Impact of leachate recirculation and recirculation volume on stabilization of municipal solid wastes in simulated anaerobic bioreactors. *Process Biochem*, **39:** 2157–2165.

Standard, I. (2006). Methods of Sampling and Test (Physical and Chemical) for Water and Wastewater. IS.

Swati, M., Karthikeyan, O., Joseph, K., Visvanathan, C. and Nagendran, R. (2011). Pilot-scale simulation of landfill bioreactor and controlled dumping of fresh and partially stabilized municipal solid waste in a tropical developing country. *J Hazard Toxic Radioactive Waste*, **15:** 321–330.

Swati, M., Karthikeyan, O.P., Joseph, K. and Nagendran, R. (2007). Landfill bioreactor: A biotechnological solution for waste management. JSIR, **66**.

Tatsi, A. and Zouboulis, A.J.A.i.E.R. (2002). A field investigation of the quantity and quality of leachate from a municipal solid waste landfill in a Mediterranean climate (Thessaloniki, Greece). Adv Environ Res, **6:** 207–219.

Tchobanoglous, G., Theisen, H., Vigil, S.A. and Alaniz, V.M. (1993). *Integrated Solid Waste Management: Engineering Principles and Management Issues*. McGraw-Hill, New York.

Tiquia, S. and Tam, N. (1998). Composting of spent pig litter in turned and forced-aerated piles. *Environ Pollut*, **99:** 329–337.

Wang, Y. and Pelkonen, M. (2009). Impacts of temperature and liquid/solid ratio on anaerobic degradation of municipal solid waste: An emission investigation of landfill simulation reactors. *Journal of Material Cycles and Waste Management*, **11:** 312.

Warith, M.J.W.M. (2002). Bioreactor landfills: experimental and field results. *Waste Manage,* **22:** 7–17.

Chapter 30
Scope of Renewable Energy Intervention for Energy Sufficiency in Nagaland

Imlisongla Aier, Anil Kumar Sakhiya, Priyanka Kaushal, and Virendra Kumar Vijay

1 Introduction

The depletion of fossil fuels and the harmful consequences of its utilisation has been the major factor contributing to the need for shift towards the use of renewable energy (Vidal-Amaro et al., 2015; Østergaard and Sperling, 2014). The exploitation of fossil fuels over the years for meeting the growing energy needs of mankind has however, left a dramatic impact on the environment. (Kaygusuz and Kaygusuz, 2002). The ever-increasing advancement in science and technologies for utilisation of renewable resources to substitute fossil fuels and meet the energy demands of the recent times require technologies which can bridge the gap between the energy demand and supply. The renewable energy sources are however unstable in nature and their applications are limited to their metrological data, geographical location of the place and specific applications depending on their availability (Debebe and Engida, 2016). While some of the renewable energy technologies are well developed, some require deeper research in the technological and social aspects for its development and implementation (Kaygusuz and Kaygusuz, 2002). To meet the ever growing demand of energy in India, the government has set up energy targets as 175 GW from Renewable energy by the year 2022, which has been a key driving force for renewable energy development in the country (National Institution for Transforming India, 2015).

About 1.1 billion of the rural population today have no access to electricity which are comprised of 95% from the African Sub-Saharan region and Asia (Kulkarni and Anil, 2015; Bailey et al., 2012). The availability of reliable energy services for electricity and cooking fuels continues to be the key driving force behind any economic development of a country as well as a major indicator of living standards in the country (Cust et al., n.d.; Zheng et al., 2010).

I. Aier (✉) · A. K. Sakhiya · P. Kaushal · V. K. Vijay
Centre for Rural Development and Technology, Indian Institute of Technology Delhi, New Delhi, Delhi, India

A. S. Kalamdhad (ed.), *Integrated Approaches Towards Solid Waste Management*, https://doi.org/10.1007/978-3-030-70463-6_30

This paper is a case study of the state of Nagaland, which is located in North East India. It is a state dominated by rural population comprising 53% of the total population (*DDUGJY& Saubhagya: Status of Rural Electrification in Nagaland,* 2019). This paper presents a study on the status of the energy demand and consumption in the state and quantifies the potential of renewable energy resource options available for electricity production to meet the growing energy demands in Nagaland.

1.1 Nagaland

Nagaland lies in the North-Eastern part of India and is the sixteenth state of the Indian union. It covers a total area of 16,579 km^2 comprising of 12 districts. It is located between 98 and 96°E longitude and 26.6 and 27.4°N latitude. The population of Nagaland as recorded in the year 2019 is estimated to be 3.27424 million ("Population of Nagaland," 2019). The climatic condition of the state ranges from 20 °C to 30 °C in summer. During winter, temperature in the state drops to a minimum below 5 °C. The annual rainfall in the state reaches up to a maximum of 2500 mm during the month of July which recorded the maximum precipitation.

1.2 Energy Status in Nagaland

The population of Nagaland comprises of a majority living in rural areas ("Population of Nagaland - StatisticsTimes.com," 2019). This Rural population constitutes a major section of communities living without the access to affordable and reliable basic fuel and energy services. There is a significant need for reliable energy service in the state if it has to improve its economic standards and living conditions.

Under the Scheme "Deendayal Upadhyaya Gram Jyoti Yojana" by the Government of India, every household in Nagaland have been provided with free electricity connections ("Overview | Government of India | Ministry of Power," 2020) which has resulted in 100% electrification of the state (DDUGJY& Saubhagya: Status of Rural Electrification in Nagaland, 2019; 24 × 7 Power For All: A joint Initiative of Government of India and Government of Nagaland, 2016). A graphical representation of the status of energy availability and demand in the state is shown in Fig. 1.

From the graph shown in Fig. 1, we observe that the energy requirement in the state is higher than the energy availability in the state (Nagaland State Action Plan on Climate Change (2021–2030), 2018).

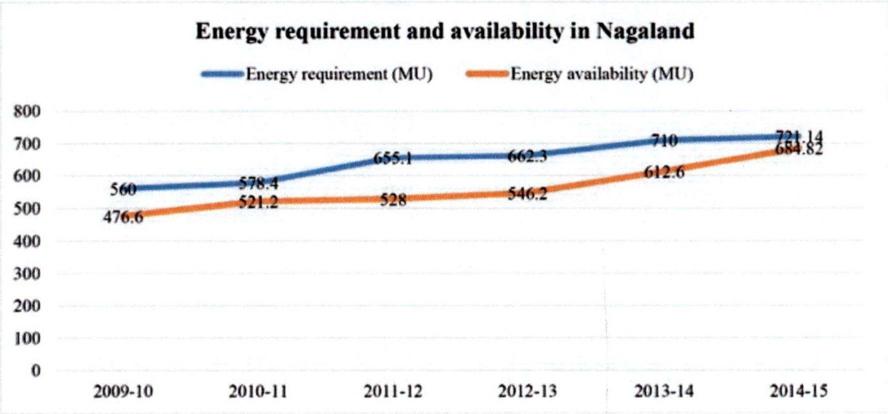

Fig. 1. Energy requirement and availability in Nagaland.

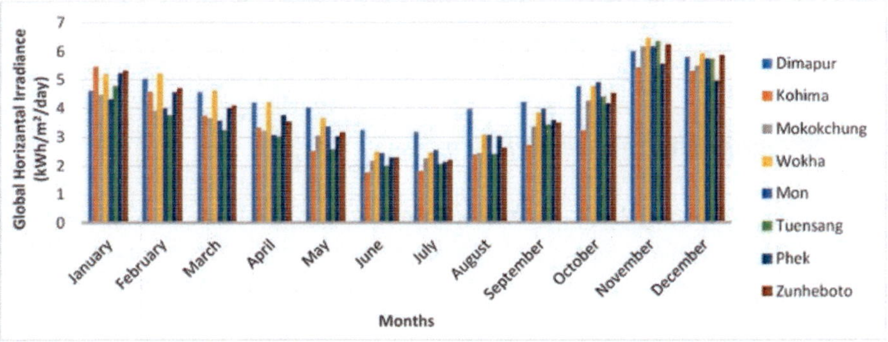

Fig. 2. Solar radiation data in Nagaland.

2 Renewable Energy Sources Available In Nagaland

2.1 Solar Energy Resource

The solar radiation received in Nagaland are shown in Fig. 2.

Figure 2 shows the global solar irradiance data throughout the year for eight districts in Nagaland. The months of November and December experiences the highest intensity of radiation. The district Dimapur and Wokha receives the highest intensity of solar radiation ("Climate-Data.org," n.d.).

2.2 Hydroenergy Resource

The hydroenergy resource is one of the most developed forms of renewable power and it contributes more than 19% of world electricity production (Debebe and Engida, 2016).

There are 11 major rivers in Nagaland, which grows larger as they collect water from more tributaries along its course ("Rivers of Nagaland," 2018). They are shown in Table 1.

2.3 Wind Energy Resource

The potential of wind energy resource is primarily dependent on the geography of the location. According to the Global Wind data information obtained for wind speed in Nagaland, the speed lies in the range of 0–5.4 m/s ("Global wind Atlas," 2019).

2.4 Biomass Energy Resource

The unique geographical location of the state has resulted in wide variety of flora and forest types. The total geographical area of the state under forest area is 52.04% while the remaining geographical area is classified as under miscellaneous grooves, trees and agricultural land, wasteland, cultivable non-forest area, etc. The Woody

Table 1. Major rivers in Nagaland and districts joining them

S. No.	*River*	*Districts to where the river flows*
1	Doyang river	Kohima, Zunheboto and Wokha district
2	Dikhu river	Zunheboto, Mokokchung, Longleng and Tuensang
3	Dhansiri river	Dimapur
4	Tizu river	Zunheboto and Phek
5	Milak river	Mokokchung
6	Zungki river	Noklak and Kiphire
7	Chathe river	Dimapur
8	Tapi river	Mon
9	Sidzu river	Pfutsero
10	Dzu-u river	Kohima
11	Miki river	Kiphire

There are so far nine small-, mini-, micro- and pico-hydropower projects installed in Nagaland.

forest makes up 20% of the total land area which is a home to rich flora and fauna (Changkija, 2012).

The biomass resources available in the state are mainly classified into agroresidue and forest and wasteland residue ("Biomass resource potential in Nagaland," 2016). Biomass potential of the agroresidue and forest and wasteland residue are shown in Tables 2 and 3.

3 Status of Energy Use In Nagaland

The state has seen a huge rise and fall in the urban and rural population in the last decade (Jamir, 2016; Aier and Anungla, 2011). Correspondingly, the households in Nagaland has also seen a huge increase in the number of urban household during the period 2015–2019 (DDUGJY& Saubhagya: Status of Rural Electrification in Nagaland, 2019). The comparison of the urban and rural households from 2014 to 2019 is shown in Fig. 3.

With the transition from rural to urban household, an increasing trend of energy requirement is observed due to the increase in the energy consumption (Phurailatpam et al., 2018). This has caused an alarming strain on the power demand in the state. The state derives power from energy sources such as hydro, coal based thermal, gas based thermal and renewable energy sources. Its major share of power is derived from hydropower plants. The contribution of power obtained from the different energy sources are shown in Fig. 4.

4 Energy Potential and Project Initiatives In Nagaland

Presently, the state has a total energy generation capacity of 72.25 MW derived from sources such as solar, and small hydropower. They are shown in Table 3.

4.1 Hydrop Plant

There is promising scope of hydropower plant in the state particularly the small and pico-hydropower plants. The potential of small-hydropower plant in Nagaland today stands at 196.98 MW while the installed projects are 102 MW. The pico-hydropower plant is a Hydroger of capacity 3–5 kW and delivers an output efficiency of 75–80%. In Nagaland, a total 55 Hydrogers units of total capacity 165 kW have been installed by the NePED (Nagaland empowerment of People through Energy Development) (All India Installed Capacity (in MW) Of Power Stations, 2019). Under NEPED, decentralised micro hydropower based Hydrogers dissemination target of 300 kW has been implemented as reported by the Nagaland State

Table 2. Potential from agroresidues and wasteland and forest ("Biomass resource potential in Nagaland," 2016)

Districts	*Area in (kha)*		*Biomass generation Potential (kT/year)*		*Potential biomass surplus (kT/year)*		*Total power potential (MW_e)*	
	Agroresidue	*Forest & waste land*	*Agroresidue*	*Forest & waste land*	*Agroresidue*	*Forest & waste land*	*Agroresidue*	*Forest & waste land*
Kohima	33	169.7	116.5	281.5	20.4	102.9	2.4	14.4
Phek	31.1	100	83.3	111.4	14.2	73.5	1.7	10.3
Wokha	28.7	67.2	73.8	70.3	12.8	46.4	1.5	6.5
Mokokchung	24.6	63.6	62.1	64.3	10.9	42.5	1.3	5.9
Mon	21.5	89.7	55.3	93	9.7	61.4	1.1	8.6
Tuensang	20.7	224.8	52.7	281.5	9.2	185.8	1.1	26
Zunheboto	20	71.3	48.5	67.3	8	44.4	0.9	6.2
Total	179.6	786.3	492.2	843.7	85.2	556.9	10	77.9

Table 3. Status of energy generation from available renewable energy sources

Renewable generating capacities (MW)	*2015–2016*	*2016–2017*	*2017–2018*	*2018–2019*	*Total*
Solar	0	20	40	0	60
Wind	0	0	0	0	0
Small hydropower	2.1	0.45	2.1	7.6	12.25
Biomass	0	0	0	0	0
Total	2.1	20.45	42.1	7.6	72.25

The power generation plants from different source of energy set up in the state are as follows.

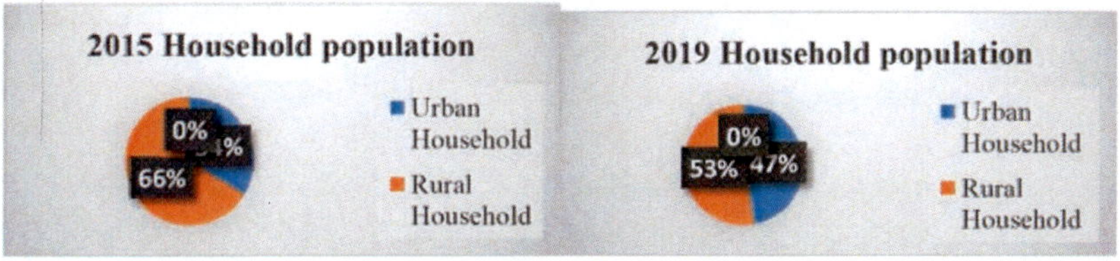

Fig. 3. Urban and rural household of Nagaland for 2015 and 2019.

Action Plan on Climate Change (Nagaland State Action Plan on Climate Change (2021–2030), 2018). The mini- and micro-hydro projects taken up in the state are listed in Table 4.

4.2 *Solar Energy*

Nagaland receives the maximum solar radiation from months March to September. The district Dimapur recorded the highest amount of solar radiation among all the districts in Nagaland with average annual solar radiation 4.93 $kWh/m^2/day$ followed by Wokha, which recorded an average of 4.79 $kWh/m^2/day$ and the districts receiving the least solar radiation were 4.49 $kWh/m^2/day$ in Mokokchung, 4.33 $kWh/m^2/day$ in Tuensang and 4.3 $kWh/m^2/day$ in Kohima. This indicates that setting up of solar photovoltaic plants in Nagaland could be a viable option. A solar power plant programme in the state with a target of 742 kW_p was planned in the year 2011–2012 (Chishi, 2011), which has now targeted 61 MW where 50 MW to be harvested from rooftop. A total 459 numbers of off-grid solar photovoltaic, five solar water pumps with a capacity of 7 HP has been disseminated in the state. The state has also targeted 3170 m^2 capacity of solar water heating systems under the planning of state department of New and Renewable Energy (Nagaland State Action Plan on Climate Change (2021–2030), 2018).

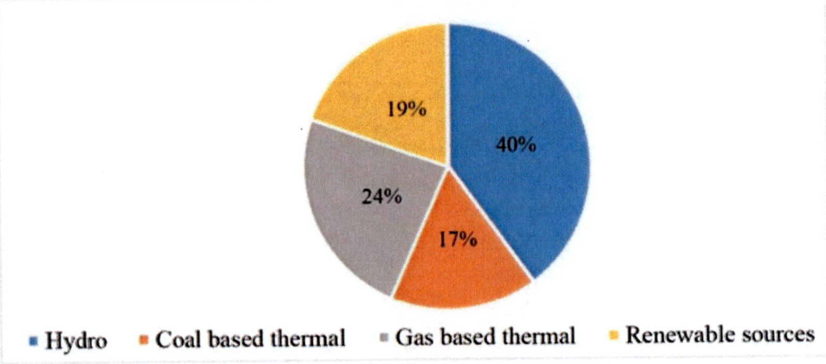

Fig. 4. The contribution of power by different energy sources.

Table 4. Mini- and micro-hydropower projects

S. No.	*Name of project*	*Capacity*	*District*
1	Duilumreu Mini Hydro Project	300 kW	Peren
2	Thivopisii Microhydel	2 × 50 kW	Phek
3	Dzüdza MicroHydel	500 kW	Kohima
4	Nguiki MicroHydel	200 kW	Peren

4.3 Wind Energy

The wind speed in Nagaland is an average of 3.5 m/s while the average minimum winds speed required for setting up wind turbine is 6 m/s. In Nagaland, a 20 kW capacity of a small wind energy hybrid system has been installed ("Global wind Atlas," 2019; Nagaland State Action Plan on Climate Change (2021–2030), 2018).

4.4 Biomass Energy

The wide availability of biomass resources in the state has led to the set-up of biomass gasifier in the state under the initiative of the state department of New and Renewable Energy and NEPeD which are reported as follows ("Report of the Task Force Nagaland Executive Summary," 2014; "MNRE Annual Report 2005–2006," 2005; "Medziphema village gets biomass gasifier power plant," 2012).

Gasifier of capacity (2 × 100 kW) in Ashukhomi village at Zunheboto District, (2 × 20 kW) in Panso village at Tuensang District, (2 × 20 kW) in Sutemi village at Zunheboto District, (2 × 50 kW) in Tizit village at Mon District, (2 × 100 kW) in Longwa village at Mon District and (2 × 100 kW) Medziphema Gasifier plant at Dimapur District.

The use of anaerobic bio-digesters to produce methane rich biogas was implemented in the state which could be used for the purpose of domestic cooking. A total of 250 biogas units have been installed so far across the state with a cumulative capacity of 500 m^3. However, the main constraint for setting up biogas plant in the state was lack of kitchen waste as they were mostly used as piggery feed (Nagaland State Action Plan on Climate Change (2021–2030), 2018).

Thus, thermochemical route of biomass conversion reported a better scope in the energy conversion process because of the following reasons:

a. Climatic condition of most of the districts cannot support the temperature required for anaerobic digestion.
b. Since most districts have cold temperature throughout the year, thermochemical conversion of biomass for electricity generation will provide greater reliable power.
c. The gasifiers have already been introduced in major cold regions of the state where solar radiation and temperature are minimum. Therefore, they are more likely to be effective and reliable as compared to other renewable energy sources.
d. There are sufficient biomass resources available in the state which can be used for power generation whenever desired.

5 Technological Intervention

5.1 Hydropower

Hydrogers can be used to convert the energy of water flowing from rivers or streams falling from a height. The force of the flowing river has the potential to develop hydropower, which can be useful for providing power supply to the villages. There is ample scope of installation at every district in the state and can be supervised by the local residents.

5.2 Solar Energy

Installation of solar photovoltaic system at rooftops, street lights and study lamps can have positive impacts on the power sufficiency in the state especially in districts like Dimapur and Wokha. However, the installation of solar concentrators in the state is not feasible in low temperature districts because solar radiation, temperature and climatic conditions are not favourable for achieving the high temperature desired. Subsidies should be provided to the buyers of solar photovoltaic system.

5.3 Wind Energy

Wind turbine structures require huge infrastructure which would be difficult to set up due to major setbacks such as low wind speed, bad road conditions, lack of skilled labour and technicians, lack of incentives and initiative by the government.

5.4 Biomass Energy

Gasifier could be the most reliable option for power generation in the state because there is abundant availability of biomass and the climatic conditions in the state is favourable for operation of biomass gasifier.

Biomass cogeneration system can also be a viable technology for providing power and heating application in the state. Besides power generation, the heat production can be useful for performing many applications related to heating thus increasing the overall efficiency of the system.

6 Conclusion

It can be concluded from the current study that there is untapped potential for renewable energy resources in the state. The solar photovoltaic energy holds a promising scope for electricity production in districts like Dimapur and Wokha. Energy from hydropower plant can also be seen as a potential source as most of the districts are connected by major river sources. In extremely remote villages far from electricity power source, the Hydroger has the potential to electrify remote hilly villages in Nagaland where source of water flowing from a height is available. Though biomass is abundantly available in the state, there is a need for technology intervention in the usage and conversion route of biomass to useful form of power. Besides gasifier which has been introduced in certain districts in the state, biomass cogeneration system could be an effective source of power as well as heat source for several applications. Besides these, government policies and incentives can also play an effective role in making the technologies accessible to the public as well as for sensitising the general public which is altogether a bigger challenge, especially in the rural areas.

References

24 × 7 Power For All: A joint Initiative of Government of India and Government of Nagaland. (2016). Retrieved from https://powermin.nic.in/sites/default/files/uploads/joint_initiative_of_govt_of_india_and_nagaland.pdf.

Aier, Anungla, T.K. (2011). Strengthening of State Plans for Human Development. Retrieved from https://www.nagaland.gov.in//Nagaland/Report/PROBLEMS OF RURAL-URBAN MIGRATION - A Thematic Report.pdf.

All India Installed Capacity (in MW) of Power Stations. (2019). Retrieved from http://www.cea.nic.in/reports/monthly/installedcapacity/2019/installed_capacity.

Bailey, M., Henriques, J., Holmes, J. and Jain, R. (2012). Providing village-level energy services in developing countries. Malaysian Commonwealth Studies Centre, (October), 2–5. https://doi.org/10.1002/ana.23759.

Biomass resource potential in Nagaland (2016). Retrieved December 17, 2019, from Biomass Portal: Ministry of New and Renewable Energy website: https://biomasspower.gov.in/nagaland.php.

Changkija, S. (2012). Forests and Floristic Diversity of Nagaland. Glimpses of Forestry Research in the Indian Himalayan Region, 67–74.

Chishi, T.S.A. (2011). Nagaland Renewable energy for the people. *Akshay Urjay*, **5(1)**. Retrieved from https://mnre.gov.in/file-manager/akshay-urja/july-august-2011/EN/Nagaland Renewable Energy.pdf.

Climate-Data.org. (n.d.). Retrieved from https://en.climate-data.org/asia/india/nagaland-787/#example1.

Cust, J., Singh, A. and Neuhoff, K. (n.d.). Rural Electrification in India: Economic and Institutional aspects of Renewables.

DDUGJY& Saubhagya: Status of Rural Electrification in Nagaland. (2019). Retrieved from http://www.ddugjy.gov.in/assets/uploads/State_wise/Nagaland.pdf.

Debebe, A. and Engida, D. (2016). Sustainable renewable energy resources utilization in rural areas. *Renewable Sustainable Energy Rev*, **66:** 1–9. doi: https://doi.org/10.1016/j.rser.2016.07.013,

Global Wind Atlas (2019). Retrieved January 8, 2019, from https://globalwindatlas.info/area/India/Nagaland.

Jamir, Y.T. (2016). A study on the changing population structure in Nagaland. *Econ Affairs*, **61(2):** 215. doi: https://doi.org/10.5958/0976-4666.2016.00029.2.

Kaygusuz, K. and Kaygusuz, A. (2002). Renewable energy and sustainable development in Turkey. *Renewable Energy*, **25(3):** 431–453. doi: https://doi.org/10.1016/S0960-1481(01)00075-1.

Kulkarni, S.H. and Anil, T.R. (2015). Strategic Planning for Energy and the Environment Status of Rural Electrification in India, Energy Scenario and People's Perception of Renewable Energy Technologies. doi: https://doi.org/10.1080/10485236.2015.11439123.

Medziphema village gets biomass gasifier power plant. (2012). Retrieved December 23, 2019, from Indian Environment Portal: Knowledge for change website: http://www.indiaenvironmentportal.org.in/category/2852/thesaurus/biomass-gasifier-programme/.

MNRE Annual Report 2005–2006. (2005). Retrieved from https://mnre.gov.in/file-manager/annual-report/2005-2006/EN/contents.htm.

Nagaland State Action Plan on Climate Change (2021–2030). (2018). Retrieved from https://www.nagaland.gov.in/Nagaland/UsefulLinks/Nagaland State Action Plan on Climate Change (NSAPCC).pdf.

National Institution for Transforming India (2015). Report of the Expert Group on 175 GW RE by 2022. Retrieved from https://niti.gov.in/writereaddata/files/175-GW-Renewable-Energy.pdf.

Østergaard, P.A. and Sperling, K. (2014). Towards sustainable energy planning and management. *Int J Sustainable Energy Plann Manage*, **1:** 1–6. https://doi.org/10.5278/ijsepm.2014.1.1

Overview I Government of India I Ministry of Power (2020). Retrieved January 14, 2020, from Government of India I Ministry of Power website: https://powermin.nic.in/en/content/overview-1.

Phurailatpam, C., Singh, B. and Wang, L. (2018). Planning and optimization of autonomous DC microgrids for rural and urban applications in India. *Renewable and Sustainable Energy Rev*, **82(September 2017):** 194–204. doi: https://doi.org/10.1016/j.rser.2017.09.022.

Population of Nagaland - StatisticsTimes.com. (2019). Retrieved December 29, 2019, from Statistics times website: http://statisticstimes.com/demographics/population-of-nagaland.php.

Population of Nagaland 2019 (2019). Retrieved December 28, 2019, from India Population 2019 website: https://indiapopulation2019.com/population-of-nagaland-2019.html.

Report of the Task Force Nagaland Executive Summary (2014). Retrieved from https://www.niti.gov.in/niti/writereaddata/files/Nagaland_Report_0.pdf.

Rivers of Nagaland (2018). Retrieved December 30, 2019, from Nagaland GK website: https://nagalandgk.com/list-of-rivers-of-nagaland-map/.

Vidal-Amaro, J.J., Østergaard, P. A. and Sheinbaum-Pardo, C. (2015). Optimal energy mix for transitioning from fossil fuels to renewable energy sources - The case of the Mexican electricity system. *Applied Energy*, **150:** 80–96. doi: https://doi.org/10.1016/j.apenergy.2015.03.133.

Zheng, Y.H., Li, Z.F., Feng, S.F., Lucas, M., Wu, G.L., Li, Y. et al. (2010). Biomass energy utilization in rural areas may contribute to alleviating energy crisis and global warming: A case study in a typical agro-village of Shandong, China. *Renewable Sustainable Energy Rev*, **14**. doi: https://doi.org/10.1016/j.rser.2010.07.052.

Index

A. S. Kalamdhad (ed.), *Integrated Approaches Towards Solid Waste Management*, https://doi.org/10.1007/978-3-030-70463-6

L

M

N

O

P